Emerging Strategies to Reduce Transmission and Thermalization Losses in Solar Cells

Jonas Sandby Lissau • Morten Madsen
Editors

Emerging Strategies to Reduce Transmission and Thermalization Losses in Solar Cells

Redefining the Limits of Solar Power Conversion Efficiency

Springer

Editors
Jonas Sandby Lissau
Mads Clausen Institute
University of Southern Denmark
Sønderborg, Denmark

Morten Madsen
Mads Clausen Institute
University of Southern Denmark
Sønderborg, Denmark

ISBN 978-3-030-70357-8 ISBN 978-3-030-70358-5 (eBook)
https://doi.org/10.1007/978-3-030-70358-5

This Springer imprint is published by the registered company Springer Nature Switzerland AG
The registered company address is: Gewerbestrasse 11, 6330 Cham, Switzerland

Preface

This book describes a number of promising emerging strategies for decreasing the substantial transmission and thermalisation losses present in current solar energy conversion. Chapter 1 has an introductory discussion of solar cell efficiency, and further provides a detailed overview of the various new strategies for going beyond current power conversion efficiency limits that are presented throughout this book.

Part I of this book presents photon upconversion via triplet fusion in organic molecules as a strategy for reducing transmission losses in solar energy conversion. Chapter 2 introduces the mechanism, while Chaps. 3 to 6 discuss potential solutions to a number of challenges associated with this particular mechanism, namely near-infrared to visible light upconversion (Chap. 3), upconversion in the solid state (Chap. 4), heavy metal-free sensitisers (Chap. 5), and plasmon-enhanced upconversion (Chap. 6).

Common to a number of the mechanistic concepts described in this book is the central roles played by triplet excitons. The strong interaction of these quasiparticles with molecular oxygen is a major challenge to tackle, and Part II of the book is dedicated to this area of research. Molecular oxygen in relation to photoresponsive organic materials is introduced in Chap. 7. Approaches to circumvent the negative effects of molecular oxygen are given in Chap. 8 (for photon upconversion) and Chap. 9 (organic optoelectronics in general).

Part III of the book concerns the implementation of photon upconversion via triplet fusion in solar cells, either by optical coupling (Chap. 10) or via electronic integration in the solar cell (Chap. 11).

Materials based on rare earth metals for photon upconversion are presented in Part IV of the book. Chapter 12 introduces the mechanism, reviews current research topics in this area, and gives an overview of solar cell implementations of this strategy. Plasmonic enhancement of rare earth ion-based photon upconversion is reviewed in Chap. 13.

Strategies dealing with the thermalisation losses in solar cells are covered in Part V of the book. Singlet fission mechanism and molecular design considerations are given in Chap. 14, and implementation of singlet fission and quantum cutting in solar cells is reviewed in Chap. 15.

We would like to acknowledge all the authors for their contributions and patience with the editing process and thank Springer Nature for publishing this book. This work was supported by a research grant (17677) from VILLUM FONDEN.

Sønderborg, Denmark Jonas Sandby Lissau

Sønderborg, Denmark Morten Madsen
April 2021

Contents

Chapter 1
Introduction: Solar Cell Efficiency and Routes Beyond Current Limits

Jonas Sandby Lissau and Morten Madsen

The energy carried by sunlight has an enormous potential to provide humanity with all the energy it needs while limiting the catastrophic consequences of climate change caused to a large extent by current combustion of fossil fuels [1, 2]. A convenient way to harvest solar energy is by making use of the photovoltaic effect of certain semiconducting materials. When a solar photon interacts with such a material, its energy can be converted into electronic energy, which can be extracted to an external electrical circuit, providing both a photovoltage and a photocurrent. The material absorbing the sunlight imposes certain boundary conditions on its electrons which allow electronic states to have only specific distinct energies. Importantly, there will always be a certain minimum energy needed to promote a "resting" electron in a material to a higher electronic state. Therefore, all materials used for the absorption of sunlight will have a requirement on the minimal energy that must be carried by a photon in order for that photon to be absorbed by the material. Photons with energies lower than this absorption threshold will be transmitted by the material. In a solar cell, this portion of solar energy will be lost. This loss of solar photons, known as a transmission loss, restricts the photocurrent achievable from sunlight (see Fig. 1.1a). In a hypothetical solar cell with an optimal absorption threshold (also known as band gap) of 1.3 eV, the fraction of incident solar energy lost by transmission is $\sim$25% [3].

For an optimized material, photons with energies *larger* than the band gap are absorbed with high efficiency. However, the fraction of the resulting electronic energy which is in excess of the band gap energy will be lost due to interactions with phonons in the material, reducing the achievable photovoltage. This is known as a thermalization loss, and in a solar cell with optimal 1.3 eV band gap, it amounts to $\sim$30% of the incident solar energy [3] (Fig. 1.1a). Taken together, transmission

J. S. Lissau (✉) · M. Madsen
Mads Clausen Institute, University of Southern Denmark, Sønderborg, Denmark
e-mail: jlissau@mci.sdu.dk; madsen@mci.sdu.dk

J. S. Lissau, M. Madsen (eds.), *Emerging Strategies to Reduce Transmission and Thermalization Losses in Solar Cells*, https://doi.org/10.1007/978-3-030-70358-5_1

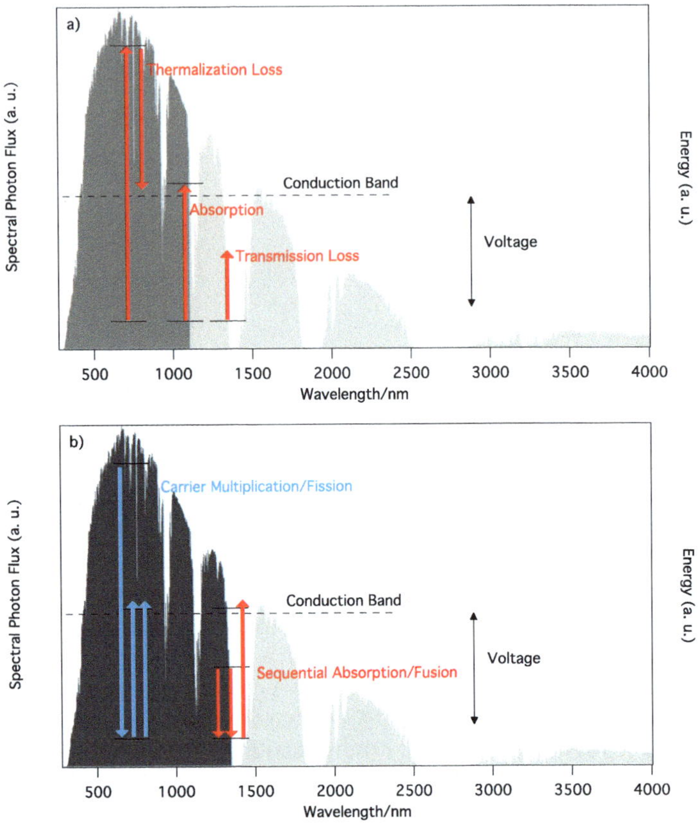

Fig. 1.1 (**a**) The two main loss mechanisms responsible for the internal power conversion efficiency limit of single threshold solar cells. (**b**) General depiction of strategies to overcome thermalization and transmission losses, respectively

and thermalization losses amount to more than 50% of the incident light on a hypothetical solar cell with optimal band gap. Mainly for this reason, the theoretical upper limit of power conversion efficiency for a conventional semiconductor single junction solar cells is 32.9%, and it is somewhat lower for emerging excitonic solar cell technologies [4]. Hence, every extra percent of power conversion efficiency that can be achieved by limiting the efficiency restrictions imposed by transmission and thermalization losses will be significant contributions for driving the transition toward a sustainably powered modern world. This book describes a number of promising emerging strategies for circumventing these substantial transmission and thermalization losses in current solar energy conversion. We would like to emphasize that the list of approaches covered in this book toward mitigating these solar energy losses is not exhaustive. For example, tandem solar cell is probably the most well-established strategy for increasing the power conversion efficiency of solar cells beyond the theoretical single junction limit, but a review of this research field is beyond the scope of this book.

To address transmission losses in solar cells, one needs to introduce an intermediate electronic energy level in the band gap of the solar cell to allow for sequential absorption of low-energy photons to take place (Fig. 1.1b). A relatively new and promising strategy toward this end is to take advantage of metastable triplet states in bimolecular systems. The mechanism was originally termed sensitized anti-Stokes delayed fluorescence or sensitized P-type delayed fluorescence, but after renewed interest from various scientific disciplines, a number of new synonyms have been introduced, all referring to the same process and many of them seen throughout this book [5, 6]. Terms like photochemical upconversion, molecular photon upconversion, photon upconversion via sensitized triplet–triplet annihilation, triplet fusion upconversion, and any combination of the above may all refer to the same mechanism and often reflect the background of the researchers and the period in which they started activities in the field. Photochemical upconversion is the theme of Part I of this book and will be further introduced in Chap. 2. The mechanism has several advantages such as strong and broad, tunable absorption bands that are decoupled from the metastable triplet states, which altogether makes the mechanism efficient under (non-concentrated) solar light conditions. However, for solar energy applications, the mechanism also has a number of challenges which are currently very active research topics. Research in these areas is presented in Chaps. 3 to 6.

A large energy loss associated with a required change of electronic spin multiplicity (from singlet to triplet), together with an exponential increase in non-radiative decay of excited states as their energy is reduced [7], has made the search for good molecular upconverters of near-infrared light a great challenge. Solutions to this problem proposed so far are outlined in Chap. 3.

The transfer of triplet state energy in molecules, which is an essential part of the photochemical upconversion mechanism, happens via an electron exchange mechanism which requires orbital overlap. However, molecular interactions can also affect the mechanism in ways that are not beneficial. In solution, photochemical upconversion works well because the involved molecules can be separate (non-interacting) while still being able to transfer triplet state energy via molecular diffusion. Condensing the system further into the solid state often poses problems caused by molecular interactions, such as aggregate formation and phase separation. For solar energy applications, using a volatile solvent as the upconverter host medium is not a durable solution. Chapter 4 reports on the progress in the development of solid-state photochemical upconverters.

To accommodate the required spin flip in photochemical upconversion, a common strategy has been to employ a metal complex where a heavy-metal ion induces a strong spin-orbit coupling. Even if the strategy is very efficient, it has its limitations in terms of applications in solar energy conversion. The heavy-metal ions commonly used, such as palladium, platinum, ruthenium, iridium, osmium, and rhenium, are all among the rarest metals on earth, and they may be environmentally hazardous when concentrated. For these reasons, it is an important task to substitute the heavy-metal complexes in photochemical upconversion with cheaper and environmentally friendly candidates. This field of research is described in Chap. 5.

To efficiently harvest all of the light at a certain wavelength in a given material, a simple method is to increase the thickness of the material. For a number of reasons, this is not a good strategy to follow for upconversion by triplet fusion. First, due to the inherent mechanism, the efficiency of the upconversion process could be very different throughout the material, with the highest efficiency attained in the first layer that the incident light penetrates. Second, depending on the specific upconversion system, inner filter effects reducing the amount of upconverted light escaping the upconverter will be more or less pronounced. These negative effects will only increase with thickness. Concentrating light in a thinner material could be a solution to the above issues. Since use of external lenses is not compatible with commercialization, the use of nanophotonic structures for light trapping seems like a more interesting approach to explore. In Chap. 6, plasmon-enhanced upconversion via sensitized triplet–triplet annihilation is reviewed.

One very central aspect of the function of photochemical upconversion is its reliance on metastable triplet states as an intermediate energy level for sequential absorption. In the mechanism, these states perform homogenous triplet–triplet annihilation among similar excited states. Meanwhile, they also perform very efficient heterogeneous triplet–triplet annihilation with the triplet ground state of molecular oxygen. Hence under atmospheric conditions, the process is often entirely quenched by the presence of oxygen. Furthermore, a product of the heterogeneous triplet–triplet annihilation is the singlet excited oxygen molecule which reacts with, and thereby degrades, a large number of organic molecules including those most often used for triplet fusion upconversion. For real-world applications of this mechanism for upconversion, the oxygen problem is therefore crucial to deal with. Part II of the book is dedicated entirely to the interaction of triplet states with molecular oxygen and how to avoid this interaction. This part of the book is relevant not only to upconversion but also to all other oxygen-sensitive organic systems, especially those where triplet states (or excitons) are formed in high yield, like singlet fission systems, which will be discussed in a later part of the book. Chapter 7 introduces the general photophysics and photochemistry of molecular oxygen in relation to photoresponsive organic materials. Chapter 8 outlines protective strategies that have been explored to stabilize upconverters based on sensitized triplet–triplet annihilation, and Chap. 9 gives a broader discussion on the use of additives for the stabilization of organic optoelectronic devices against photooxidation.

Part III of the book concerns the implementation of photochemical upconversion in solar cells. The studies done so far in this area can be divided into two conceptually different approaches which have been dedicated a separate chapter each. Chapter 10 discusses optically coupled photochemical upconversion solar cells. In this approach, the upconverter is external to the solar cell, and upconverted photons are emitted into the solar cell where they are absorbed. In the second approach reviewed in Chap. 11, the sensitized triplet–triplet annihilation mechanism is electronically integrated internally in the solar cell. In this scheme, no upconverted photons are emitted; instead, the high-energy singlet state produced in the triplet fusion process is harvested via direct charge separation. Due to the absence of upconverted photons, this strategy is a molecular analogue of the intermediate band solar cell [8]. Chapter 11 also contains a comparison between the two different approaches to photochemical upconversion in solar cells presented in Chaps. 10 and 11.

In Part IV of the book, an alternative class of materials is presented which also facilitate upconversion via sequential absorption. The photophysical mechanism for upconversion in rare-earth ions is presented in Chap. 12, together with a discussion of current research topics in the area and a review of attempts to implement this upconversion mechanism in solar cells. In comparison to the photochemical upconversion mechanism discussed in the preceding chapters, upconversion in rare-earth ions has the advantage that it upconverts from wavelengths far into the near-infrared region. Furthermore, the process runs in solid-state materials well-suited for solar cell integration and less prone to degradation than organic molecules. The efficiency of the mechanism is however much lower than the corresponding efficiencies seen for photochemical upconversion under simulated sunlight conditions. For this reason, nanophotonic enhancement of rare-earth ion-based upconversion is an important topic covered separately in Chap. 13.

Approaches that address thermalization losses are discussed in Part V (Fig. 1.1b). Some of the more promising strategies are based on a photophysical mechanism very similar to the photophysics responsible for upconversion via sensitized triplet fusion as discussed in Part I. Whereas the latter relies on the fusion of two triplet states to form a singlet state, singlet fission which is discussed in Part V is the same scheme run in reverse, with the fission of one singlet state into two triplet states. Therefore, the two fields can learn important lessons from each other. Often what is needed to run the scheme efficiently in one direction should be avoided for the reverse reaction, to minimize loss processes. The mechanism of singlet fission and considerations on molecular design are presented in Chap. 14, while implementations in solar cells are reviewed in Chap. 15. Like for solar cell integration of triplet fusion upconversion as discussed in Part III, also for singlet fission, there exist two categories of implementations in solar cells, optically and electronically coupled, respectively. Whereas the product state in triplet fusion is an emissive singlet state, in singlet fission, the product is two non-radiative triplet states. The singlet fission mechanism can therefore be electronically coupled by internal implementation in solar cells where the triplet exciton energy is harvested by charge separation. Alternatively the mechanism can be implemented externally by optical coupling via triplet energy transfer to emissive species. The latter approach is an example of quantum cutting. Both types of implementations are discussed in Chap. 15.

Summarized, this book provides a thorough introduction to current research strategies that are set out to mitigate part of the transmission and thermalization losses present in current photovoltaic technologies. This includes overview of state-of-the-art, basic scientific principles and also detailed elaboration of latest research progress made in these topics. Clearly, given the current challenges on climate changes, the need for efficient photovoltaic technologies has not been more urgent, providing a worldwide focus on these research areas. While the research strategies and efforts outlined in this book mainly remain at a research stage at rather low technology readiness level, the constant efforts from research groups are expected to change that in the near future, providing further support to the green energy transition worldwide.

References

1. R.E. Blankenship, D.M. Tiede, J. Barber, G.W. Brudvig, G. Fleming, M. Ghirardi, M.R. Gunner, W. Junge, D.M. Kramer, A. Melis, T.A. Moore, C.C. Moser, D.G. Nocera, A.J. Nozik, D.R. Ort, W.W. Parson, R.C. Prince, R.T. Sayre, Science **332**(6031), 805 (2011). http://www.sciencemag.org/content/332/6031/805.abstract
2. J. Hansen, P. Kharecha, M. Sato, V. Masson-Delmotte, F. Ackerman, D.J. Beerling, P.J. Hearty, O. Hoegh-Guldberg, S.L. Hsu, C. Parmesan, J. Rockstrom, E.J. Rohling, J. Sachs, P. Smith, K. Steffen, L. Van Susteren, K. von Schuckmann, J.C. Zachos, PLoS One **8**(12), e81648 (2013). http://dx.doi.org/10.1371%2Fjournal.pone.0081648
3. L.C. Hirst, N.J. Ekins-Daukes, Proc. SPIE: Next Generation (nano) Photonic and Cell Technologies For Solar Energy Conversion **7772**, 777211 (2010)
4. M.J.Y. Tayebjee, L.C. Hirst, N.J. Ekins-Daukes, T.W. Schmidt, J. Appl. Phys. **108**(12), 124506 (2010)
5. C.A. Parker, C.G. Hatchard, Proc. Chem Soc., 386 (1962)
6. C.A. Parker, C.G. Hatchard, T.A. Joyce, Nature **205**(4978), 1282 (1965). https://doi.org/10.1038/2051282a0
7. R. Englman, J. Jortner, Mol. Phys. **18**(2), 145 (1970). https://doi.org/10.1080/00268977000100171
8. A. Luque, A. Martí, Phys. Rev. Lett. **78**(26), 5014 (1997)

Part I
Addressing Transmission Losses: Sequential Absorption via Triplet Fusion in Organic Materials

Chapter 2
Photophysics

Nancy Awwad, Mo Yang, and Felix N. Castellano

Abstract Sensitized triplet-triplet annihilation-based photochemical upconversion (TTA-UC) or sensitized triplet fusion (TF) is a process by which low-energy photons are converted into high-energy light through a sequence of bimolecular triplet-triplet energy-transfer (TTET) reactions. TTA-UC attracts interest due to its potential applications in solar energy conversion and storage, biological imaging, photochemical synthesis, photocatalysis, photopolymerization, and photoresponsive devices. In this chapter, we present the principles and photophysics governing the observation, mechanism, and chemical kinetics of TTA-UC to establish general energetic and kinetic rules which dictate the conditions under which the phenomenon can be observed.

Keywords Molecular photophysics · Upconversion · Energy transfer · Photosensitization · Sensitized triplet-triplet annihilation

2.1 Original Experimental Observations of Delayed Fluorescence in Solution

TTA-UC is an anti-Stokes fluorescence process where the energy of two low-energy photons of incident light is pooled to produce one high-energy photon [1]. The process is denoted "anti-Stokes" because the result is an emission spectrum markedly blue-shifted with respect to the absorption spectrum of the sensitizer. This phenomenon is promising because of its potential applications in oxygen sensing [1], biological imaging [2], photocatalyst systems [3], photovoltaics [4], and photochromic displays [5]. Focusing specifically on solar energy conversion, UC represents an attractive method for overcoming theoretical power efficiency limits in operational solar cells. Certain highly efficient upconversion systems can

N. Awwad · M. Yang · F. N. Castellano (✉)
Department of Chemistry, North Carolina State University, Raleigh, NC, USA
e-mail: nawwad@ncsu.edu; myang19@ncsu.edu; fncastel@ncsu.edu

J. S. Lissau, M. Madsen (eds.), *Emerging Strategies to Reduce Transmission and Thermalization Losses in Solar Cells*, https://doi.org/10.1007/978-3-030-70358-5_2

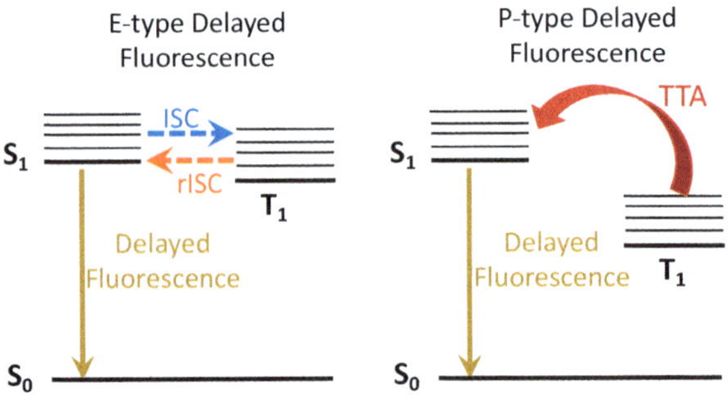

Fig. 2.1 Oversimplified Jablonski diagrams illustrating the differences between E-type and P-type delayed fluorescence. ISC and rISC represent intersystem crossing and reverse intersystem crossing, respectively, where the latter generates the S_1 excited state via thermal activation of the T_1 excited state in the same molecule. TTA is triplet-triplet annihilation, representing a bimolecular reaction between two T_1 excited states ultimately forming one S_1 excited state. Both processes form the S_1 excited state well after the excitation event and are collectively termed delayed fluorescence

operate efficiently even with noncoherent light sources, such as sunlight [6–8]. By upconverting sub-bandgap photons, a UC-equipped solar cell could potentially increase efficiency from 33% (as set by the Shockley–Queisser limit in Si) [9] to greater than 43% [10–12].

In the early 1960s, Parker and Hatchard first observed what they referred to as "sensitized anti-Stokes delayed fluorescence" in donor/acceptor solutions of proflavine hydrochloride and anthracene or phenanthrene and naphthalene [13]. Excitation of the blue-absorbing proflavine hydrochloride resulted in near-UV emission from anthracene. This form of delayed fluorescence was denoted pyrene-type (or P-type) to distinguish it from eosin-type (E-type fluorescence), denoted in Fig. 2.1. In E-type delayed fluorescence, conversion from triplet excited states to singlet excited states occurs thermally, by reverse intersystem crossing. However, in P-type delayed fluorescence the energy gap between the singlet fluorescent state (S_1) and the lowest-lying triplet state (T_1) is sufficiently large that thermal activation to the excited singlet state at room temperature is not possible [14]. It can only occur through the pooling of the energy of two triplet excited states to produce one singlet excited state, and is therefore designated triplet-triplet annihilation upconversion (TTA-UC). In the 1960s, the phenomenon of delayed fluorescence from an excited singlet state due to a TTA-UC mechanism had been reported in solutions of pyrene, phenanthrene, and other aromatic hydrocarbons [13, 15, 16].

However, the low efficiency of TTA-UC, due to the low yield of intersystem crossing (ISC) of the organic sensitizer, belied the utility of this phenomenon, and TTA-UC was infrequently investigated over the subsequent decades following this result. More recently, several research groups [6, 17–20] have surmounted the obstacle that limited the efficiency of Parker and Hatchard's systems by integrating metal-organic chromophores containing Ru(II) [21], Ir(III) [22], Zn(II) [23], Pt(II) [24, 25], Pd(II) [26, 27], and others as sensitizers, which exhibit much higher ISC efficiencies. However, before discussing the factors that dictate the overall efficiency of the TTA-UC, an explanation of the mechanism by which it occurs is necessary.

2.2 The Mechanism of TTA-UC

TTA-UC typically draws on two species (though not always, see Sect. 2.6), a sensitizer and an annihilator. The sensitizer is responsible for absorbing light and transferring triplet energy to the annihilator, which in turn generates the upconverted light. Beginning with all species in their ground state, a sensitizer absorbs a photon and is promoted to an electronically excited singlet state ($^1S^*$). The excited sensitizer then undergoes ISC, resulting in the lowest excited triplet state ($^3S^*$). Subsequently, a sensitizer in the $^3S^*$ engages in TTET with an energetically appropriate acceptor/annihilator chromophore, thereby producing the triplet excited state of the acceptor ($^3A^*$). When two acceptor triplets collide, they may take part in TTA, resulting in one acceptor molecule in the excited singlet state ($^1A^*$) and another in the ground state (^1A). Ultimately, $^1A^*$ returns to the ground state through radiative decay in most instances. The Jablonski diagram presented in Fig. 2.2 illustrates the reaction sequence by which TTA-UC occurs along with the relevant energy levels of the participating excited states.

2.3 Requirements for TTA-UC

Figure 2.2 illustrates the generalized energy levels of sensitizer (S) and annihilator (A) necessary for achieving TTA-UC. Because this is a multi-chromophore,

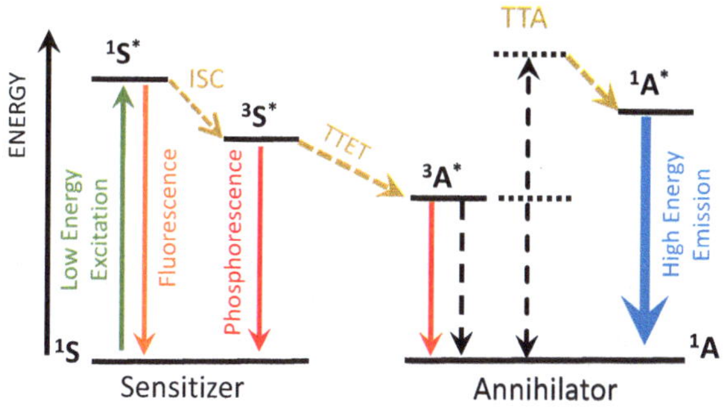

Fig. 2.2 Simplified Jablonski diagram depicting the energy levels of the sensitizer (S) and annihilator (A) species that may lead to photochemical upconversion. Note that two $^3A^*$ species must be generated in order to produce the desired $^1A^*$ species from TTA-UC. Solid and dashed lines denote radiative and non-radiative transitions, respectively

multiphoton process, several key parameters must be addressed. These include the energetics of relevant excited states and the kinetics governing all of the first- and second-order processes necessary to achieve this phenomenon as described below.

Chromophores and Energetic Requirements

When selecting a potential sensitizer-acceptor system for upconversion, the first consideration must be the relative excited-state energies as they relate to the favorability of TTET and TTA. First, in order for the acceptor to quench the triplet excited state of the sensitizer, the triplet excited state of the acceptor must be lower than the triplet excited state of the sensitizer: $E(^3A^*) \leq E(^3S^*)$. The larger the energy difference between the triplet sensitizer and triplet acceptor, the more thermodynamically favorable the TTET process, but the smaller the potential anti-Stokes shift [5]. Second, the merged triplet energy from two acceptor molecules must be greater than or equal to the acceptor's singlet excited-state energy, $2E(^3A^*) \geq E(^1S^*)$. This second energetic requirement ensures that TTA is favorable with increasing driving force at the expense of the anti-Stokes shift. In addition to these energetic requirements, the sensitizer should be a strong light absorber and the acceptor should be a strong emitter in cases where light production is desired. This manifests in a high oscillator strength for the absorption into the sensitizer's singlet excited state, and a high fluorescence quantum yield for the emission of the acceptor. Given the appropriate energies of the relevant excited states, the rates and yields of their interconversion must also be considered.

Rules That Govern ISC

As mentioned in Sect. 2.1, the efficiency of Parker and Hatchard's initial systems for TTA-UC was hampered by a low yield of ISC in their organic photosensitizers. However, sensitizers containing heavy atoms, more specifically transition metals, display high ISC efficiencies by virtue of their large spin-orbit coupling constants. This makes them effective sensitizers for TTA-UC. Although ISC is a spin-forbidden process, it is still possible due to the spin-orbit coupling. Spin-orbit coupling arises when the spin of an electron is influenced by the electron's motion about a nucleus (i.e., an interaction between the spin angular momentum and the orbital angular momentum). This interaction allows change in the spin angular momentum (i.e., S_1 to T_1) to coincide with a change in orbital angular momentum, thereby compensating for the change in the former.

A simplified Hamiltonian for spin-orbit coupling (H_{SO}) is described according to Eq. (2.1):

$$H_{SO} = -\frac{Z^4 e^2}{8\pi \varepsilon_0 m_e^2 c^2} \, l \bullet s \tag{2.1}$$

There are two main components to this factor: the coefficient and the dot product. The latter portion is the dot product of two vectors, the electron's angular momentum, l, and spin angular momentum, s. This determines that spin-orbit coupling rigorously depends on both the motion of the electron and its spin. The coefficient consists of a collection of constants where c is the speed of light in vacuum, m_e is the mass of an electron, e is the elementary charge of an electron, ε_0 is the permittivity of the vacuum, and Z is the atomic number of the nucleus [28].

The proportionality of H_{SO} to Z^4 implies that, in general, molecules containing heavier atoms exhibit significantly higher ISC rates, leading to the so-called heavy atom effect. Typical species incorporated into sensitizers to enhance the heavy atom effect include transition metals (e.g., Pd, Pt, or Zn) and/or large halogens (e.g., I, Br, or Cl) in porphyrins, phthalocyanines, or other transition metal complexes, which helps realize singlet-triplet ISC efficiencies near unity [29, 30]. This arises from providing a heavy/highly charged nucleus to increase the factor Z from Eq. (2.1) while also providing a possible improvement to the vector dot product, $l \bullet s$.

Energy Transfer Mechanisms: Förster and Dexter

Following intersystem crossing, the next step in the TTA-UC mechanism is triplet-triplet energy transfer (TTET), with the sensitizer acting as donor and the annihilator as acceptor. Ignoring trivial energy transfer, there are two fundamental mechanisms by which TTET can occur, Förster [31] and Dexter [32] energy transfer. Because the kinetics of these processes depend differently on fundamental parameters, it is important to highlight two significant differences between the Dexter and Förster mechanisms. First, the Dexter mechanism is a spin-conserved process that depends on electron interactions [33], whereas the Förster mechanism is based on the Coulombic interaction between transition dipole moments [34]. The implications of this will be discussed in the following paragraph. Second, both mechanisms depend on distance (denoted R) and operate most efficiently at short range; however, the rate at which they fall off at long-range differs. Because Förster energy transfer scales as $(1/R)^6$ and Dexter energy transfer scales as $\exp(-2R/L)$, Förster energy transfer can occur at much longer distances with respect to Dexter. The reaction rate constant of Dexter energy becomes negligible once the distance between the sensitizer and acceptor (R) increases beyond 10 Å (>1 nm).

These transfer mechanisms are shown schematically in Figs. 2.3 and 2.4 and are described below.

Förster Theory Förster, or fluorescence resonance energy transfer (FRET), represents non-radiative electronic energy transfer resulting from Coulombic interactions between the oscillating electric dipoles of an excited sensitizer molecule and a

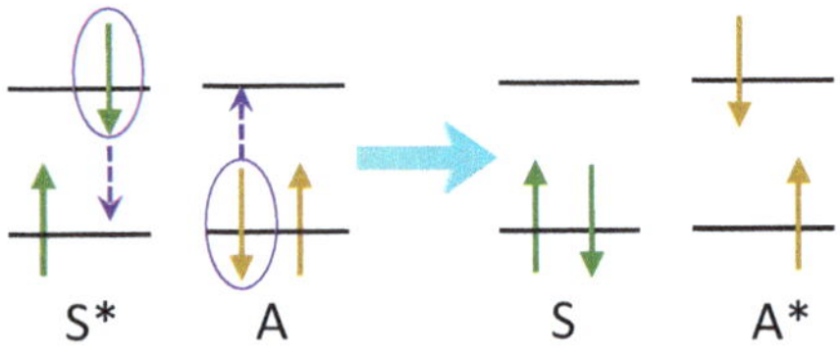

Fig. 2.3 Schematic representation of the Förster-type energy transfer mechanism occurring between two molecular singlets: the sensitizer is excited (S*) and the acceptor is in its ground state (A) prior to energy transfer

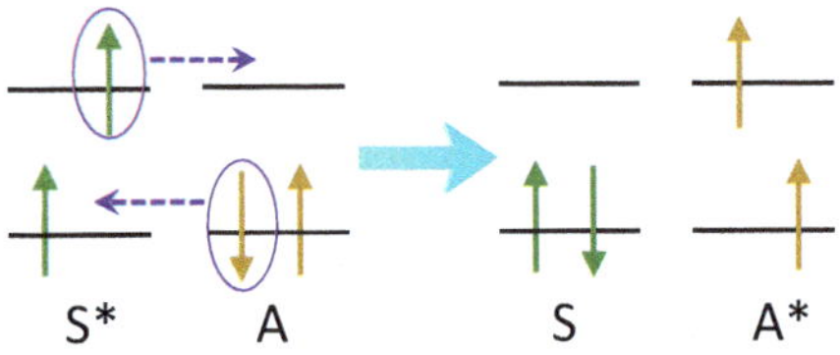

Fig. 2.4 Schematic representation of the Dexter-type energy transfer mechanism using an excited triplet photosensitizer (S*) and a ground-state acceptor producing a ground-state singlet photosensitizer (S) and an excited triplet acceptor (A*). Note how the electron exchange results in conservation of spin

ground-state acceptor molecule, with energy transfer occurring through a classical resonant interaction between these two dipoles [35]. Since the transfer rate is governed by dipolar interactions, it has been most commonly observed in singlet-singlet energy transfer processes, Fig. 2.3.

Dexter Theory The Dexter energy transfer (DET) mechanism is a quantum mechanical effect which is sometimes described as a double-electron-exchange mechanism, as illustrated in Fig. 2.4 for a triplet excited sensitizer.

In this mechanism, an excited sensitizer molecule transfers an electron to the acceptor concurrently with an electron transfer from the acceptor back to the sensitizer, leaving the acceptor in an excited state [36]. Although both FRET and Dexter ET depend on the spectral overlap integral, the exchange contribution to ET in the Dexter mechanism ensures that ET can still occur in the absence of good spectral overlap. However, in Dexter ET, overlap of the sensitizer and acceptor molecular orbitals is necessary [32], rendering it a short-range phenomenon. As a result, Dexter energy transfer typically occurs following intermolecular collisions, so the rate of ET is, at its fastest, diffusion controlled [32]. Here, the dependence of the ET rate constant on the distance between the sensitizer and acceptor, R, is exponential and is calculated by Eq. (2.2) [37]:

$$k_{\mathrm{ET}} = K J^{\mathrm{D}} \exp\left(-2R/L\right) \tag{2.2}$$

where L is the effective average Bohr radius for the sensitizer and the acceptor molecules. K is a constant given by the equation, $K = (2\pi/\hbar)\kappa^2$, where κ is a constant in units of energy. In this case, J^D is the Dexter overlap integral evaluated from the absorption spectrum of the acceptor (ε_A) and the fluorescence spectrum of the sensitizers (f_S) over wave number $\widetilde{\upsilon}$ according to Eq. (2.3) [37–39]:

$$J^D = \frac{\int f_S(\widetilde{\upsilon})\,\varepsilon_A(\widetilde{\upsilon})\,d\widetilde{\upsilon}}{\int f_S(\widetilde{\upsilon})\,d(\widetilde{\upsilon})\,d\widetilde{\upsilon} \int \varepsilon_A(\widetilde{\upsilon})\,d\widetilde{\upsilon}} \tag{2.3}$$

In the overall TTA mechanism, there are two cases of Dexter ET. First is the TTET step where a triplet excited state $^3S^*$ and a ground-state annihilator encounter each other, producing $^3A^*$ and regenerating the ground state of the sensitizer. The second case is the TTA step, where two $^3A^*$ molecules combine to produce one $^1A^*$ molecule along with one ground-state A species. Both are bimolecular Dexter processes necessitating physical contact of the two components, but the efficiency of the TTA step is also governed by the nature of the species generated in the bimolecular collision.

Spin Statistics of TTA

The penultimate step in the overall TTA-UC mechanism is TTA, in which two triplet excited state annihilators interact, resulting in simultaneous, spin-allowed transitions on both molecules. While Nickel [40] determined that annihilation results in a highest energetically available singlet excited-state product, early studies suggested that the efficiency of TTA would be limited by spin statistics and that the singlet is only one of several possible product states [41–45]. Because the number of possible encounter-pair spin states is the product of the spin multiplicities of the two species, and each $^3A_1{}^*$ has a spin multiplicity, M, of 3 (total spin quantum number, $S = 1$), there are 9 (3×3) possible encounter pair sub-states. According to the rules for combining spin quantum numbers, where S ($M = 2S + 1$), which is the magnitude of the vector sum of the electron spins +1/2 or −1/2, can have a minimum value of $|s_1 - s_2|$, a maximum value of $s_1 + s_2$, and an intermediate value incremented by 1, those nine possible sub-states are either singlet $^1(^3A_1{}^3A_1)_0$, triplet $^3(^3A_1{}^3A_1)_1$, or quintet $^5(^3A_1{}^3A_1)_2$ states. When considering the multiplicity of the spin states, it can be seen that out of the nine sub-states, five are of quintet ($S = 2$) character, three are of triplet ($S = 1$) character, and one is of singlet ($S = 0$) character (calculated from $(2S + 1)/9$) as presented in Eq. (2.4) [14, 46]:

$$
{}^{3}A_1{}^{*} + {}^{3}A_1{}^{*} \begin{cases} \xrightarrow{(5/9)} {}^{5}(AA)_2{}^{*} \leftrightarrow {}^{5}A_2{}^{*} + {}^{1}A_0 \\ \xrightarrow{(3/9)} {}^{3}(AA)_1{}^{*} \leftrightarrow {}^{3}A_1{}^{*} + {}^{1}A_0 \\ \xrightarrow{(1/9)} {}^{1}(AA)_0{}^{*} \leftrightarrow {}^{1}A_0{}^{*} + {}^{1}A_0 \end{cases} \quad (2.4)
$$

Based purely on spin statistics, the probability that an annihilation event should yield an excited singlet molecule ($S = 0$) is therefore 1/9 or 11.1% [41–45]. However, TTA quantum efficiency in excess of 40% is frequently measured [30, 47]. The efficiency in excess of the statistical limit can be understood when considering the fate of the triplet and quintet encounter complex in the other pathways. Given that the excited quintet state is always at extremely high energy (four parallel spins need to be generated) with respect to the other relevant excited states, the energy of the annihilating triplets is therefore insufficient for the formation of an excited quintet molecule, Eq. (2.4) [43, 44]. Accordingly, it is more likely that the quintet encounter complex reverts back into two triplet excited states ($2 \times {}^{3}A_1{}^{*}$). Similarly, the ${}^{3}A_1{}^{*}$ upper triplet excited state (such as T_2 or higher) is frequently inaccessible given the energetic input, so the triplet encounter complex would likely dissociate back to two ${}^{3}A_1{}^{*}$ states, becoming available for additional annihilation events. If the reaction did indeed produce an upper ${}^{3}A_1{}^{*}$ excited state, the product of Eq. (2.4), internal conversion to the lowest ${}^{3}A_1{}^{*}$ state is possible (T_1) in addition to ISC to form the ${}^{1}A_0{}^{*}$ fluorescent singlet. Thus, the quintet and triplet encounter complexes are recycled back into triplets, enabling TTA quantum efficiencies well in excess of 11.1%, approaching an upper limit surpassing 40% [30, 43, 47, 48].

2.4 The Importance of Kinetics

Since the phenomenon of sensitized TTA was first reported by Parker and Hatchard in the 1960s [13], there has been extensive research on TTA-UC and its application in solar cells [18, 49–51], photocatalysis [52–54], biological imaging [55, 56], and photodynamic therapy [57, 58]. Despite the extensive exploration of this topic, the chemical kinetics that control the efficiency of the upconversion process remain underappreciated. Here we provide a tutorial to illustrate how reaction kinetics ultimately control most facets of photochemical upconversion and guide efficient design for established and newly conceived upconversion compositions.

Stern-Volmer Analysis

Quenching refers to any process that decreases photoluminescence intensity. Photoluminescence quenching is experienced in various photochemical events, including excited-state reactions, molecular rearrangements, and energy transfer, and can occur through a variety of pathways, including ground-state complex formation and collisional quenching [36]. The TTET process typically occurs via a bimolecular collisional quenching mechanism; therefore collisional quenching will be the focus of this section. In collisional quenching, the molecule in its excited state collides with a quencher through a diffusion process within the lifetime of the excited chromophore. Upon molecular contact, the originally excited molecule can return to the ground state through a non-radiative photophysical process. Collisional quenching of photoluminescence is described by the Stern-Volmer equation:

$$\frac{I_0}{I} = \frac{\tau_0}{\tau} = 1 + k_q\tau_0\,[Q] = 1 + K_{SV}\,[Q] \tag{2.5}$$

In Eq. (2.5), I_0 and I are the photoluminescence intensities in the absence and presence of quencher, respectively; k_q is the bimolecular quenching rate constant; τ_0 is the lifetime of the excited-state fluorophore in the absence of quencher; [Q] is the concentration of the quencher molecule; and K_{sv} is the Stern-Volmer constant, where K_{sv} is defined in Eq. (2.6):

$$K_{SV} = k_q\tau_0 \tag{2.6}$$

Note that Eq. (2.5) is not applicable to static quenching where a non-emissive complex is formed between the excited molecule and the quencher. Static quenching does not decrease the measured photoluminescence lifetime because only the uncomplexed photoluminescent molecules can be observed, and they retain their unquenched lifetime, τ_0. Dynamic Stern-Volmer plots are usually presented as the ratio of I_0/I or τ_0/τ as function of [Q], which is expected to display linear behavior as shown in Fig. 2.5. The slope of a Stern-Volmer plot represents K_{sv}, which allows for the determination of the quenching rate constant k_q by Eq. (2.6), if the τ_0 is known. Since TTET occurs through a diffusion-controlled Dexter mechanism, this energy transfer process can be formally regarded as sensitizer quenching resulting from direct collision with the acceptor. Therefore, k_q is equivalent to the triplet energy transfer rate constant (k_{TTET}), which will be further discussed in the following section.

TTET Rate Constants

The TTET from triplet sensitizer/donor to the triplet state of an acceptor involves a collisional Dexter-type mechanism [59, 60]. When the decay of triplet sensitizer

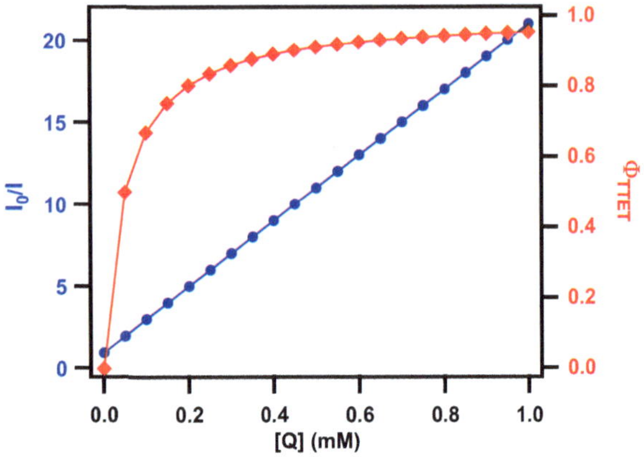

Fig. 2.5 Stern-Volmer plot (in blue dotted points) and TTET efficiency plot (in red diamond points) for quenching of a hypothetical sensitizer with an unquenched lifetime of 20 μs. A linear Stern-Volmer plot yields an intercept of 1 on the *y*-axis and a slope equal to the Stern-Volmer constant $K_{sv} = 2.0 \times 10^4$ M^{-1}, giving $k_{TTET} = 1.0 \times 10^9$ M^{-1} s^{-1} ($K_{sv} = k_{TTET}\tau_0$)

follows first-order kinetics, k_{TTET} can be considered equivalent to the quenching rate constant k_q. The triplet energy transfer efficiency (Φ_{TTET}) can be estimated using the Stern-Volmer equation and varies systematically with the concentration of the acceptor as shown in Eq. (2.7) and Fig. 2.5:

$$\Phi_{TTET} = \left(1 - \frac{I}{I_0}\right) = \left(1 - \frac{\tau}{\tau_0}\right) = \frac{k_{TTET}\tau_0\,[Q]}{1 + k_{TTET}\tau_0\,[Q]} \tag{2.7}$$

Ideally, TTET will occur with near-unity efficiency under strong quenching conditions, meaning the entire population of $^3S^*$ is converted into $^3A^*$ states [59]. To ensure this, all other species which can quench the triplet state of the sensitizer, particularly ground-state dioxygen, must be removed.

Since the quenching of the sensitizer triplet excited state is due to a second-order bimolecular reaction, k_{TTET} is ultimately limited by the diffusion limit k_d, which can be estimated by Eq. (2.8):

$$k_d = \frac{8RT}{3\eta} \tag{2.8}$$

where R is the gas constant, T is the temperature, and η is the solvent viscosity. If quenching is diffusion limited, $k_q \approx k_d$. It should be noted that, to the first-order approximation in Eq. (2.8), k_d is only dependent on the properties of the solvent and no molecular parameters associated with the reacting species are incorporated [61].

The value of the diffusion limit in the solvent of interest provides a useful benchmark for researchers to compare measured rate constants for TTET and a performance target. Because the overall rate of TTET needs to be significantly larger (usually by more than 2 orders of magnitude [59]) than the triplet decay in order to have an effective energy transfer, the yield of TTET can be enhanced by increasing k_q, extending the lifetime of the triplet, or increasing the concentration of quencher. Because the former two are fundamental properties, the latter tends to be the most accessible parameter to change. For example, a solvent viscosity of 1 mPa·s corresponds to $k_d = 7 \times 10^9\ M^{-1}\ s^{-1}$, whereas the rate constant for TTET (k_{TTET}) is typically more than one order of magnitude lower than this value. For a diffusion-controlled $k_{TTET} = 1 \times 10^9\ M^{-1}\ s^{-1}$, a quencher concentration of 0.1 mM enables quenching to become effective for a sensitizer having an excited-state lifetime of 10 μs. Generally, it is difficult to achieve efficient (nearly quantitative) TTET for sensitizers with lifetimes shorter than 1 μs without employing massive corrections for inner filtering and trivial energy transfer, whereas sensitizers possessing long excited-state lifetimes (longer than tens to hundreds of μs) are more suitable for supporting TTA-UC processes where light is produced [59].

TTA Rate Constants

Following TTET, the acceptor triplets can take one of the two competing pathways. This manifests as parallel first- and second-order decay components. The underlying kinetics can be expressed as Eq. (2.9):

$$\frac{d[^3A^*]_t}{dt} = -k_T\left[^3A^*\right]_t - k_{TT}\left[^3A^*\right]_t^2 \tag{2.9}$$

The first-order decay term (k_T) incorporates all of the first-order processes by which the triplet state can decay, including the intrinsic lifetime of the acceptor and pseudo-first-order quenching by dioxygen and other species. The second-order component ($k_{TT}[^3A^*]_t$) represents the triplet quenching by the bimolecular TTA process. The competition between these two kinetic processes results in two distinct kinetic regimes, which will be further discussed in Sect. 2.6. The analytical solution to Eq. (2.9) from Bachilo and Weisman is presented here [43]:

$$\frac{[^3A^*]_t}{[^3A^*]_0} = \frac{1-\beta}{e^{k_T t} - \beta} \tag{2.10}$$

where

$$\beta = \frac{k_{TT}[^3A^*]_0}{k_T + k_{TT}[^3A^*]_0} \tag{2.11}$$

In Eqs. (2.10) and (2.11) above, the dimensionless parameter β runs from 0 to 1 and equals the fraction of $^3A^*$ which decays through the second-order TTA channel. In some studies, α is introduced to represent the product of the TTA rate and the initial concentration of the acceptor triplets, i.e., $\alpha = k_{TT}[^3A^*]_0$; thus $\beta = \alpha/(k_T + \alpha)$.

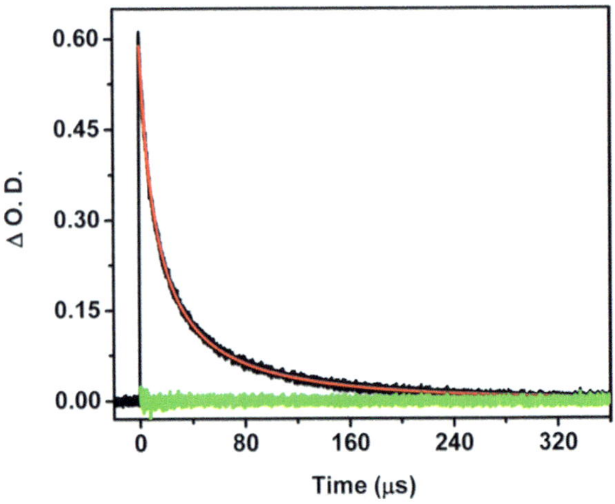

Fig. 2.6 Representative transient absorption kinetic decay of triplet anthracene at 430 nm produced under 547 nm excitation in a mixture of PdOEP and anthracene in degassed toluene. The red line represents the fit to the data in black and the residuals of the fit are provided in green. The figure was adapted from [30]

In order to relate these kinetic components to experimental observables, Eq. (2.10) can be written in terms of transient absorption kinetic parameters. The initial concentration of acceptor triplets $[^3A^*]_0$ can be readily calculated from the triplet decay kinetics, if one has measured the extinction coefficient of the $^3A^*$ state (Eq. 2.12):

$$\Delta A = \frac{\Delta A_0 (1 - \beta)}{e^{k_T t} - \beta} \tag{2.12}$$

Calculation of triplet extinction coefficients will not be discussed here but has been covered elsewhere [62]. A representative example where a transient absorption decay kinetic trace was fit using Eq. (2.12) is given in Fig. 2.6. In this case, anthracene was sensitized by the green-absorbing palladium(II) octaethylporphyrin (PdOEP). The transient absorption kinetic decay at 430 nm, which corresponds to the characteristic absorption of the triplet excited state of anthracene, was acquired to extract the kinetic parameters. ΔA_0 is readily converted into molar concentration, $[^3A^*]_0$, using the extinction coefficient for triplet anthracene (4.2×10^4 M^{-1} cm^{-1} [63]). Fitting the collected kinetic data using Eq. (2.12) yields values for β and k_T, and thus calculation of α and k_{TT}.

2.5 Quadratic-to-Linear Power Dependence

One of the conventional methods to determine if the observed photoluminescence originates from upconversion is to measure the dependence of photoluminescence intensity on the incident light intensity. This power dependence experiment is typically evaluated using steady-state photoluminescence, executed using a wide variety of coherent or noncoherent excitation sources that permit scanning a range of irradiance intensities over many orders of magnitude [64]. In the weak annihilation regime, the upconverted photoluminescence intensity generally shows a quadratic dependence on incident light power [13]. In this regime, the kinetics of TTA-UC are mainly governed by the first-order decay of acceptor triplets. However, the incident power dependence changes from quadratic to linear as the incident power increases above an onset threshold (denoted I_{th}), which is a kinetic regime known as the strong annihilation limit [47, 60, 65]. Above I_{th}, the steady-state population of $^3A^*$ is sufficiently large that bimolecular TTA becomes the dominant pathway and upconversion is linearly proportional to $^3A^*$, and therefore linearly proportional to incident light intensity. The highest possible upconversion quantum efficiency for a specific sensitizer/acceptor pair necessarily occurs in this strong annihilation regime [64, 65]. It should be noted that the I_{th} crossover point for a given donor-acceptor pair is markedly affected by the sensitizer absorption at the excitation wavelength as the overall kinetics can only be as good as the light absorption rate departure point.

The quadratic-to-linear power dependence is represented in Fig. 2.7 and quantitively illustrated in Eq. (2.9). By evaluating Eq. (2.9), two kinetic limits of the $^3A^*$ decay can be determined, since the associated upconverted fluorescence intensity (N_F) is proportional to the square of the $[^3A^*]$ population as shown in Eq. (2.13) [64]:

$$N_F = \int_0^{\infty} I_F(t)\mathrm{d}t = \int_0^{\infty} \Phi_F k_{TT}\left[{}^3A^*\right]_t^2 \tag{2.13}$$

where Φ_F represents the fluorescence quantum yield of the acceptor. Recall from Eqs. (2.10) and (2.11) that there are two competing kinetic processes, k_T, all of the

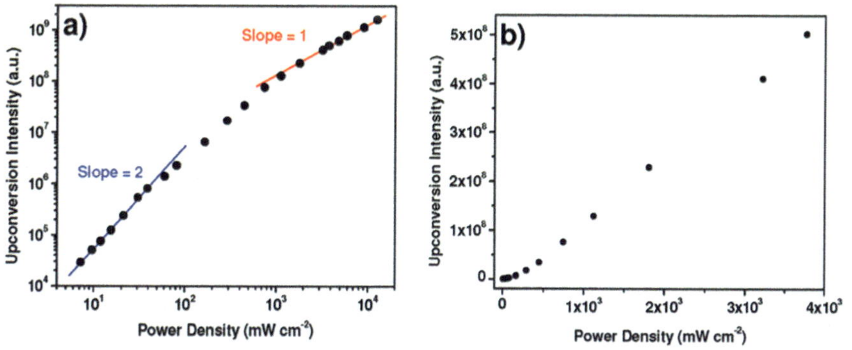

Fig. 2.7 (**a**) Double-logarithmic plot of the upconversion emission signal at 420 ± 6 nm measured as a function of 514.5 nm incident laser power in a mixture of PdOEP and DPA in argon-degassed toluene. The colored lines are linear fits with slopes of 1.0 (red, linear response) and 2.0 (blue, quadratic response) in the high- and low-power regimes, respectively. (**b**) Linear plot of the first 19 data points presented in panel a. The figure was adapted from [64]

unimolecular and pseudo-first-order decay pathways for $[^3A^*]$, and $k_{TT}[^3A^*]$, the second-order TTA of the acceptor triplets [59, 64, 66].

In the weak annihilation limit, where $k_T > k_{TT}[^3A^*]$, Eq. (2.10) reduces to a single exponential decay presented in Eq. (2.14). Upon substitution of Eq. (2.14) into Eq. (2.13) and integration with respect to time, the result is Eq. (2.15):

$$\left[{}^3A^*\right]_t = \left[{}^3A^*\right]_0 e^{-k_T t} \tag{2.14}$$

$$N_F = \frac{\Phi_F k_{TT}\left[{}^3A^*\right]_0^2}{2k_T} \tag{2.15}$$

According to Eq. (2.15), N_F is proportional to $[^3A^*]^2$ and expected to be quadratic with respect to the absorbed optical power. In the strong annihilation limit, where $k_{TT}[^3A^*] > k_T$, Eq. (2.10) can be simplified to Eq. (2.16) and integration with respect to time results in Eq. (2.17):

$$\left[{}^3A^*\right]_t = \frac{\left[{}^3A^*\right]_0}{1 + k_{TT}\left[{}^3A^*\right]_0 t} \tag{2.16}$$

$$N_F = \Phi_F\left[{}^3A^*\right]_0 \tag{2.17}$$

The result in Eq. (2.17) indicates that the total intensity of upconverted emission is linearly proportional to $[^3A^*]$ and thus exhibits a linear dependence on incident light power. This kinetic analysis is consistent with the kinetic treatments of conventional delayed fluorescence performed by Birks and Parker more than 50 years ago [14, 67].

Based on the above discussion about the kinetic of TTA, several general considerations should be made to achieve the highest efficiency. First, the maximum upconversion quantum efficiencies in samples are expected to lie in the strong annihilation limit since the second-order bimolecular TTA is no longer the rate-limiting step with respect to first-order decay of acceptor triplets. Second, any energy-wasting, pseudo-first-order quenching process of the acceptor triplets should be minimized. Therefore, using degassed solvent or oxygen-impermeable host media in the upconversion experiments is key. Third, the excitation wavelengths and the optical densities of the associated sensitizers should be optimized to reach the strong annihilation regime at the lowest possible optical power [64]. The upconversion quantum yield is an important parameter used to evaluate the overall efficiency of each upconversion composition. Photochemical upconversion quantum yield measurements have been detailed in a recently published review article [68] along with recommended best practices, and will not be detailed here.

2.6 Heteromolecular vs. Homomolecular TTA

Up until this point, the discussion has focused on upconverted fluorescence using heteromolecular TTA, in which one molecule serves as a sensitizer and another as the acceptor. A second reported scheme invokes homomolecular TTA, illustrated in the Jablonski diagram shown in Fig. 2.8, in which the sensitizer and acceptor are the same compound. Both mechanisms result in emission at a higher energy than the light that was absorbed, but the main difference between the hetero- and homomolecular TTA mechanism is that TTA emission in the one-component system occurs from an upper excited singlet state (S_n, where $n > 1$). This represents a violation of Kasha's rule, which states that molecules radiate with considerable intensity only from the lowest excited electronic state of a given multiplicity [69].

Homomolecular TTA is rarely observed, owing to several unique challenges relative to heteromolecular TTA. These challenges arise from the fact that singlet excited states above the S_1 manifold typically have very short lifetimes ($\leq 10^{-12}$ s), resulting in low quantum yields of emission from these states ($\leq 10^{-3}$), so it is essential to use a powerful excitation source and a sensitive detection system [70].

Homomolecular TTA processes are rare, and have been observed in a handful of systems such as aromatic thiones, metalloporphyrins, and non-alternant aromatic hydrocarbons (including azulene) [71, 72]. Azulene is the first-described and best-known exception to Kasha's rule [73]; it is a closed-shell hydrocarbon molecule which exhibits its strongest fluorescence from its second excited singlet electronic state S_2 in condensed media at room temperature [74]. As another example, photon upconversion resulting in delayed S_2 fluorescence by homomolecular TTA, $2T_1 \rightarrow S_0 + S_2$, was first observed in metalloporphyrins by Tsvirko, Stel'makh, and coworkers [70, 75], and has also been observed in other molecular families that exhibit appropriate electronic energy gaps and triplet lifetimes [76, 77]. We have recently harnessed the power of homomolecular TTA from Zn(II) tetraphenylporphyrin to activate the free radical polymerization of acrylates using long-wavelength visible light [78].

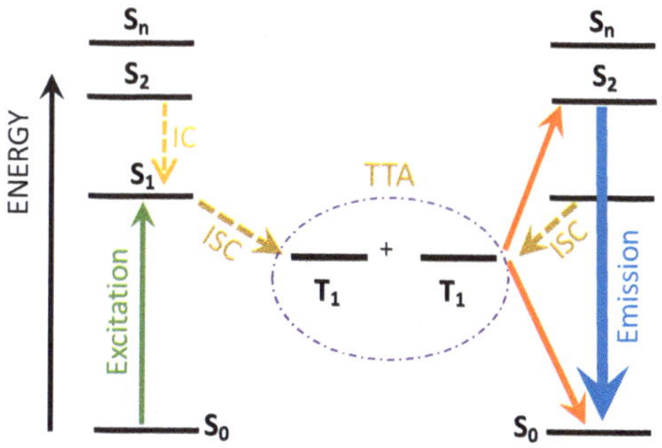

Fig. 2.8 Qualitative Jablonski diagram illustrating homomolecular triplet-triplet annihilation. Singlet states are labeled S_0 (ground state), S_1 (first excited state), and S_2 (second excited state). The lowest excited triplet state is denoted as T_1

2.7 Conclusions

Photon upconversion is a process of converting lower energy incident light into higher energy photons. TTA-UC is a generalized phenomenon as long as all the proper energetic and kinetic requirements are met, and in some ideal cases it can be achieved with low-intensity light and incoherent irradiation. Herein, we have described in detail the rules which govern this process. First, the generation of sensitizer triplets via ISC can be facilitated by spin-orbit coupling due to heavy metal center of sensitizer molecules. This process is followed by two main diffusional bimolecular processes, TTET and TTA. TTET from triplet sensitizers to triplet state of an acceptor occurs through a collisional intermolecular Dexter-type mechanism, and the rate constant k_{TTET} can be extracted from Stern-Volmer analysis. The last process, TTA, is the photophysical process that involves interactions between two excited triplet states of acceptor molecules that undergo two competing quenching pathways: first-order decay and second-order bimolecular TTA process. This dual first-order and second-order decay pathways for acceptor triplets result in a quadratic-to-linear incident light power dependence. Because the overall efficiency of TTA-UC will be the product of the efficiencies of each step, judicious choice of sensitizer and annihilator and careful optimization of conditions are necessary to achieve high-efficiency TTA. The rules laid out in this chapter help to provide a framework by which this can be accomplished in a thoughtful and systematic manner.

References

1. P. Zhang, W. Steelant, M. Kumar, M. Scholfield, Versatile photosensitizers for photodynamic therapy at infrared excitation. J. Am. Chem. Soc. **129**(15), 4526–4527 (2007). https://doi.org/10.1021/ja0700707
2. Q. Liu, T.S. Yang, W. Feng, F.Y. Li, Blue-emissive upconversion nanoparticles for low-power-excited bioimaging in vivo. J. Am. Chem. Soc. **134**(11), 5390–5397 (2012). https://doi.org/10.1021/ja3003638
3. Z.J. Zhang, W.Z. Wang, W.Z. Yin, M. Shang, L. Wang, S.M. Sun, Inducing photocatalysis by visible light beyond the absorption edge: effect of upconversion agent on the photocatalytic activity of Bi_2WO_6. Appl. Catal. B Environ. **101**(1–2), 68–73 (2010). https://doi.org/10.1016/j.apcatb.2010.09.008
4. T.N. Singh-Rachford, J. Lott, C. Weder, F.N. Castellano, Influence of temperature on low-power upconversion in rubbery polymer blends. J. Am. Chem. Soc. **131**(33), 12007–12014 (2009). https://doi.org/10.1021/ja904696n
5. T. Miteva, V. Yakutkin, G. Nelles, S. Baluschev, Annihilation assisted upconversion: all-organic, flexible and transparent multicolour display. New J. Phys. **10**, 103002 (2008). https://doi.org/10.1088/1367-2630/10/10/103002
6. T.N. Singh-Rachford, F.N. Castellano, Photon upconversion based on sensitized triplet-triplet annihilation. Coord. Chem. Rev. **254**(21–22), 2560–2573 (2010). https://doi.org/10.1016/j.ccr.2010.01.003
7. A. Monguzzi, R. Tubino, S. Hoseinkhani, M. Campione, F. Meinardi, Low power, non-coherent sensitized photon up-conversion: modelling and perspectives. Phys. Chem. Chem. Phys. **14**(13), 4322–4332 (2012). https://doi.org/10.1039/c2cp23900k

8. R.R. Islangulov, F.N. Castellano, Photochemical upconversion: anthracene dimerization sensitized to visible light by a Ru-II chromophore. Angew. Chem. Int. Ed. **45**(36), 5957–5959 (2006). https://doi.org/10.1002/anie.200601615
9. W. Shockley, H.J. Queisser, Detailed balance limit of efficiency of p-*n* junction solar cells. J. Appl. Phys. **32**(3), 510–519 (1961). https://doi.org/10.1063/1.1736034
10. M.J.Y. Tayebjee, D.R. McCamey, T.W. Schmidt, Beyond Shockley–Queisser: molecular approaches to high-efficiency photovoltaics. J. Phys. Chem. Lett. **6**(12), 2367–2378 (2015). https://doi.org/10.1021/acs.jpclett.5b00716
11. T. Trupke, M.A. Green, P. Wurfel, Improving solar cell efficiencies by up-conversion of sub-band-gap light. J. Appl. Phys. **92**(7), 4117–4122 (2002). https://doi.org/10.1063/1.1505677
12. Y.Y. Cheng, A. Nattestad, T.F. Schulze, R.W. MacQueen, B. Fuckel, K. Lips, G.G. Wallace, T. Khoury, M.J. Crossley, T.W. Schmidt, Increased upconversion performance for thin film solar cells: a trimolecular composition. Chem. Sci. **7**(1), 559–568 (2016). https://doi.org/10.1039/c5sc03215f
13. C.A. Parker, C.G. Hatchard, Sensitized anti-stokes delayed fluorescence, in *Proceedings of the Chemical Society of London (DEC)* (1962), pp. 386–387
14. J.B. Birks, *Photophysics of Aromatic Molecules* (Wiley-Interscience, London; New York, 1970)
15. C.A. Parker, Sensitized P-type delayed fluorescence. Proc. R. Soc. Lond. Ser. A Math. Phys. Sci. **276**(1364), 125–135 (1963). https://doi.org/10.1098/rspa.1963.0197
16. C.A. Parker, T.A. Joyce, Delayed fluorescence and some properties of the chlorophyll triplets. Photochem. Photobiol. **6**(6), 395–406 (1967). https://doi.org/10.1111/j.1751-1097.1967.tb08886.x
17. M.K. Manna, S. Shokri, G.P. Wiederrecht, D.J. Gosztola, A.J.L. Ayitou, New perspectives for triplet-triplet annihilation based photon upconversion using all-organic energy donor & acceptor chromophores. Chem. Commun. **54**(46), 5809–5818 (2018). https://doi.org/10.1039/c8cc01553h
18. T.F. Schulze, T.W. Schmidt, Photochemical upconversion: present status and prospects for its application to solar energy conversion. Energy Environ. Sci. **8**(1), 103–125 (2015). https://doi.org/10.1039/c4ee02481h
19. A. Nattestad, Y.Y. Cheng, R.W. MacQueen, G.G. Wallace, T.W. Schmidt, Integrating a triplet-triplet annihilation up-conversion system to enhance dye-sensitized solar cell response to sub-bandgap light. J. Vis. Exp. (91), 52028 (2014). https://doi.org/10.3791/52028
20. F.N. Castellano, C.E. McCusker, MLCT sensitizers in photochemical upconversion: past, present, and potential future directions. Dalton Trans. **44**(41), 17906–17910 (2015). https://doi.org/10.1039/c5dt03212a
21. T.N. Singh-Rachford, F.N. Castellano, Low power visible-to-UV upconversion. J. Phys. Chem. A **113**(20), 5912–5917 (2009). https://doi.org/10.1021/jp9021163
22. W. Zhao, F.N. Castellano, Upconverted emission from pyrene and di-tert-butylpyrene using Ir(ppy)(3) as triplet sensitizer. J. Phys. Chem. A **110**(40), 11440–11445 (2006). https://doi.org/10.1021/jp064261s
23. S.K. Sugunan, U. Tripathy, S.M.K. Brunet, M.F. Paige, R.P. Steer, Mechanisms of low-power noncoherent photon upconversion in metalloporphyrin-organic blue emitter systems in solution. J. Phys. Chem. A **113**(30), 8548–8556 (2009). https://doi.org/10.1021/jp9034776
24. T.N. Singh-Rachford, F.N. Castellano, Triplet sensitized red-to-blue photon upconversion. J. Phys. Chem. Lett. **1**(1), 195–200 (2010). https://doi.org/10.1021/jz900170m
25. P.B. Merkel, J.P. Dinnocenzo, Low-power green-to-blue and blue-to-UV upconversion in rigid polymer films. J. Lumin. **129**(3), 303–306 (2009). https://doi.org/10.1016/j.jlumin.2008.10.013
26. R.R. Islangulov, J. Lott, C. Weder, F.N. Castellano, Noncoherent low-power upconversion in solid polymer films. J. Am. Chem. Soc. **129**(42), 12652–12653 (2007). https://doi.org/10.1021/ja075014k
27. A. Monguzzi, R. Tubino, F. Meinardi, Multicomponent polymeric film for red to green low power sensitized up-conversion. J. Phys. Chem. A **113**(7), 1171–1174 (2009). https://doi.org/10.1021/jp809971u

28. P.W. Atkins, R.S. Friedman, *Molecular Quantum Mechanics* (OUP, Oxford, 2011)
29. E.U. Condon, G.H. Shortley, C.U. Press, *The Theory of Atomic Spectra* (Cambridge University Press, Cambridge, 1951)
30. F. Deng, J. Blumhoff, F.N. Castellano, Annihilation limit of a visible-to-UV photon upconversion composition ascertained from transient absorption kinetics. J. Phys. Chem. A **117**(21), 4412–4419 (2013). https://doi.org/10.1021/jp4022618
31. T. Förster, Energy migration and fluorescence. J. Biomed. Opt. **17**(1), 011002 (2012)
32. D.L. Dexter, A theory of sensitized luminescence in solids. J. Chem. Phys. **21**(5), 836–850 (1953). https://doi.org/10.1063/1.1699044
33. S.M. Bai, P. Zhang, P. Antoniou, S.S. Skourtis, D.N. Beratan, Quantum interferences among Dexter energy transfer pathways. Faraday Discuss. **216**, 301–318 (2019). https://doi.org/10.1039/c9fd00007k
34. T. Forster, *ZWISCHENMOLEKULARE ENERGIEWANDERUNG UND FLUORESZENZ. Ann. Phys. **2**(1–2), 55–75 (1948). https://doi.org/10.1002/andp.19484370105
35. S.J. Jang, K.J. Shin, S.Y. Lee, Effects of excitation migration and translational diffusion in the luminescence quenching dynamics. J. Chem. Phys. **102**(2), 815–827 (1995). https://doi.org/10.1063/1.469196
36. J. Lakowicz, *Principles of Fluorescence Spectroscopy*, 3rd edn. (Springer, Boston, 2006)
37. C.B. Murphy, Y. Zhang, T. Troxler, V. Ferry, J.J. Martin, W.E. Jones, Probing Forster and Dexter energy-transfer mechanisms in fluorescent conjugated polymer chemosensors. J. Phys. Chem. B **108**(5), 1537–1543 (2004). https://doi.org/10.1021/jp0301406
38. K. Sudha, S. Sundharamurthi, S. Karthikaikumar, K. Abinaya, P. Kalimuthu, Switching of Forster to Dexter mechanism of short-range energy transfer in meso-anthrylporphyrin. J. Phys. Chem. C **121**(11), 5941–5948 (2017). https://doi.org/10.1021/acs.jpcc.6b13042
39. B. Ventura, F. Barigelletti, F. Lodato, D.L. Officer, L. Flamigni, Energy transfer processes in electronically coupled porphyrin hetero-dyads connected at the beta position. Phys. Chem. Chem. Phys. **11**(13), 2166–2176 (2009). https://doi.org/10.1039/b819138g
40. B. Nickel, Delayed fluorescence from upper excited singlet-states SN(n>1) of aromatic-hydrocarbons 1,2-benzathracene, fluoranthene, pyrene, and chrysene in methylcyclohexane. Helv. Chim. Acta **61**(1), 198–222 (1978). https://doi.org/10.1002/hlca.19780610118
41. J.L. Charlton, R. Dabestani, J. Saltiel, Role of triplet-triplet annihilation in anthracene dimerization. J. Am. Chem. Soc. **105**(11), 3473–3476 (1983). https://doi.org/10.1021/ja00349a017
42. A.J. McLean, T.G. Truscott, Efficiency of triplet-photosensitised singlet oxygen generation in benzene. J. Chem. Soc. Faraday Trans. **86**(14), 2671–2672 (1990). https://doi.org/10.1039/ft9908602671
43. S.M. Bachilo, R.B. Weisman, Determination of triplet quantum yields from triplet-triplet annihilation fluorescence. J. Phys. Chem. A **104**(33), 7711–7714 (2000). https://doi.org/10.1021/jp001877n
44. P.P. Levin, S.M.B. Costa, T.G. Nunes, L.F.V. Ferreira, L.M. Ilharco, A.M.B. do Rego, Kinetics of triplet-triplet annihilation of tetraphenylporphyrin in liquid and frozen films of decanol on the external surface of zeolite. Fast probe diffusion in monolayers and polycrystals. J. Phys. Chem. A **107**(3), 328–336 (2003). https://doi.org/10.1021/jp021635i
45. V. Jankus, E.W. Snedden, D.W. Bright, V.L. Whittle, J.A.G. Williams, A. Monkman, Energy upconversion via triplet fusion in super yellow PPV films doped with palladium tetraphenyltetrabenzoporphyrin: a comprehensive investigation of exciton dynamics. Adv. Funct. Mater. **23**(3), 384–393 (2013). https://doi.org/10.1002/adfm.201201284
46. J. Saltiel, B.W. Atwater, Spin-statistical factors in diffusion-controlled reactions. Adv. Photochem. **14**(14), 1–90 (1988). https://doi.org/10.1002/9780470133446.ch1
47. Y.Y. Cheng, T. Khoury, R. Clady, M.J.Y. Tayebjee, N.J. Ekins-Daukes, M.J. Crossley, T.W. Schmidt, On the efficiency limit of triplet-triplet annihilation for photochemical upconversion. Phys. Chem. Chem. Phys. **12**(1), 66–71 (2010). https://doi.org/10.1039/b913243k
48. Y.Y. Cheng, B. Fuckel, T. Khoury, R. Clady, M.J.Y. Tayebjee, N.J. Ekins-Daukes, M.J. Crossley, T.W. Schmidt, Kinetic analysis of photochemical upconversion by triplet-triplet annihilation: beyond any spin statistical limit. J. Phys. Chem. Lett. **1**(12), 1795–1799 (2010). https://doi.org/10.1021/jz100566u

49. R.S. Khnayzer, J. Blumhoff, J.A. Harrington, A. Haefele, F. Deng, F.N. Castellano, Upconversion-powered photoelectrochemistry. Chem. Commun. **48**(2), 209–211 (2012). https://doi.org/10.1039/c1cc16015j
50. L. Nienhaus, J.P. Correa-Baena, S. Wieghold, M. Einzinger, T.A. Lin, K.E. Shulenberger, N.D. Klein, M.F. Wu, V. Bulovic, T. Buonassisi, M.A. Baldo, M.G. Bawendi, Triplet-sensitization by lead halide perovskite thin films for near-infrared-to-visible upconversion. ACS Energy Lett. **4**(4), 888–895 (2019). https://doi.org/10.1021/acsenergylett.9b00283
51. D. Beery, J.P. Wheeler, A. Arcidiacono, K. Hanson, CdSe quantum dot sensitized molecular photon upconversion solar cells. ACS Appl. Energy Mater. **3**(1), 29–37 (2020). https://doi.org/10.1021/acsaem.9b01765
52. B.D. Ravetz, A.B. Pun, E.M. Churchill, D.N. Congreve, T. Rovis, L.M. Campos, Photoredox catalysis using infrared light via triplet fusion upconversion. Nature **565**(7739), 343–346 (2019). https://doi.org/10.1038/s41586-018-0835-2
53. B. Pfund, D.M. Steffen, M.R. Schreier, M.S. Bertrams, C. Ye, K. Borjesson, O.S. Wenger, C. Kerzig, UV light generation and challenging photoreactions enabled by upconversion in water. J. Am. Chem. Soc. **142**(23), 10468–10476 (2020). https://doi.org/10.1021/jacs.0c02835
54. M. Majek, U. Faltermeier, B. Dick, R. Perez-Ruiz, A. Jacobi von Wangelin, Application of visible-to-UV photon upconversion to photoredox catalysis: the activation of aryl bromides. Chem Eur J **21**(44), 15496–15501 (2015). https://doi.org/10.1002/chem.201502698
55. Y.I. Park, K.T. Lee, Y.D. Suh, T. Hyeon, Upconverting nanoparticles: a versatile platform for wide-field two-photon microscopy and multi-modal in vivo imaging. Chem. Soc. Rev. **44**(6), 1302–1317 (2015). https://doi.org/10.1039/c4cs00173g
56. Y.J. Liu, Y.Q. Lu, X.S. Yang, X.L. Zheng, S.H. Wen, F. Wang, X. Vidal, J.B. Zhao, D.M. Liu, Z.G. Zhou, C.S. Ma, J.J. Zhou, J.A. Piper, P. Xi, D.Y. Jin, Amplified stimulated emission in upconversion nanoparticles for super-resolution nanoscopy. Nature **543**(7644), 229–232 (2017). https://doi.org/10.1038/nature21366
57. A. Punjabi, X. Wu, A. Tokatli-Apollon, M. El-Rifai, H. Lee, Y.W. Zhang, C. Wang, Z. Liu, E.M. Chan, C.Y. Duan, G. Han, Amplifying the red-emission of upconverting nanoparticles for biocompatible clinically used prodrug-induced photodynamic therapy. ACS Nano **8**(10), 10621–10630 (2014). https://doi.org/10.1021/nn505051d
58. S.H.C. Askes, A. Bahreman, S. Bonnet, Activation of a photodissociative ruthenium complex by triplet-triplet annihilation upconversion in liposomes. Angew. Chem. Int. Ed. **53**(4), 1029–1033 (2014). https://doi.org/10.1002/anie.201309389
59. T.W. Schmidt, F.N. Castellano, Photochemical upconversion: the primacy of kinetics. J. Phys. Chem. Lett. **5**(22), 4062–4072 (2014). https://doi.org/10.1021/jz501799m
60. A. Monguzzi, J. Mezyk, F. Scotognella, R. Tubino, F. Meinardi, Upconversion-induced fluorescence in multicomponent systems: steady-state excitation power threshold. Phys. Rev. B **78**, 195112 (2008). https://doi.org/10.1103/PhysRevB.78.195112
61. K.S. Schmitz, Chapter 15: Reaction rates and mechanisms, in *Physical Chemistry*, ed. by K. S. Schmitz, (Elsevier, Boston, 2017), pp. 695–760. https://doi.org/10.1016/B978-0-12-800514-9.00015-8
62. T.N. Singh-Rachford, F.N. Castellano, Supra-nanosecond dynamics of a red-to-blue photon upconversion system. Inorg. Chem. **48**(6), 2541–2548 (2009). https://doi.org/10.1021/ic802114d
63. T.N. Singh-Rachford, A. Nayak, M.L. Muro-Small, S. Goeb, M.J. Therien, F.N. Castellano, Supermolecular-chromophore-sensitized near-infrared-to-visible photon upconversion. J. Am. Chem. Soc. **132**(40), 14203–14211 (2010). https://doi.org/10.1021/ja105510k
64. A. Haefele, J. Blumhoff, R.S. Khnayzer, F.N. Castellano, Getting to the (square) root of the problem: how to make noncoherent pumped upconversion linear. J. Phys. Chem. Lett. **3**(3), 299–303 (2012). https://doi.org/10.1021/jz300012u
65. J.E. Auckett, Y.Y. Chen, T. Khoury, R.G.C.R. Clady, N.J. Ekins-Daukes, M.J. Crossley, T.W. Schmidt, Efficient up-conversion by triplet-triplet annihilation. J. Phys. Conf. Ser. **185**, 012002 (2009). https://doi.org/10.1088/1742-6596/185/1/012002

66. C.E. McCusker, F.N. Castellano, Efficient visible to near-UV photochemical upconversion sensitized by a long lifetime Cu(I) MLCT complex. Inorg. Chem. **54**(12), 6035–6042 (2015). https://doi.org/10.1021/acs.inorgchem.5b00907
67. H. Stegemeyer, Book review: photoluminescence of solutions. By C. A. Parker. Angew. Chem. Int. Ed. Engl. **8**(12), 996–997 (1969). https://doi.org/10.1002/anie.196909963
68. Y. Zhou, F.N. Castellano, T.W. Schmidt, K. Hanson, On the quantum yield of photon upconversion via triplet-triplet annihilation. ACS Energy Lett. **5**(7), 2322–2326 (2020). https://doi.org/10.1021/acsenergylett.0c01150
69. A.P. Demchenko, V.I. Tomin, P.T. Chou, Breaking the Kasha rule for more efficient photochemistry. Chem. Rev. **117**(21), 13353–13381 (2017). https://doi.org/10.1021/acs.chemrev.7b00110
70. G.F. Stelmakh, M.P. Tsvirko, Delayed fluorescence from the upper excited electronic states of metalloporphyrins. Opt. Spectrosc. **49**, 278–281 (1980)
71. A. Maciejewski, R.P. Steer, The photophysics, physical photochemistry, and related spectroscopy of thiocarbonyls. Chem. Rev. **93**(1), 67–98 (1993). https://doi.org/10.1021/cr00017a005
72. T. Itoh, Fluorescence and phosphorescence from higher excited states of organic molecules. Chem. Rev. **112**(8), 4541–4568 (2012). https://doi.org/10.1021/cr200166m
73. M. Kasha, Characterization of electronic transitions in complex molecules. Discuss. Faraday Soc. **9**, 14–19 (1950). https://doi.org/10.1039/df9500900014
74. N. Tetreault, R.S. Muthyala, R.S.H. Liu, R.P. Steer, Control of the photophysical properties of polyatomic molecules by substitution and solvation: the second excited singlet state of azulene. J. Phys. Chem. A **103**(15), 2524–2531 (1999). https://doi.org/10.1021/jp984407q
75. G.F. Stelmakh, M.P. Tsvirko, Metalloporphyrin dimeric S_2 emission resulting from triplet-triplet annihilation. ACS Symp. Ser. **321**, 118–127 (1986)
76. H. Eisenberger, B. Nickel, A.A. Ruth, R.P. Steer, Photophysical triplet state processes of 4-H-1-benzopyrane-4-thione in a perfluoroalkane. 2. Delayed S-2->S-0 fluorescence due to triplet-triplet annihilation. J. Chem. Soc. Faraday Trans. **92**(5), 741–746 (1996). https://doi.org/10.1039/ft9969200741
77. R.P. Steer, Electronic energy-pooling processes in polyatomic-molecules. Res. Chem. Intermed. **12**(1), 81–101 (1989). https://doi.org/10.1163/156856789x00186
78. N. Awwad, A.T. Bui, E.O. Danilov, F.N. Castellano, Visible-light-initiated free-radical polymerization by homomolecular triplet-triplet annihilation. Chem **6**(11), 1–15 (2020). https://doi.org/10.1016/j.chempr.2020.08.019

Chapter 3
Near-Infrared-to-Visible Photon Upconversion

Yoichi Sasaki, Nobuhiro Yanai, and Nobuo Kimizuka

Abstract One of the promising methods to overcome the Shockley-Queisser limit in solar energy conversion is triplet-triplet annihilation-based photon upconversion (TTA-UC) from near-infrared (NIR, λ > 700 nm) light to visible (Vis, λ < 700 nm) light. However, it had been difficult to achieve efficient NIR-to-Vis TTA-UC mainly due to the absence of appropriate triplet sensitizers with less or no energy loss associated with intersystem crossing (ISC). In this chapter, we overview recent successful examples of NIR-to-Vis TTA-UC based on the developments of new NIR-absorbing triplet sensitizers, such as semiconductor nanocrystals with small singlet-triplet exchange splitting and Os complexes with direct singlet-to-triplet (S-T) absorption.

Keywords Quantum dots · Perovskites · Os complex · S-T absorption · Optogenetics

3.1 Introduction

To exceed the Shockley-Queisser limit in solar energy conversion, it is imperative to develop near-infrared (NIR, λ > 700 nm)-to-visible (Vis, λ < 700 nm) photon upconversion (UC) materials [1–3]. As a means to perform efficient UC under solar irradiance (~mW/cm^2), triplet-triplet annihilation-based UC (TTA-UC) has attracted growing attention [4–10]. However, there exist outstanding problems in NIR-to-Vis TTA-UC, notably the absence of appropriate triplet sensitizers with NIR absorption. Conventional sensitizers such as porphyrins and phthalocyanines lose hundreds of meV during the intersystem crossing (ISC) (Fig. 3.1a) [11]. This energy loss significantly reduces the anti-Stokes shift, which limits the performance

Y. Sasaki · N. Yanai (✉) · N. Kimizuka (✉)
Department of Chemistry and Biochemistry, Kyushu University, Fukuoka, Japan
e-mail: yanai@mail.cstm.kyushu-u.ac.jp; n-kimi@mail.cstm.kyushu-u.ac.jp

J. S. Lissau, M. Madsen (eds.), *Emerging Strategies to Reduce Transmission and Thermalization Losses in Solar Cells*, https://doi.org/10.1007/978-3-030-70358-5_3

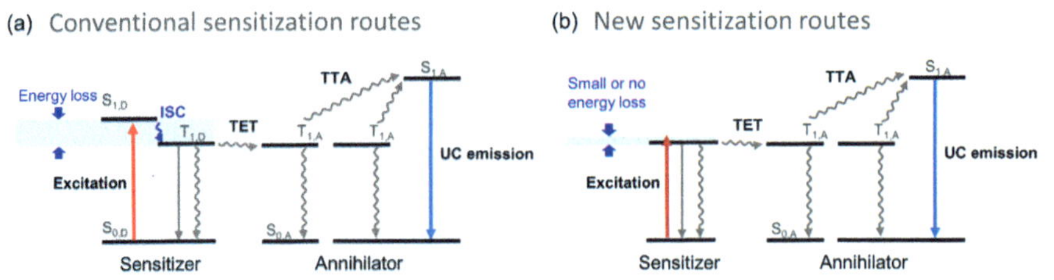

Fig. 3.1 (**a**) Conventional TTA-UC scheme that involves energy loss in the course of ISC from the excited singlet state (S_1) to triplet state (T_1). (**b**) New TTA-UC scheme with small or no energy loss due to ISC

of the TTA-UC system. Therefore, the development of NIR sensitizers with less or no energy loss due to ISC is indispensable.

In recent years, two major approaches have been developed to harvest and upconvert the NIR light. One is to reduce the S_1-T_1 energy gap, and the other is to permit a direct singlet-to-triplet (S-T) transition (Fig. 3.1b) [12–16]. These sensitizers are required to satisfy wide tunability of the absorption wavelength from visible to NIR, and also applicable to NIR-to-Vis TTA-UC in solutions as well as in solid state [12, 15–17]. As it turned out that the relatively faster excited-state deactivation of these sensitizers reduces the UC efficiency, recent efforts have been focused on the control of aggregation structures and sensitizer-emitter conjugation to solve the problem [18–24]. In this chapter, we introduce newly developed NIR-absorptive sensitizers for NIR-to-Vis TTA-UC and discuss the design principles for sensitizing the emitter triplets efficiently. It provides a new perspective to create efficient solid-state NIR-to-Vis TTA-UC that would boost the efficiency of sunlight-powered devices.

3.2 Semiconductor Nanocrystals as Triplet Sensitizers

Colloidal semiconductor nanocrystals (NCs) show unique optical features derived from quantum confinement effect—when the crystal size becomes comparable to the exciton Bohr radius, the bandgap becomes discrete and larger than that of the bulk semiconductors. Since the bandgap depends on the size of NCs, the absorption range can be synthetically tuned by controlling the growth of NCs (Fig. 3.2a) [25].

The idea of colloidal NC triplet sensitizers comes from the NC photoluminescence sensitized by dark triplets. Rao et al. and Bawendi and Baldo et al. observed efficient triplet energy transfer from thermally deposited acenes to semiconductor NCs [26, 27]. Transient absorption and photoluminescence studies revealed that the triplet energy produced via singlet fission in the acene films could be transferred to the NCs (Fig. 3.2b), which enhanced their emission intensity. They found that the triplet energy transfer (TET) process depends on the length of ligands covering the NC surfaces and on the energy gap between the bandgap of NCs and acene triplets,

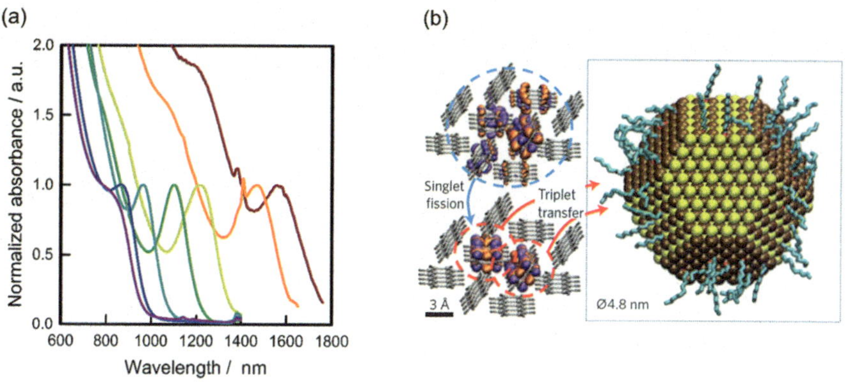

Fig. 3.2 (**a**) Absorption spectra of PbS NCs with different sizes, demonstrating tunable exciton peak absorption wavelengths. Adopted with permission from [25]. Copyright 2017, Royal Society of Chemistry. (**b**) Schematic of triplet exciton transfer from tetracene to a PbS NC with aliphatic acid ligands. Adopted with permission from [27]. Copyright 2014, Springer Nature

which is typical for the Dexter-type energy transfer mechanism. These reports drove the interest of other researchers to the opposite process, i.e., TET from NCs to organic dyes. Importantly, the singlet-triplet exchange splitting at the band edge of the NCs is as small as 1–25 meV because of the small electron-hole exchange interactions as compared with the value in molecules (~1 eV) [20, 28]. Therefore, when the NCs are used as triplet sensitizers, quite small energy loss is expected during ISC.

The triplet sensitization with NIR-absorptive NCs has been recently utilized for NIR-to-Vis TTA-UC. Tang, Bardeen, and co-workers sensitized rubrene triplets by PbSe NCs in degassed toluene solution (Fig. 3.3) [12]. Upon excitation with NIR light ($\lambda_{ex} = 980$ nm), a yellow UC emission was observed ($\lambda_{em} = 568$ nm). This large anti-Stokes shift of 0.92 eV was obtained thanks to the small energy loss during the triplet sensitization. The UC emission intensity showed a quadratic dependence on excitation intensity and a lifetime of delayed fluorescence that reflected the decay of the rubrene triplet, as observed in conventional TTA-UC systems. However, the efficiency of NIR-to-Vis TTA-UC was low ($\eta_{UC} = 0.01\%$ among 100% maximum). Deactivation pathways in NCs such as defects on the NC surface, thermal relaxation, and radiative recombination seem to have contributed to dropping the sensitization efficiency [29].

To improve the efficiency of TET from the NCs to acceptors in solution, it is important to modify the NC surface with triplet-accepting chromophores since the NC surface is usually covered with aliphatic ligands and the excited-state lifetime of NCs ($\tau < 1$ μs) [30] is shorter than that of benchmark porphyrin-based sensitizers (e.g., τ_p ~ 42 μs for platinum(II) tetraphenyltetrabenzoporphyrin) [31]. Tang, Bardeen, and co-workers used a combination of CdSe NCs and 9,10-diphenyl anthracene (DPA) as a model sensitizer-emitter system. They introduced 9-anthracene carboxylic acid (9-ACA) to the surface of CdSe NCs, which served

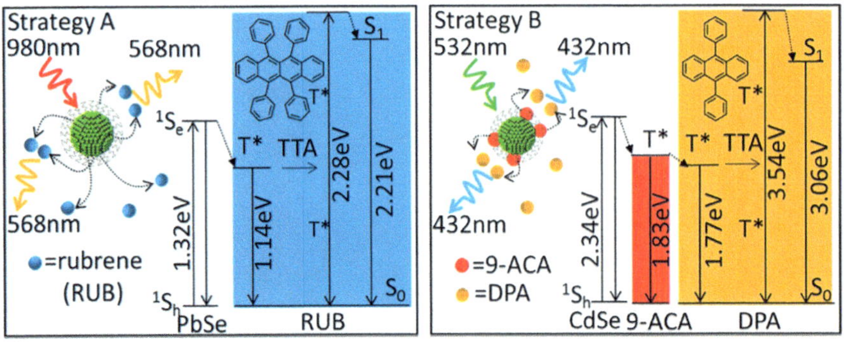

Fig. 3.3 Schematic illustrations for TTA-UC combining PbSe NCs with rubrene and 9-ACA-modified CdSe NCs with DPA. Adopted with permission from [12]. Copyright 2015, American Chemical Society

as a triplet transmitting unit to DPA in solution (Fig. 3.3) [12]. The attachment of 9-ACA dramatically enhanced UC emission from DPA, and a high UC efficiency of $9 \pm 2\%$ was obtained.

The direct evidence of the TET process from CdSe to 9-ACA was reported by Castellano and co-workers [18]. Their transient absorption results confirmed the direct TET from the selectively excited CdSe NCs to the surface-anchored ACA chromophores without any intermediates. The interfacial TET underwent with rate constants between 2.0×10^9 s^{-1} and 2.8×10^9 s^{-1}, which were much faster than the excited CdSe decay rate ($k \sim 3.0 \times 10^7$ s^{-1}) and resulted in a high TET efficiency ($\Phi_{\mathrm{TET}} = 0.92$).

The distance between the NC and the triplet transmitter is also an important factor in obtaining good TET efficiencies. Tang and co-workers changed the distance from CdSe NCs to attached anthracene derivatives by introducing *p*-phenylene bridges and evaluated the TET rate [32]. When the length of the *p*-phenylene spacer was increased, the TET rate decreased exponentially, and the data showed a good agreement with the Dexter-type TET model ($k_{\mathrm{et}} = k_0 \exp(-\beta d)$, $\beta = 0.43 \pm 0.07$ Å^{-1}) (Fig. 3.4a, b).

It is to note that, when the energy levels of the attached ligands are close to those of NCs, the net excited-state emission decay profiles are affected by the back TET processes from the ligands to NCs. Castellano and co-workers attached 1-pyrenecarboxylic acid (PCA) on CdSe NCs with different sizes and investigated the reverse TET from PCA to CdSe NCs [33]. Interestingly, in the case of CdSe-PCA with the energy gap below 0.54 eV, the reverse TET occurred, and the NCs showed thermally activated delayed photoluminescence. This feature was explained by the excited-state thermal equilibrium, as observed for purely molecular based systems [34–37]. To promote TTA-UC, it is necessary to optimize the energy gap to make the back TET rate slow enough and to achieve efficient TET to annihilator molecules. Also, the lifetime of the excited state can be elongated over tens of microsecond timescale (>10 μs), which is beneficial for quantitative TET in solution [33, 38].

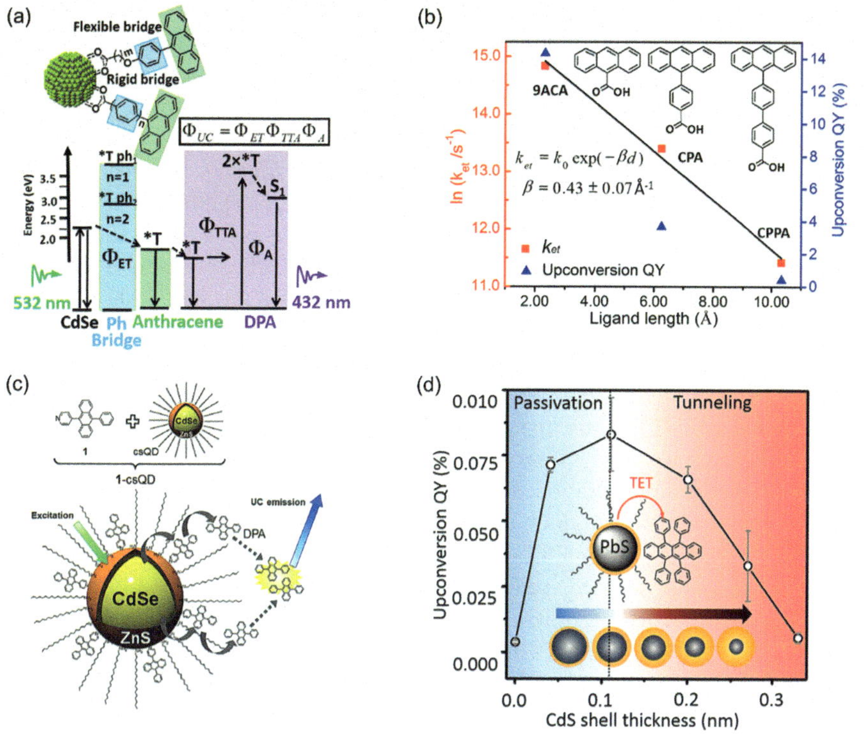

Fig. 3.4 (**a**) Schematic illustration of TTA-UC with CdSe-bridge-anthracene sensitizers. (**b**) Distance dependence of the TET rate from CdSe NC to anthracene (red squares) and the maximum UC efficiency (blue triangles). Adopted with permission from [32]. Copyright 2016, American Chemical Society. (**c**) Schematic illustration of the TTA-UC system using transmitter-modified CdSe/ZnS core-shell NCs. Adopted with permission from [41]. Copyright 2016, John Wiley & Sons. (**d**) Dependence of the UC efficiency on CdS shell thickness for TTA-UC with rubrene-PbS/CdS in toluene. Adopted with permission from [19]. Copyright 2016, American Chemical Society

Another possible deactivation in the NC sensitizers is due to the trap sites at the NC surface. In general, this deactivation can be suppressed by covering the NCs with other semiconductors having larger bandgaps [39, 40]. The Yanai and Kimizuka group employed CdSe/ZnS core-shell NCs to passivate the trap and modified the surface with transmitter ligand 4-(10-phenylanthracene-9-yl)pyridine (PAPy) (Fig. 3.4c) [41]. When the core-shell sensitizer was combined with DPA, UC efficiencies improved compared with that with core-only CdSe NCs. The excitation intensity threshold I_{th}, where the excitation intensity dependence of TTA-UC emission intensity undergoes a transition from quadratic to linear, also showed a decrease for the CdSe/ZnS core-shell NCs compared with the core-only CdSe NCs. The detailed investigations on shell structures for efficient TET and TTA have been reported by Tang and co-workers (Fig. 3.4d) [19, 42]. They synthesized PbS/CdS core-shell NCs with varied thickness of the CdS shell. When the shell thickness was increased, the passivation of the surface trap states became more effective, thus leading to the improved UC efficiency. Since TET depends on the wave function overlap between the NC and triplet transmitters, a further increase of shell thickness caused the reduction of TET efficiency.

Synthetic protocols for NCs also exert considerable impacts on the surface states and the resultant photochemical properties. Tang and Lian synthesized PbS NCs by using highly purified lead and thiourea precursors according to the method reported by Owen et al. [43] and modified the surface with 5-carboxylic acid tetracene (5-CT) [44]. The PbS NCs synthesized from highly purified precursors showed a long triplet lifetime of 2.86 μs for the surface-anchored 5-CT, which is comparable to that observed for 5-CT on PbS/CdS core-shell NCs (2.70 μs). Meanwhile, the PbS/5-CT NCs showed 2.5 times faster TET ($k_{et} = 2.56 \times 10^9\ s^{-1}$) as compared to the core-shell PbS/CdS/5-CT. Thus, the elimination of the insulating shell layer has been shown to improve the TET process. When the PbS/5-CT NCs were mixed with rubrene, continuous excitation at NIR ($\lambda_{ex} = 781$ nm) resulted in a yellow emission ($\lambda_{em} = 560$ nm) with a high UC efficiency η_{UC} of $11.8 \pm 1.1\%$.

Solid-state NIR-to-Vis TTA-UC, which is vital for solar device applications, has been achieved with NC sensitizers. Bulović, Bawendi, Baldo, and co-workers developed solid-state TTA-UC devices based on PbS NCs and rubrene (Fig. 3.5a) [17]. They spread the PbS NCs on a substrate and deposited rubrene through thermal evaporation. Rubrene has a small energy difference between the involved states ($S_1 = 2.23$ eV, $2 \times T_1 = 2.28$ eV), and their aggregation causes the decrease of the fluorescence quantum yield due to singlet fission which converts one excited singlet (S_1) to two excited triplets ($2 \times T_1$) [45]. To circumvent this process, they doped 0.5 vol% dibenzotetraphenylperiflanthene (DBP) into rubrene films to collect singlet energy to emissive DBP. Upon NIR excitation ($\lambda_{ex} = 808$ nm), UC emission from DBP ($\lambda_{em} = 612$ nm) was observed with an intrinsic UC quantum efficiency

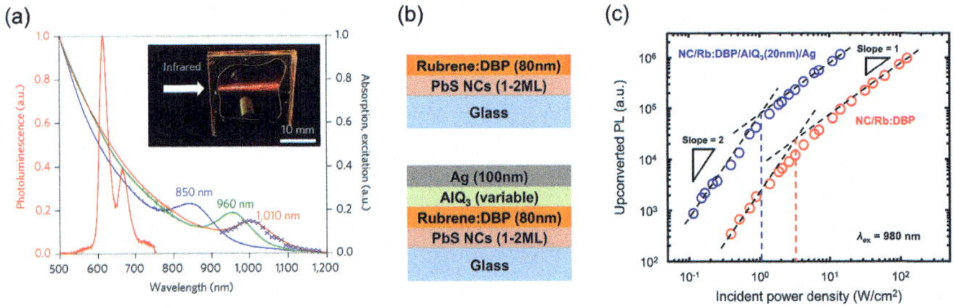

Fig. 3.5 (**a**) Absorption spectra of PbS NCs and emission spectrum of DBP at 0.5 vol% in rubrene. Inset: Photograph showing DBP photoluminescence sensitized by PbS NCs ($\lambda_{ex} = 808$ nm). Adopted with permission from [17]. Copyright 2015, Springer Nature. (**b**) Device structures for upconversion without (top) and with (bottom) optical interference. (**c**) Dependence of upconverted photoluminescence on excitation light intensity ($\lambda_{ex} = 980$ nm) for devices with (blue) and without (red) optical interference. Adopted with permission from [46]. Copyright 2017, AIP Publishing

η_{UC}/Φ_{PL} of 1.2 ± 0.2%. Excitation intensity threshold I_{th} was high (12 W/cm^2) because of the low absorption of the thin NC layer. The optical interference was used to enhance the net excitation intensity of the UC devices. Baldo and co-workers deposited a thin layer of tris-(8-hydroxyquinoline)aluminum (AlQ_3) of various thicknesses and a 100-nm-thick film of silver by thermal evaporation to maximize the optical field at the position of the PbS NC layer (Fig. 3.5b, c) [46]. Under the optimized condition, the I_{th} value decreased to 1.1 W/cm^2 ($\lambda_{ex} = 980$ nm).

In the solid-state, sensitizers are close to annihilator molecules and accordingly efficient TET would be expected without the use of transmitters. Bawendi and co-workers conducted a theoretical and experimental study on the optimal length of ligand aliphatic acid that allows efficient TET in the solid state [20]. They found that hexanoic acid ligands give sufficient distance to show near-unity TET efficiency, which led to TTA-UC with high intrinsic UC quantum efficiency η_{UC}/Φ_{PL} of 7 ± 1% ($\lambda_{ex} = 808$ nm, $\lambda_{em} = 610$ nm).

Solid-state upconverters have also been constructed by hybridizing NCs with metal-organic frameworks (MOFs). MOFs offer a promising platform for TTA-UC because of the ordered arrangements of chromophores with fast triplet exciton diffusion [47–50]. Rao, Yanai, and Kimizuka et al. explored the NC-MOF hybrids by combining PbS NCs with diphenyltetracene-based MOF (PbS@T-MOF) through post-modification of the crystal surface (Fig. 3.6a) [51]. Under excitation at 785 nm, the PbS@T-MOF showed an upconverted emission at around 600 nm (Fig. 3.6b). However, a high I_{th} value of 13 W/cm^2 was observed for PbS@T-MOF (Fig. 3.6c). The triplet lifetime of the annihilator was short ($\tau = 28$ μs), possibly due to the presence of defect sites in the crystal that deactivate the excited triplets. Besides, some aggregations of NCs were observed in scanning transmission electron microscopy (STEM) images. Therefore, simultaneous pursuit of the improved nanocrystal dispersibility and MOF crystallinity would be required to achieve efficient TTA-UC at low excitation intensity.

3.3 Triplet Sensitization with Lead Halide Perovskites

Organic-inorganic halide perovskite materials have attracted increasing attention because of their excellent optoelectronic properties suitable for photovoltaics [52]. Typically they are composed of metal halides (MX_2) and amine halides (RNH_3X), which gives MX_6 octahedra intercalated with organic cations as structural units. One of the distinct features of the perovskites is its structural variety. The dimensionality and optoelectronic properties can be tuned by exchanging metals, halogens, and amine cations [53]. The photoexcitation of perovskite materials produces singlet and triplet excited states, and the triplet energy can be transferred to organic dyes. The first Dexter-type triplet energy transfer from a two-dimensional (2D) perovskite to an organic dye has been reported by Era and co-workers [54, 55]. They observed phosphorescence of naphthalene chromophores electrostatically introduced in the

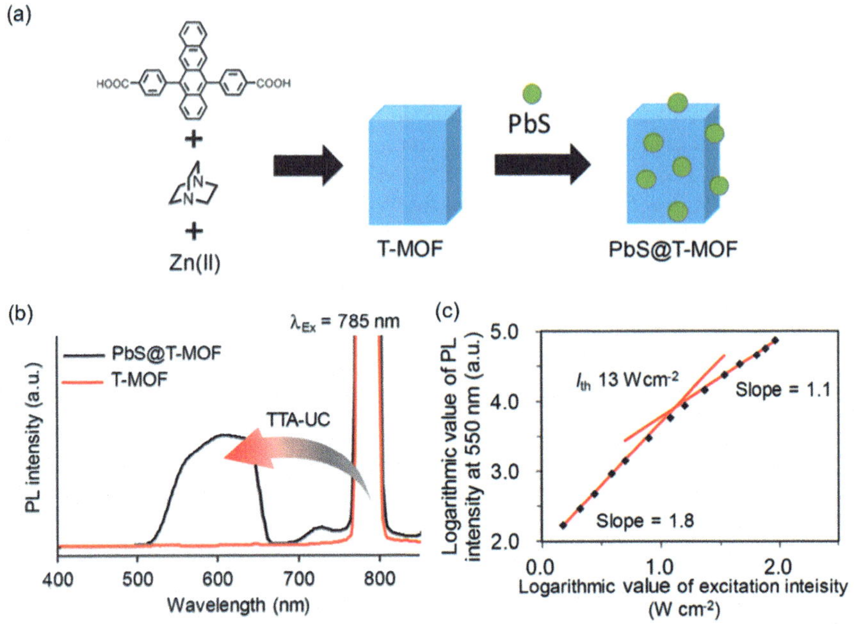

Fig. 3.6 (**a**) Schematic illustration of the preparation of PbS@T-MOF. (**b**) The UC emission spectra of the T-MOF with (black) and without (red) PbS (λ_{ex} = 785 nm). (**c**) Dependence of upconverted photoluminescence on excitation light intensity for PbS@T-MOF (λ_{ex} = 785 nm). Adopted with permission from [51]. Copyright 2018, Royal Society of Chemistry

PbBr-based layered perovskites, even at room temperature. The singlet-triplet gap was around 20 meV, and a large number of singlet excitons quickly relaxed to the triplet states within picoseconds. This small energy loss is useful for TTA-UC. Mitzi and Gundogdu et al. have reported TTA-UC sensitized by a 2D halide perovskite $(MA)_2Pb(SCN)_2I_2$ (MA = methylammonium) at room temperature [13]. The phosphorescence lifetime of $(MA)_2Pb(SCN)_2I_2$ dropped from 181 to 20.7 ns by overlying rubrene, indicating the efficient TET (Fig. 3.7; Φ_{TET} = 89%). Based on the PL dynamics and 1D diffusion model, the diffusion length of triplet excitons in a 213-nm-thick perovskite film was calculated to be 152 ± 12 nm, which is long enough to transport the triplet energy to the interface. When the perovskite was submerged in a chlorobenzene solution of perylene (10 mM) and excited at 570 nm, UC emission around 520 nm was observed with Φ_{TET} of 26%.

Three-dimensional (3D) metal-halide perovskites are promising as triplet sensitizers due to their strong absorption ability and tunable bandgap. The Yanai and Kimizuka group reported the triplet sensitization with 3D perovskites for the first time and demonstrated its application in TTA-UC [14]. NCs of cesium lead halide perovskite $CsPb(Br/I)_3$ were employed as a novel class of inorganic triplet sensitizer (Fig. 3.8a). The bandgap of perovskite was tuned for 9,10-diphenylanthracene (DPA) by exchanging halide ions. Since the short excited-state lifetime (τ = 16.1 ns)

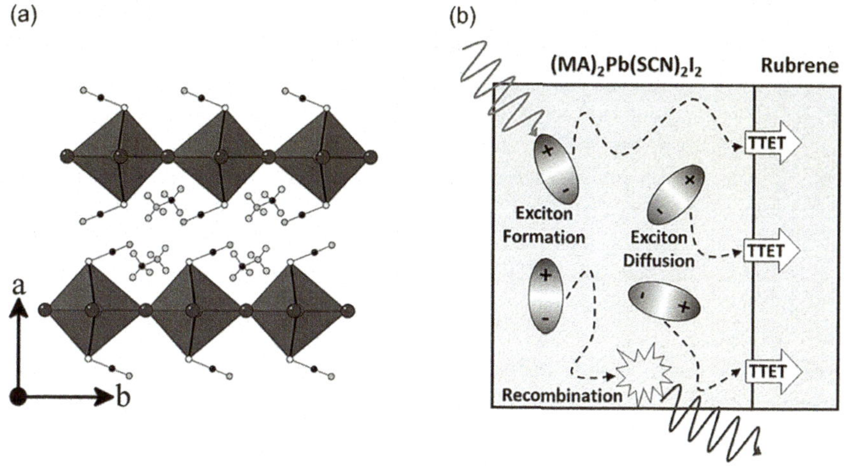

Fig. 3.7 (**a**) Crystal structure of $(MA)_2Pb(SCN)_2I_2$ perovskite. (**b**) Schematic illustration of exciton diffusion and the energy transfer at the $(MA)_2Pb(SCN)_2I_2$/rubrene interface. Adopted with permission from [13]. Copyright 2017, John Wiley & Sons

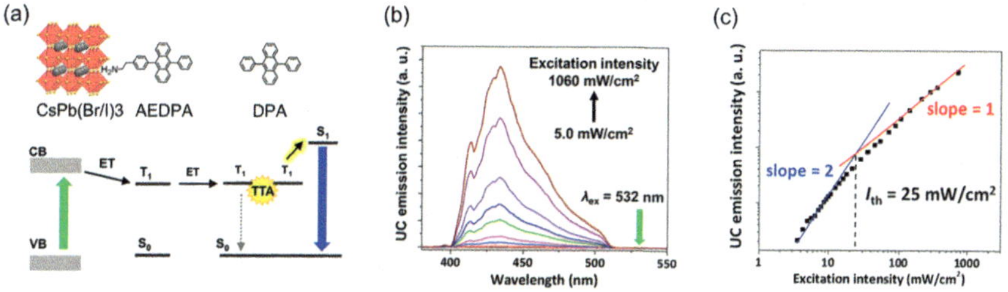

Fig. 3.8 (**a**) Scheme for TTA-UC sensitized by inorganic perovskite nanocrystals $CsPb(Br/I)_3$. (**b**) UC emission spectra of $CsPb(Br/I)_3$-AEDPA and DPA in toluene under various excitation intensities at 532 nm. (**c**) Excitation intensity dependence of UC emission intensity for $CsPb(Br/I)_3$-AEDPA and DPA in toluene. Adopted with permission from [14]. Copyright 2017, Royal Society of Chemistry

and a thick surface passivation layer interrupt TET, 2-(4-(10-phenylanthracene-9-yl)phenyl)ethan-1amine (AEDPA) was attached to the $CsPb(Br/I)_3$ surface as a triplet transmitter unit. The photoluminescence lifetime of $CsPb(Br/I)_3$ dropped to 10.0 ns, indicating TET from $CsPb(Br/I)_3$ to the surface AEDPA. Under excitation at 532 nm or 635 nm, UC emission at 434 nm was observed from a toluene solution containing $CsPb(Br/I)_3$-AEDPA and DPA (Fig. 3.8b). Considering that the conventional porphyrin-based UC systems cannot sensitize DPA under excitation at 635 nm due to the energy loss during ISC, $CsPb(Br/I)_3$ sensitizer works as a superior sensitizer for TTA-UC with a larger anti-Stokes shift. Also, a low I_{th} value of 25 mW/cm^2 illustrates the high sensitization ability of the surface-modified perovskite nanocrystals (Fig. 3.8c). By generalizing this concept, the Yanai and

Kimizuka group achieved the first example of Vis-to-UV TTA-UC sensitized by lead halide perovskite NCs [56]. $CsPb(Cl/Br)_3$ NCs modified with 1-naphthoic acid (NCA) were combined with 2,5-diphenyloxazole (PPO) to achieve the Vis-to-UV TTA-UC with a good efficiency exceeding 4%.

The TET from perovskite nanocrystals to surface triplet transmitters can be enhanced by decreasing the size of the perovskite NCs. Wu and co-workers synthesized various sizes of $CsPbBr_3$ NCs (3.5–11.2 nm) and investigated the dynamics of TET rates from the perovskite NCs to surface-bound 1-pyrenecarboxylic acid [21]. When the size of $CsPbBr_3$ NCs was decreased, the exciton bleaching lifetime became shortened, and TET became more favorable (Fig. 3.9). The TET rate increased up to 9.1×10^9 s^{-1}, which enabled almost quantitative TET (Φ_{TET} ~ 99%). The linear dependence of the TET rate on the carrier probability density at the NC surface was consistent with the Dexter-type TET mechanism.

The bulk-perovskite film has also been successfully used for solid-state NIR-to-Vis TTA-UC. Nienhaus, Baldo, Bawendi, and co-workers reported NIR-to-Vis TTA-UC in the solid state by employing perovskite thin films as triplet sensitizers [15]. They synthesized perovskite thin films based on methylammonium (MA) formamidinium (FA) lead triiodide ($MA_{0.15}FA_{0.85}PbI_3$; MAFA) on glass substrates and deposited rubrene and 1mol% DBP as an annihilator-emitter layer (Fig. 3.10). Under the excitation at 785 nm, the device showed UC emission at 610 nm with an intrinsic UC quantum efficiency of 3.1% ($\Phi_{PL} = 9.3 \pm 0.8\%$, I_{ex} ~ 88 W/cm^2).

Thanks to the long-range exciton transporting feature, a thick perovskite layer showing strong NIR absorption can also be used for solid-state triplet sensitization. In the case of semiconductor NCs such as PbS, the inefficient exciton diffusion has limited the thickness of the NC layer to one or two monolayers, which significantly

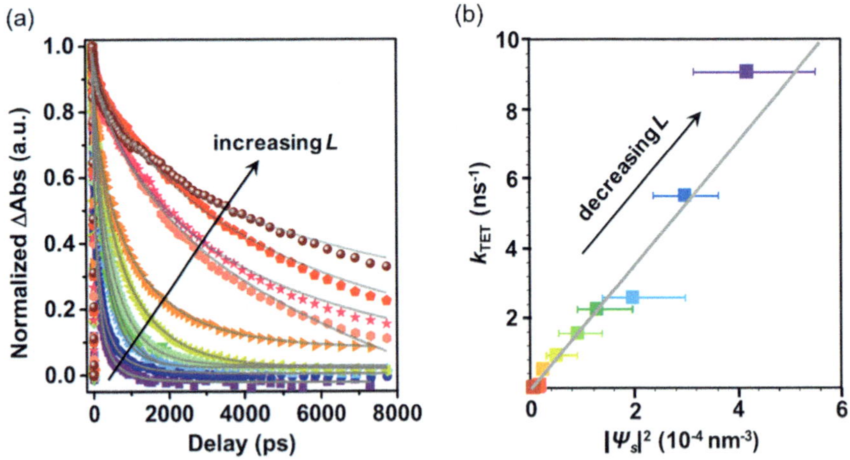

Fig. 3.9 (**a**) Normalized TA kinetics probed at the exciton bleaching of $CsPbBr_3$-PCA with various edge lengths (L) (colored symbols). (**b**) Carrier probability density dependence of the TET rate. Adopted with permission from [21]. Copyright 2019, American Chemical Society

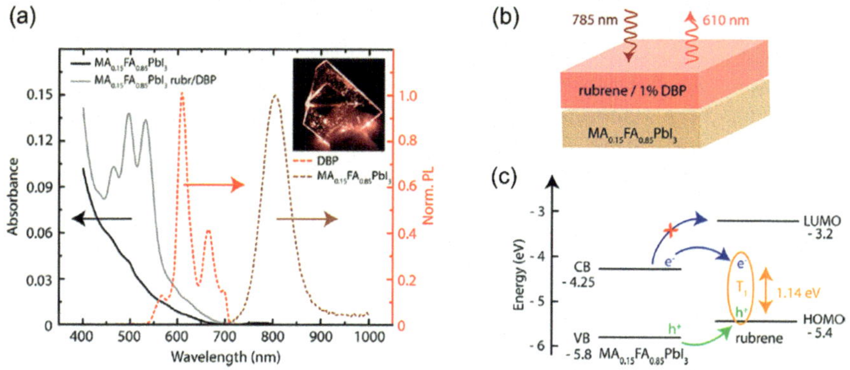

Fig. 3.10 (**a**) Normalized absorption spectra of a neat $MA_{0.15}FA_{0.85}PbI_3$ perovskite film (black), $MA_{0.15}FA_{0.85}PbI_3$/rubrene/1mol% DBP bilayer device (gray), the UC emission from rubrene/DBP (red), and the NIR emission of the $MA_{0.15}FA_{0.85}PbI_3$ perovskite (brown). Inset: Photograph showing DBP photoluminescence sensitized by $MA_{0.15}FA_{0.85}PbI_3$ ($\lambda_{ex} = 785$ nm). (**b**) Schematic and (**c**) band diagram of the UC device. Adopted with permission from [15]. Copyright 2019, American Chemical Society

decreases the NIR absorption to less than 1%. On the other hand, the bulk perovskite films show long triplet diffusion length that allows increasing the film thickness. Nienhaus and co-workers varied the thicknesses of the $MA_{0.15}FA_{0.85}PbI_3$ film from 20 to 380 nm and studied the TTA-UC properties by attaching a rubrene/1mol% DBP annihilator-emitter layer [22]. A low excitation intensity threshold I_{th} of 7.1 mW/cm^2 was observed for the sample with the thickest MAFA film of 380 nm. As the thicker layer increases the chance of singlet back energy transfer from rubrene/DBP to $MA_{0.15}FA_{0.85}PbI_3$ or reabsorption by $MA_{0.15}FA_{0.85}PbI_3$ in the bilayer device, it remains a challenge how to overcome the trade-off between the I_{th} value and the UC efficiency.

3.4 Molecular Sensitizers Showing a Singlet-to-Triplet Direct Transition

Molecular based NIR-absorbing triplet sensitizers have been desired to minimize the dissipative energy and optimize the sensitization process based on the molecular design. However, the development of NIR-absorptive molecules with small S_1-T_1 energy gaps has been challenging. To circumvent the energy loss during S_1-to-T_1 ISC, the Yanai and Kimizuka group proposed to utilize molecular sensitizers with singlet-to-triplet (S-T, S_0-to-T_1) absorption in the NIR range. Although the direct S-T transition is spin-forbidden, several osmium complexes have been known to show S-T absorption with a sufficiently high absorption coefficient ($\varepsilon > 1000\ M^{-1}\ cm^{-1}$) thanks to the strong spin-orbit coupling [57–60]. This absorption is attributed to be the triplet metal-to-ligand charge transfer (^{3}MLCT) transition, and the absorption wavelength can be tuned by the judicious selection of metals and ligands.

Based on the S-T absorption, the NIR-to-Vis TTA-UC under excitation over 900 nm has been achieved by Yanai and Kimizuka et al. [16]. An Os complex, Os(atpy)(tbbpy)Cl$^+$ (Fig. 3.11a, atpy = tris(2-ethylhexyl)-[2,2′:6′,2″-terpyridine]-4,4′,4″-tricarboxamide; tbbpy = 4,4′-di-tert-butyl-2,2′-bipyridine), was developed by functionalizing a previously reported structure [59]. The branched alkyl chains were attached to obtain sufficient solubility in an organic solvent and compatibility with the nonpolar emitter assemblies. Os(atpy)(tbbpy)Cl$^+$ exhibited a broad singlet-to-triplet (S-T) absorption at 888 nm (Fig. 3.11b, $\varepsilon = 3200\ M^{-1}\ cm^{-1}$). Excitation at the ^{3}MLCT band gave a phosphorescence with a small Stokes shift (λ_{em} ~ 1000 nm), illustrating the absence of the ISC process. Under excitation at 938 nm, a mixed CH_2Cl_2 solution of rubrene and Os(atpy)(tbbpy)Cl$^+$ showed an UC emission at 570 nm with a large anti-Stokes shift of 0.86 eV. However, the UC efficiency η_{UC} was less than 0.01% in solution because the triplet lifetime of Os(atpy)(tbbpy)Cl$^+$ is short ($\tau_p = 12$ ns).

When Os(atpy)(tbbpy)Cl$^+$ was doped into rubrene nanoparticles (NPs), TET efficiency was dramatically improved. Rapid injection of Os(atpy)(tbbpy)Cl$^+$-rubrene mixed tetrahydrofuran (THF) solution into an aqueous solution of sodium dodecyl sulfate (SDS) resulted in the formation of amorphous rubrene NPs in which the sensitizer molecules were homogeneously doped. The sensitizer-doped rubrene

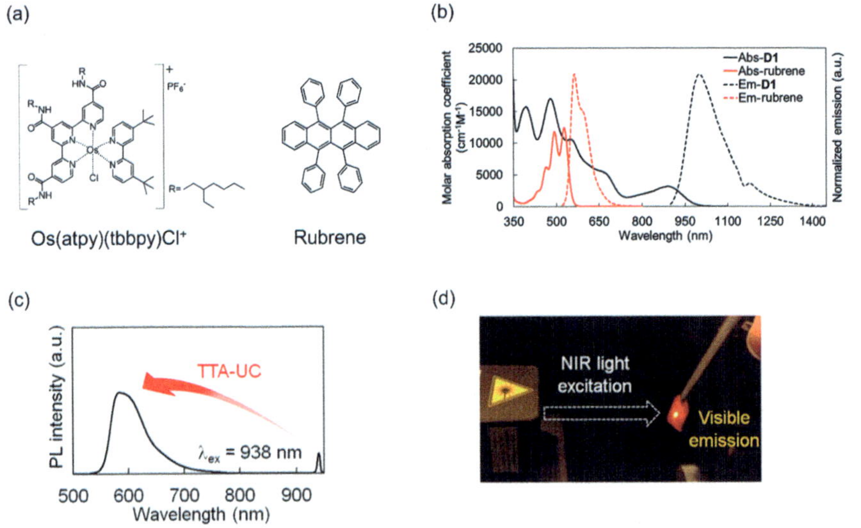

Fig. 3.11 (**a**). Molecular structures of Os(atpy)(tbbpy)Cl$^+$ and rubrene. (**b**) Absorption and emission spectra of Os(atpy)(tbbpy)Cl$^+$ and rubrene in CH_2Cl_2. (**c**) UC emission spectrum of Os(atpy)(tbbpy)Cl$^+$-doped rubrene NPs dispersed in PVA ($\lambda_{ex} = 938$ nm, in air). (**d**) Photograph showing UC emission of rubrene NPs sensitized by Os(atpy)(tbbpy)Cl$^+$ in PVA. Adopted with permission from [16]. Copyright 2016, American Chemical Society

NPs were further dispersed in poly(vinyl alcohol) (PVA) to suppress the access of molecular oxygen, which quenches excited triplets. Even under ambient conditions, the solid film showed UC emission at around 580 nm upon excitation at 938 nm (Fig. 3.11c, d). An improved I_{th} (10 W/cm^2) was observed thanks to the enhanced TET in aggregated states. The combination of molecular based S-T sensitizers with the triplet energy migration-based UC scheme is a powerful strategy to realize efficient NIR-to-Vis TTA-UC and high air stability.

By tuning the S-T absorption wavelength to sensitize a blue-emissive annihilator, the Yanai and Kimizuka group realized a NIR (excitation wavelength >700 nm)-to-blue (emission peak <500 nm) TTA-UC (Fig. 3.12a) [61]. The complex $Os(bptpy)_2^{2+}$ was designed with terpyridine ligands with 4-bromophenyl substituents. $Os(bptpy)_2^{2+}$ showed a ^{3}MLCT absorption beyond 700 nm with an absorption coefficient of 1600 M^{-1} cm^{-1} at 724 nm. When 2,5,8,11-tetra-*tert*-butylperylene (TTBP) was used as an annihilator, the $Os(bptpy)_2^{2+}$-TTBP mixed DMF solution showed an upconverted blue emission at 462 nm under NIR excitation at 724 nm (Fig. 3.12b). The large anti-Stokes shift of 0.97 eV was obtained thanks to the reduced energy loss in the triplet sensitization using the S-T absorption. The mixed solution showed a TTA-UC efficiency η_{UC} of 2.7% and a threshold excitation intensity of 320 mW/cm^2 under an optimized condition. However, TET efficiency was not maximized ($\Phi_{TET} = 47\%$) due to the short excited-state lifetime of $Os(bptpy)_2^{2+}$($\tau_p = 207$ ns).

The excited-state lifetime of heavy metal complexes in solution media can be improved by attaching triplet accepting moieties. The covalent conjugation of metal complexes with chromophoric groups having a slightly lower triplet energy level than the parent metal complexes has been reported to show elongated excited-state lifetimes [34–37]. For example, Zhao and co-workers combined tris(2,2′-bipyridine)osmium(II) complex $Os(bpy)_3^{2+}$ with BODIPY and observed 31-fold increase of the triplet lifetime from 56 ns to 1.73 μs [62]. In those works, the excited-state lifetime is elongated via intramolecular energy transfer (IMET) processes

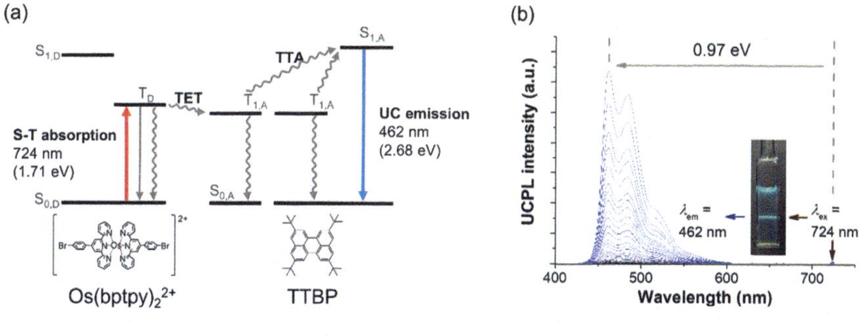

Fig. 3.12 (**a**) Energy diagram of singlet-to-triplet (S-T) absorption-based TTA-UC processes with $Os(bptpy)_2^{2+}$ and TTBP. (**b**) UC emission spectra of $Os(bptpy)_2^{2+}$ (20 μM) and TTBP (2 mM) in deaerated DMF with various excitation intensities ($\lambda_{ex} = 724$ nm). Adopted with permission from [61]. Copyright 2017, Royal Society of Chemistry

and succeeding excited-state equilibration. The equilibrium constant K_{eq} for the reversible processes in one-chromophore-combined metal complexes is estimated by the following relationship:

$$\Delta E \approx -RT \ln\left(K_{eq}\right)$$

$$K_{eq} = k_{IMET}/k_{bIMET} = \alpha_P/(1 - \alpha_P)$$

where k_{IMET} and k_{bIMET} represent the rate constants of IMET and back IMET, respectively, and α_P corresponds to the fraction of excited triplet pooling units. The Yanai and Kimizuka group designed a new chromophore-combined osmium complex $Os(peptpy)_2{}^{2+}$ by optimizing the energy gap and the triplet lifetime for efficient NIR-to-blue TTA-UC (Fig. 3.13a) [23]. In order to elongate the triplet lifetime of donor osmium(II) bis(4′-phenyl-2,2′:6′,2″-terpyridine) ($Os(ptpy)_2$, $T_1 = 1.63$ eV), a phenyl-perylene (pPe) unit ($T_1 \sim 1.50$ eV) was employed as a counterpart which gives a small ΔE of −0.13 eV for a high IMET efficiency ($\alpha_P > 99\%$). Two pPe units were attached to the $Os(ptpy)_2$ unit so that the IMET process becomes entropically

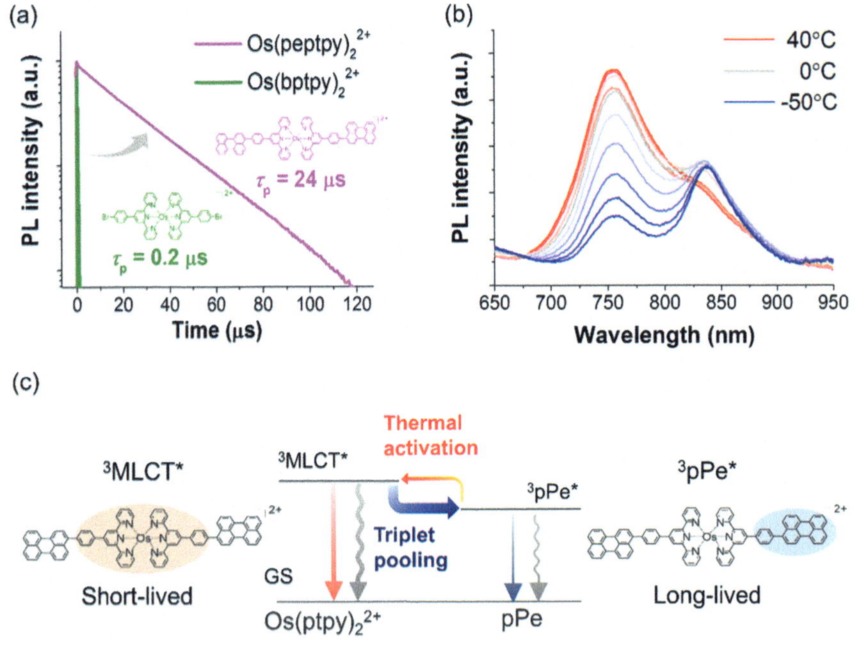

Fig. 3.13 (**a**) Phosphorescence decay profiles of $Os(peptpy)_2{}^{2+}$ (magenta) and $Os(bptpy)_2{}^{2+}$ (green). (**b**) Temperature-dependent emission spectra of $Os(peptpy)_2{}^{2+}$. (**c**) The mechanism of the extension of an excited-state lifetime and thermally activated delayed phosphorescence of $Os(peptpy)_2{}^{2+}$. Adopted with permission from [23]. Copyright 2019, John Wiley & Sons

more favorable [63]. $Os(peptpy)_2^{2+}$ showed a 120-fold longer excited-state lifetime of 24 μs than $Os(bptpy)_2^{2+}$. The emission from ^{3}MLCT state of $Os(peptpy)_2^{2+}$ ($\lambda_{em} = 760$ nm) showed a thermally activated feature, which indicates the excited-state thermal equilibrium (Fig. 3.13b, c). When $Os(peptpy)_2^{2+}$ was used as a triplet sensitizer for TTBP, almost unity IMET and TET efficiencies (Φ_{IMET} ~ 100%, Φ_{TET} ~ 100%) were obtained. As a result, η_{UC} increased by a factor of two from 2.7 to 5.9%. The extended lifetime of $Os(peptpy)_2^{2+}$ enabled sensitization even in a viscous hydrogel, which resulted in the exploration of NIR optogenetic genome engineering based on TTA-UC.

Highly efficient TET in the solid state has been achieved by introducing a S-T absorber into an annihilator-based MOF. The Yanai and Kimizuka group constructed a new zirconium-based UiO-type MOF, which is constructed with an annihilator linker, [4-((10-(4-carboxyphenyl)anthracene-9-yl)ethynyl)benzoic acid] (H_2CPAEBA) [24]. A new S-T absorption sensitizer $Os(tpyCOOH)_2^{2+}$ was designed to have coordination sites along with an appropriate length close to H_2CPAEBA. The sensitizer was successfully introduced through in situ coordination to Zr_6 clusters during the MOF crystallization without disturbing the framework structure (Fig. 3.14a, b). Energy-dispersive X-ray spectrometry (EDX) mapping in scanning transmission electron microscopy (STEM) images clarified

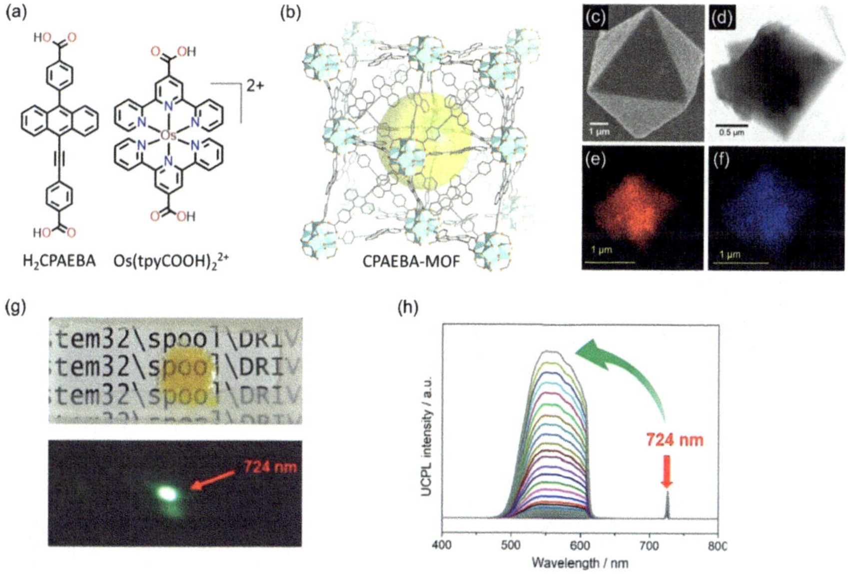

Fig. 3.14 (**a**) Molecular structures of H_2CPAEBA and $Os(tpyCOOH)_2^{2+}$. (**b**) Structure of CPAEBA-MOF with UiO-69-type topology. (**c**) SEM and (**d**) STEM images of Os-CPAEBA-MOF. STEM-EDX mapping images for (**e**) Zr and (**f**) Os atoms. (**g**) Photograph of Os-CPAEBA-MOF in TNB under white light (above) and 724 nm laser light (below). (**h**) UC emission spectra under various excitation intensities of 724 nm laser with 610 nm short-pass filter. Adopted with permission from [24]. Copyright 2020, John Wiley & Sons

that the Os sensitizer was homogeneously dispersed into the annihilator MOF structure (Fig. 3.14c–f). Under the excitation at 724 nm, the Os-CPAEBA-MOF in a glassy 1,3,5-tri(1-naphthyl)benzene (TNB) host showed an UC emission at around 540 nm (Fig. 3.14g, h). The phosphorescence lifetime of the sensitizer was much shorter in MOF ($\tau_{\mathrm{p}} < 10$ ns) than that in DMF ($\tau_{\mathrm{p},0} = 299$ ns), indicating a high triplet energy transfer efficiency ($\Phi_{\mathrm{TET}} > 95\%$). This result highlights the advantage of the S-T absorption sensitizer to produce upconverting MOFs compared with the previous work where the semiconductor NC sensitizers aggregated in MOFs [51]. The TTA-UC efficiency (0.012%) was not high as expected from the Φ_{TET} and fluorescence quantum yield of the MOF ($\Phi_{\mathrm{F}} \sim 5.5\%$). Triplet might be trapped in the defect sites such as locally disordered structures with the lower triplet energy levels in the framework. Further study on the optimization of assembly structure would be necessary to develop ideal solid-state upconverters.

3.5 Conclusion and Outlook

In this chapter, we discussed the recent advancement of NIR-absorptive sensitizers for TTA-UC. The energy loss during ISC—a bottleneck achieving NIR-to-Vis TTA-UC—has been circumvented by developing new types of triplet sensitizers such as semiconductor NCs, lead halide perovskites, and S-T absorption metal complexes. These triplet sensitizers enabled the upconversion of NIR light over 700 nm with large anti-Stokes shifts. While these new sensitizers have faced the problem of low TET efficiency due to the short excited-state lifetimes, the control of the distance between sensitizers and triplet transmitting moieties allows the efficient triplet energy transfer via the Dexter mechanism. When the sensitizers are functionalized with triplet transmitter moieties with long triplet lifetimes, the triplet lifetime of the whole sensitizer-transmitter conjugate becomes long enough to transfer the triplet energy to free emitter molecules.

Whereas newly developed sensitizers successfully realized NIR-to-Vis TTA-UC even in the solid state, there are remaining challenges towards the practical applications. First, the excitation intensity threshold I_{th} tends to be much higher than solar irradiance (~mW/cm^2) partially because of the weak absorption of UC devices. The developed new sensitizers have strong and broad visible light absorption, which partly quenches the upconverted emission. Therefore, it is difficult to increase the sensitizer concentration and material thickness. It would be necessary to develop sensitizers with an optical window in the visible region and/or waveguide materials that selectively trap the NIR light. Besides, for reducing I_{th}, it is also important to optimize the chromophore assembly structure for efficient energy transfer and long triplet lifetimes. Second, the obtained efficiencies of NIR-to-Vis TTA-UC are still low ($\eta_{\mathrm{UC}} < 10\%$), which is partly due to the formation of triplet trap sites as well as the strong reabsorption/back energy transfer from the upconverted singlet state to the sensitizers. The formation of high-quality crystals with fewer defects would be beneficial to improve TTA-UC efficiency. The problem of reabsorption/back energy transfer would be solved by controlling the energy flow and harvesting the upconverted energy to singlet energy collectors [64]. Given the currently expanded absorption wavelength over 800 nm, the future advancement both in the optical device structure and in the molecular/hybrid material design should provide a route to overcome the Shockley-Queisser limit by realizing the highly efficient NIR-to-Vis TTA-UC under 1 sun.

References

1. T. Trupke, M.A. Green, P. Würfel, Improving solar cell efficiencies by up-conversion of sub-band-gap light. J. Appl. Phys. **92**(7), 4117–4122 (2002)
2. S. Baluschev, V. Yakutkin, T. Miteva, G. Wegner, T. Roberts, G. Nelles, A. Yasuda, S. Chernov, S. Aleshchenkov, A. Cheprakov, A general approach for non-coherently excited annihilation up-conversion: transforming the solar-spectrum. New J. Phys. **10**(1), 013007 (2008)
3. N.J. Ekins-Daukes, T.W. Schmidt, A molecular approach to the intermediate band solar cell: the symmetric case. Appl. Phys. Lett. **93**, 063507 (2008)
4. S. Baluschev, T. Miteva, V. Yakutkin, G. Nelles, A. Yasuda, G. Wegner, Up-conversion fluorescence: noncoherent excitation by sunlight. Phys. Rev. Lett. **97**(14), 143903 (2006)
5. T.N. Singh-Rachford, F.N. Castellano, Photon upconversion based on sensitized triplet–triplet annihilation. Coord. Chem. Rev. **254**(21–22), 2560–2573 (2010)
6. J. Zhao, S. Ji, H. Guo, Triplet–triplet annihilation based upconversion: from triplet sensitizers and triplet acceptors to upconversion quantum yields. RSC Adv. **1**(6), 937–950 (2011)
7. Y.C. Simon, C. Weder, Low-power photon upconversion through triplet-triplet annihilation in polymers. J. Mater. Chem. **22**(39), 20817–20830 (2012)
8. A. Monguzzi, R. Tubino, S. Hoseinkhani, M. Campione, F. Meinardi, Low power, non-coherent sensitized photon up-conversion: modelling and perspectives. Phys. Chem. Chem. Phys. **14**(13), 4322–4332 (2012)
9. V. Gray, K. Moth-Poulsen, B. Albinsson, M. Abrahamsson, Towards efficient solid-state triplet–triplet annihilation based photon upconversion: supramolecular, macromolecular and self-assembled systems. Coord. Chem. Rev. **362**(1), 54–71 (2018)
10. J.C. Wang, S.P. Hill, T. Dilbeck, O.O. Ogunsolu, T. Banerjee, K. Hanson, Multimolecular assemblies on high surface area metal oxides and their role in interfacial energy and electron transfer. Chem. Soc. Rev. **47**(1), 104–148 (2018)
11. T.F. Schulze, T.W. Schmidt, Photochemical upconversion: present status and prospects for its application to solar energy conversion. Energy Environ. Sci. **8**(1), 103–125 (2015)
12. Z. Huang, X. Li, M. Mahboub, K.M. Hanson, V.M. Nichols, H. Le, M.L. Tang, C.J. Bardeen, Hybrid molecule-nanocrystal photon upconversion across the visible and near-infrared. Nano Lett. **15**(8), 5552–5557 (2015)
13. R. Younts, H.-S. Duan, B. Gautam, B. Saparov, J. Liu, C. Mongin, F.N. Castellano, D.B. Mitzi, K. Gundogdu, Efficient generation of long-lived triplet excitons in 2D hybrid perovskite. Adv. Mater. **29**(9), 1604278 (2017)
14. K. Mase, K. Okumura, N. Yanai, N. Kimizuka, Triplet sensitization by perovskite nanocrystals for photon upconversion. Chem. Commun. **53**(59), 8261–8264 (2017)
15. L. Nienhaus, J.-P. Correa-Baena, S. Wieghold, M. Einzinger, T.-A. Lin, K.E. Shulenberger, N.D. Klein, M. Wu, V. Bulović, T. Buonassisi, M.A. Baldo, M.G. Bawendi, Triplet-sensitization by lead halide perovskite thin films for near-infrared-to-visible upconversion. ACS Energy Lett. **4**(4), 888–895 (2019)
16. S. Amemori, Y. Sasaki, N. Yanai, N. Kimizuka, Near-infrared-to-visible photon upconversion sensitized by a metal complex with spin-forbidden yet strong S_0-T_1 absorption. J. Am. Chem. Soc. **138**(28), 8702–8705 (2016)

17. M. Wu, D.N. Congreve, M.W.B. Wilson, J. Jean, N. Geva, M. Welborn, T.V. Voorhis, V. Bulović, M.G. Bawendi, M.A. Baldo, Solid-state infrared-to-visible upconversion sensitized by colloidal nanocrystals. Nat. Photon. **10**(1), 31–34 (2015)
18. C. Mongin, S. Garakyaraghi, N. Razgoniaeva, M. Zamkov, F.N. Castellano, Direct observation of triplet energy transfer from semiconductor nanocrystals. Science **351**(6271), 369–372 (2016)
19. M. Mahboub, Z. Huang, M.L. Tang, Efficient infrared-to-visible upconversion with subsolar irradiance. Nano Lett. **16**(11), 7169–7175 (2016)
20. L. Nienhaus, M. Wu, N. Geva, J.J. Shepherd, M.W.B. Wilson, V. Bulović, T.V. Voorhis, M.A. Baldo, M.G. Bawendi, Speed limit for triplet-exciton transfer in solid-state PbS nanocrystal-sensitized photon upconversion. ACS Nano **11**(8), 7848–7857 (2017)
21. X. Luo, R. Lai, Y. Li, Y. Han, G. Liang, X. Liu, T. Ding, J. Wang, K. Wu, Triplet energy transfer from CsPbBr3 nanocrystals enabled by quantum confinement. J. Am. Chem. Soc. **141**(10), 4186–4190 (2019)
22. S. Wieghold, A.S. Bieber, Z.A. VanOrman, L. Daley, M. Leger, J.-P. Correa-Baena, L. Nienhaus, Triplet sensitization by lead halide perovskite thin films for efficient solid-state photon upconversion at subsolar fluxes. Matter **1**(3), 705–719 (2019)
23. Y. Sasaki, M. Oshikawa, P. Bharmoria, H. Kouno, A. Hayashi-Takagi, M. Sato, I. Ajioka, N. Yanai, N. Kimizuka, Near-infrared optogenetic genome engineering based on photon-upconversion hydrogels. Angew. Chem. Int. Ed. **58**(49), 17827–17833 (2019)
24. B. Joarder, A. Mallick, Y. Sasaki, M. Kinoshita, R. Haruki, Y. Kawashima, N. Yanai, N. Kimizuka, Near-infrared-to-visible photon upconversion by introducing an S-T absorption sensitizer into a metal-organic framework. ChemNanoMat **6**(6), 1–5 (2020)
25. S. Chan, M. Liu, K. Latham, M. Haruta, H. Kurata, T. Teranishi, Y. Tachibana, Monodisperse and size-tunable PbS colloidal quantum dots via heterogeneous precursors. J. Mater. Chem. C **5**(8), 2182–2187 (2017)
26. M. Tabachnyk, B. Ehrler, S. Gélinas, M.L. Böhm, B.J. Walker, K.P. Musselman, N.C. Greenham, R.H. Friend, A. Rao, Resonant energy transfer of triplet excitons from pentacene to PbSe nanocrystals. Nat. Mater. **13**(11), 1033–1038 (2014)
27. N.J. Thompson, M.W.B. Wilson, D.N. Congreve, P.R. Brown, J.M. Scherer, T.S. Bischof, M. Wu, N. Geva, M. Welborn, T.V. Voorhis, V. Bulović, M.G. Bawendi, M.A. Baldo, Energy harvesting of non-emissive triplet excitons in tetracene by emissive PbS nanocrystals. Nat. Mater. **13**(11), 1039–1043 (2014)
28. J.H. Kim, C.Y. Wong, G.D. Scholes, Exciton fine structure and spin relaxation in semiconductor colloidal quantum dots. Acc. Chem. Res. **42**(8), 1037–1046 (2009)
29. Z. Huang, M.L. Tang, Semiconductor nanocrystal light absorbers for photon upconversion. J. Phys. Chem. Lett. **9**(21), 6198–6206 (2018)
30. B.L. Wehrenberg, C. Wang, P. Guyot-Sionnest, Interband and intraband optical studies of PbSe colloidal quantum dots. J. Phys. Chem. B **106**(41), 10634–10640 (2002)
31. T.N. Singh-Rachford, F.N. Castellano, Triplet sensitized red-to-blue photon upconversion. J. Phys. Chem. Lett. **1**(1), 195–200 (2010)
32. X. Li, Z. Huang, R. Zavala, M.L. Tang, Distance-dependent triplet energy transfer between CdSe nanocrystals and surface bound anthracene. J. Phys. Chem. Lett. **7**(11), 1955–1959 (2016)
33. C. Mongin, P. Moroz, M. Zamkov, F.N. Castellano, Thermally activated delayed photoluminescence from pyrenyl-functionalized CdSe quantum dots. Nat. Chem. **10**(2), 225–230 (2018)
34. W.E. Ford, M.A.J. Rodgers, Reversible triplet-triplet energy transfer within a covalently linked bichromophoric molecule. J. Phys. Chem. **96**(7), 2917–2920 (1992)
35. D.S. Tyson, F.N. Castellano, Intramolecular singlet and triplet energy transfer in a ruthenium(II) diimine complex containing multiple pyrenyl chromophores. J. Phys. Chem. A **103**, 10955–10960 (1999)
36. X.-y. Wang, A.D. Guerzo, R.H. Schmehl, Photophysical behavior of transition metal complexes having interacting ligand localized and metal-to-ligand charge transfer states. J. Photochem. Photobiol. C **5**(1), 55–77 (2004)

37. N.D. McClenaghan, Y. Leydet, B. Maubert, M.T. Indelli, S. Campagna, Excited-state equilibration: a process leading to long-lived metal-to-ligand charge transfer luminescence in supramolecular systems. Coord. Chem. Rev. **249**(13–14), 1336–1350 (2005)
38. M.L. Rosa, S.A. Denisov, G. Jonusauskas, N.D. McClenaghan, A. Credi, Designed long-lived emission from CdSe quantum dots through reversible electronic energy transfer with a surface-bound chromophore. Angew. Chem. Int. Ed. **57**(12), 3104–3107 (2018)
39. X. Peng, M.C. Schlamp, A.V. Kadavanich, A.P. Alivisatos, Epitaxial growth of highly luminescent CdSe/CdS core/shell nanocrystals with photostability and electronic accessibility. J. Am. Chem. Soc. **119**(30), 7019–7029 (1997)
40. J.J. Li, Y.A. Wang, W. Guo, J.C. Keay, T.D. Mishima, M.B. Johnson, X. Peng, Large-scale synthesis of nearly monodisperse CdSe/CdS core/shell nanocrystals using air-stable reagents via successive ion layer adsorption and reaction. J. Am. Chem. Soc. **125**(41), 12567–12575 (2003)
41. K. Okumura, K. Mase, N. Yanai, N. Kimizuka, Employing core-shell quantum dots as triplet sensitizers for photon upconversion. Chem. Eur. J. **22**(23), 7721–7726 (2016)
42. M. Mahboub, P. Xia, J.V. Baren, X. Li, C.H. Lui, M.L. Tang, Midgap states in PbS quantum dots induced by Cd and Zn enhance photon upconversion. ACS Energy Lett. **3**(4), 767–772 (2018)
43. M.P. Hendricks, M.P. Campos, G.T. Cleveland, I.J.-L. Plante, J.S. Owen, A tunable library of substituted thiourea precursors to metal sulfide nanocrystals. Science **348**(6240), 1226–1230 (2015)
44. Z. Huang, Z. Xu, M. Mahboub, Z. Liang, P. Jaimes, P. Xia, K.R. Graham, M.L. Tang, T. Lian, Enhanced near-infrared-to-visible upconversion by synthetic control of PbS nanocrystal triplet photosensitizers. J. Am. Chem. Soc. **141**(25), 9769–9772 (2019)
45. K. Miyata, Y. Kurashige, K. Watanabe, T. Sugimoto, S. Takahashi, S. Tanaka, J. Takeya, T. Yanai, Y. Matsumoto, Coherent singlet fission activated by symmetry breaking. Nat. Chem. **9**(10), 983 (2017)
46. M. Wu, J. Jean, V. Bulović, M.A. Baldo, Interference-enhanced infrared-to-visible upconversion in solid-state thin films sensitized by colloidal nanocrystals. Appl. Phys. Lett. **110**(21), 211101 (2017)
47. M. Oldenburg, A. Turshatov, D. Busko, M. Jakoby, R. Haldar, K. Chen, G. Emandi, M.O. Senge, C. Wöll, J.M. Hodgkiss, B.S. Richards, I.A. Howard, Enhancing the photoluminescence of surface anchored metal-organic frameworks: mixed linkers and efficient acceptors. Phys. Chem. Chem. Phys. **20**(17), 11564–11576 (2018)
48. J. Park, M. Xu, F. Li, H.-C. Zhou, 3D long-range triplet migration in a water-stable metal–organic framework for upconversion-based ultralow-power in vivo imaging. J. Am. Chem. Soc. **140**(16), 5493–5499 (2018)
49. J.M. Rowe, J. Zhu, E.M. Soderstrom, W. Xu, A. Yakovenko, A.J. Morris, Sensitized photon upconversion in anthracene-based zirconium metal–organic frameworks. Chem. Commun. **54**(56), 7798–7801 (2018)
50. F. Meinardi, M. Ballabio, N. Yanai, N. Kimizuka, A. Bianchi, M. Mauri, R. Simonutti, A. Ronchi, M. Campione, A. Monguzzi, Quasi-thresholdless photon upconversion in metal–organic framework nanocrystals. Nano Lett. **19**(3), 2169–2177 (2019)
51. S. Amemori, R.K. Gupta, M.L. Böhm, J. Xiao, U. Huynh, T. Oyama, K. Kaneko, A. Rao, N. Yanai, N. Kimizuka, Hybridizing semiconductor nanocrystals with metal–organic frameworks for visible and near-infrared photon upconversion. Dalton Trans. **47**(26), 8590–8594 (2018)
52. A. Kojima, K. Teshima, Y. Shirai, T. Miyasaka, Organometal halide perovskites as visible-light sensitizers for photovoltaic cells. J. Am. Chem. Soc. **131**(17), 6050–6051 (2009)
53. J. Calabrese, N.L. Jones, R.L. Harlow, N. Herron, D.L. Thorn, Y. Wang, Preparation and characterization of layered lead halide compounds. J. Am. Chem. Soc. **113**(6), 2328–2330 (1991)
54. M. Era, K. Maeda, T. Tsutsui, Enhanced phosphorescence from naphthalene-chromophore incorporated into lead bromide-based layered perovskite having organic–inorganic superlattice structure. Chem. Phys. Lett. **296**(3–4), 417–420 (1998)

55. K. Ema, M. Inomata, Y. Kato, H. Kunugita, M. Era, Nearly perfect triplet-triplet energy transfer from Wannier excitons to naphthalene in organic-inorganic hybrid quantum-well materials. Phys. Rev. Lett. **100**(25), 257401 (2008)
56. K. Okumura, N. Yanai, N. Kimizuka, Visible-to-UV photon upconversion sensitized by Lead halide perovskite nanocrystals. Chem. Lett. **48**(11), 1347–1350 (2019)
57. J.-P. Sauvage, J.-P. Collin, J.-C. Chambron, S. Guillerez, C. Coudret, V. Balzani, F. Barigelletti, L.D. Cola, L. Flamigni, Ruthenium (II) and osmium (II) bis (terpyridine) complexes in covalently-linked multicomponent systems: synthesis, electrochemical behavior, absorption spectra, and photochemical and photophysical properties. Chem. Rev. **94**(4), 993–1019 (1994)
58. S. Altobello, R. Argazzi, S. Caramori, C. Contado, S.D. Fré, P. Rubino, C. Choné, G. Larramona, C.A. Bignozzi, Sensitization of nanocrystalline TiO2 with black absorbers based on Os and Ru polypyridine complexes. J. Am. Chem. Soc. **127**(44), 15342–15343 (2005)
59. T. Kinoshita, J. Fujisawa, J. Nakazaki, S. Uchida, T. Kubo, H. Segawa, Enhancement of near-IR photoelectric conversion in dye-sensitized solar cells using an osmium sensitizer with strong spin-forbidden transition. J. Phys. Chem. Lett. **3**(3), 394–398 (2012)
60. X. Zhang, S.E. Canton, G. Smolentsev, C.-J. Wallentin, Y. Liu, Q. Kong, K. Attenkofer, A.B. Stickrath, M.W. Mara, L.X. Chen, K. Wärnmark, V. Sundström, Highly accurate excited-state structure of $[Os(bpy)_2dcbpy]^{2+}$ determined by X-ray transient absorption spectroscopy. J. Am. Chem. Soc. **136**(24), 8804–8809 (2014)
61. Y. Sasaki, S. Amemori, H. Kouno, N. Yanai, N. Kimizuka, Near infrared-to-blue photon upconversion by exploiting direct S-T absorption of a molecular sensitizer. J. Mater. Chem. C **5**(21), 5063–5067 (2017)
62. D. Liu, Y. Zhao, Z. Wang, K. Xu, J. Zhao, Exploiting the benefit of $S_0 \rightarrow T_1$ excitation in triplet-triplet annihilation upconversion to attain large anti-stokes shifts: tuning the triplet state lifetime of a tris(2,2'-bipyridine) osmium(II) complex. Dalton Trans. **47**(26), 8619–8628 (2018)
63. Y.Y. Cheng, B. Fückel, T. Khoury, R.G.C.R. Clady, N.J. Ekins-Daukes, M.J. Crossley, T.W. Schmidt, Entropically driven photochemical upconversion. J. Phys. Chem. A **115**(6), 1047–1053 (2011)
64. N. Yanai, N. Kimizuka, New triplet sensitization routes for photon upconversion: thermally activated delayed fluorescence molecules, inorganic nanocrystals, and singlet-to-triplet absorption. Acc. Chem. Res. **50**(10), 2487–2495 (2017)

Chapter 4
Photon Upconversion Based on Sensitized Triplet-Triplet Annihilation (*s*TTA) in Solids

Angelo Monguzzi

Abstract The conversion of low-energy photons into radiation of higher energy is useful for bioimaging, 3D displays and other applications. In particular, upconversion of the infrared portion of the solar spectrum, which is typically not absorbed by the light-harvesting materials used in solar technologies, allows additional photons to be harnessed and boosts the efficiency of photovoltaic and photocatalytic devices. Therefore, low power photon upconversion of non-coherent light based on sensitized triplet-triplet annihilation (sTTA) has been recently recognized as a potential viable approach towards enhancing the efficiency of sunlight-powered devices through sub-bandgap photon harvesting.

The sTTA permits the conversion of light into radiation of higher energy involving a sequence of photophysical processes between two moieties, respectively a light harvester/triplet sensitizer and an annihilator/emitter. High up-conversion yields under solar irradiance can be observed in low viscosity solutions of dyes, but in solid materials, which are better suited for integration in devices, the process is usually less efficient. The ability to control triplet excitons in the solid state is therefore crucial to obtain high performance solid upconverters. In this chapter, we will discuss the results obtained in several systems, such as dye-doped polymers/nanoparticles, amorphous/crystalline supramolecular structures and many others, highlighting the materials design guidelines to obtain efficient upconverters at the solid state that can match the strict requirements of solar technologies.

Keywords Photon managing · Photon upconversion · Triplet-triplet annihilation · Nanostructured materials

A. Monguzzi (✉)
Scienza dei Materiali, Università degli Studi Milano-Bicocca, Milan, Italy
e-mail: angelo.monguzzi@unimib.it

J. S. Lissau, M. Madsen (eds.), *Emerging Strategies to Reduce Transmission and Thermalization Losses in Solar Cells*, https://doi.org/10.1007/978-3-030-70358-5_4

4.1 Diffusion-Limited sTTA-UC

As detailed in the previous chapters, photon upconversion based on sensitized triplet-triplet annihilation (*s*TTA-UC) is a multistep process that involves two bimolecular interactions, namely the sensitizer-to-emitter energy transfer (ET) and the fusion of two emitter triplets by TTA that generates the high-energy singlet responsible for the ultimate photon emission [1]. Both these *bimolecular* mechanisms are short-range interactions. The ET usually occurs indeed between the triplet state of the sensitizer moiety towards the optically dark triplet state of the emitter through Dexter-type interaction mechanism, i.e., a simultaneous hole and electron transfer between two moieties driven by exchange interactions [2–4]. The spatial overlap of the energy donor (the sensitizer) and the energy acceptor (the emitter) molecular orbital wave functions is therefore required to enable the charge transfer-mediated ET. By considering the extent in space of molecular orbitals, typical interaction radii for Dexter ET are assessed on the order of 1–2 nm, which means that molecule-sized donor and acceptors must collide in order to allow ET. Similarly, on the other hand, the TTA requires the collision of two triplet excitons to form a collisional complex <T-T> that contains simultaneously singlet, triplet, and quintet characters, according to the spin combination statistic of the triplet states involved. A possible result of the <T-T> complex fission is therefore formation of a singlet-state molecule, a triplet-state molecule, or a quintet-state molecule with a branching ratio determined by the excited-state properties of the involved species, their mutual arrangement, and local environment [5, 6].

Several approximations can be made in order to simplify the analytical discussion of the problem of bimolecular interaction modelling. One of the most used is the Perrin model [7]. Considering s as the average intermolecular distance between donor and acceptor, this model considers the interaction rate constant $k_{d,\,a}$ between an energy *donor* (d) and an energy *acceptor* (a) system as

$$k_{d,a} = \begin{cases} \infty & if\ s < r_0 \\ 0 & if\ s > r_0 \end{cases} \tag{4.1}$$

where r_0 is the characteristic interaction radius, i. e. , the distance at which the (d, a) interaction efficiency is 50 % . In the case of the Dexter ET in sTTA − UC systems, r_0 takes the value of the Dexter radius r_{Dx}, i. e. , the distance at which the ET rate k_{ET} equals to the rate of the spontaneous decay k_{D} of the energy donor in the absence of energy acceptors. From an experimental point of view, in a system where sensitizers and emitters are blocked in fixed positions the Dexter radius can be obtained from measurements on the sensitizer relative photoluminescence (PL) quantum yield, which is defined as the ratio between the sensitizer PL intensity I in the presence of a defined acceptor concentration C_{a} and that without it I_0. By using Eq. 4.1, the ET yield can be written as [8]

$$\phi_{\mathrm{ET}} = 1 - (-I/I_0) = \exp\left(-C_{\mathrm{a}}/C_0\right) \tag{4.2}$$

where C_0 is a constant related to the Dexter parameters, which is called critical transfer concentration. By using units of moles per liter for C_0, the r_{Dx} value expressed in Å equals 7.346 (C_0) –1/3. In other words, Eq. 4.2 says that the ET yield is set by how many acceptors are included in a sphere of radius r_{Dx} around the energy donor.

These considerations point out that the eventual diffusion of energy, by translation motion of excited molecules or by exciton diffusion in solid matrixes, is crucial in determining the yield of the two fundamental steps of the *s*TTA-UC mechanism. In particular, in the case of short-range processes where r_0 is ~1 nm, according to the Perrin model, it is the collision between the two species involved that activates the interaction with a yield of 100%. Therefore, it is the collisional rate, i.e., the number of collisions with acceptors per second during the donor excited state lifetime, that finally determines the efficiency of the ET and TTA. Diffusion-limited processes have been studied extensively in the last century and several analytical methods have been developed in order to quantitatively face their phenomenology. In general, the diffusion length of a particle moving by random Brownian motion is calculated as the mean square displacement in any possible isotropic direction [9, 10]:

$$L = \sqrt{2ZD\tau}, \tag{4.3}$$

where D [cm^2 s^{-1}] is the diffusivity coefficient of the system considered and τ is the lifetime of the donor excited state considered. The parameter Z takes the value of 1, 2, or 3 in the case of 1D, 2D, or 3D isotropic diffusion environments. In the 3D case, Eq. 4.3 allows to calculate the diameter of the sphere explored by the donor excited molecules/exciton before spontaneous recombination. The collisional rate with the acceptor species is set by the acceptor concentration C_a contained in the volume explored by the donor in the lifetime τ. For low acceptor concentrations the collisional rate $k_{d,\,a}$ is not proportional to C_a, but a useful simplification of the problem can be used when the so-called *rapid diffusion* limit is reached [11]. The values of C_a or D can be indeed high enough to verify the condition:

$$L \gg s \text{ or } \frac{D\tau}{s^2} \gg 1, \tag{4.4}$$

where s is the average intermolecular donor/acceptor distance in the system. If Eq. 4.4 is verified, it means that the donor is able to encounter all the acceptor available in the unit volume before spontaneous deactivation. Consequently, the collisional/interaction rate becomes proportional to C_a by

$$k_{d,a} = 4\pi D_{\text{tot}} r_0 C_a, \tag{4.5}$$

where the total diffusivity is calculated as $D_{\text{tot}} = D_1 + D_2$. In the case of *s*TTA-UC, the rapid diffusion condition allows to write a simple expression for both the ET and TTA rates as

$$k_{\mathrm{ET}} = 4\pi D_{\mathrm{tot}} r_{\mathrm{ET}} C_{\mathrm{em}}, \tag{4.6}$$

$$k_{\mathrm{TTA}} = 8\pi D_{\mathrm{T}} r_{\mathrm{TT}} [T] = \gamma_{\mathrm{TT}} [T]. \tag{4.7}$$

Equations 4.6 and 4.7 point out two possible strategies to enhance the short-range interaction rate. First, increase as high as possible the density of interacting species (the acceptor moiety for Dexter ET and the density of triplet for TTA). Second, enhance the energy diffusion, in order to minimize the average intermolecular distance between interacting systems and therefore maximize the interaction rate. Considering the solubility and size of typical molecular and semiconductor nanocrystal systems employed, the rapid diffusion regime is actually easy to obtain in low-viscosity solvents, where energy diffusion relies on molecular translational diffusion by Brownian motion [12]. In this case the diffusivity of the involved moiety i can be estimated using the Stokes-Einstein equation:

$$D_i = \frac{k_{\mathrm{B}} T}{6\pi \eta R_i}, \tag{4.8}$$

where η is the solvent viscosity at temperature T and R_i is the radius of the sphere that approximates the considered molecule/nanocrystal diffusion and rotation in the liquid environment. It is worth noting that for *s*TTA-UC the energy diffusivity parameter can be crucial. For imaging or optical sensing applications where no limits to the excitation fluence, the energy diffusivity can be considered a minor aspect, since the triplet density $[T]$ can be raised without limitation by increasing the excitation power. Conversely, in solar applications the amount of triplets that can be generated has an upper limit that is the solar irradiance. According to Eq. 4.7, k_{TTA} depends directly on the triplet diffusivity D_{T} through the second-order rate constant γ_{TT}. Consequently, D_{T} must be high enough to have a k_{TTA} that surpasses the spontaneous recombination triplet decay rate k_{T} at low $[T]$ available. In this *high*-efficiency regime ($k_{\mathrm{TTA}} \gg k_{\mathrm{T}}$) energy losses by non-radiative recombination channels are avoided, and the system shows its best maximum upconversion quantum yield $\mathrm{QY}_{\mathrm{uc}}$ which is independent from the excitation intensity (Fig. 4.1) [13]. At very low excitation power usually the triplet density $[T]$ is so low that the annihilation results to be negligible with respect to k_{T}. In this *low*-efficiency regime, the *s*TTA-UC efficiency shows the expected linear dependency on the excitation fluence, as expected for bimolecular mechanisms. The switch from the *low*- to the *high*-efficiency regime is marked by the so-called excitation intensity threshold I_{th} ($k_{\mathrm{TTA}} = k_{\mathrm{T}}$). It corresponds to the excitation intensity at which the $\mathrm{QY}_{\mathrm{uc}}$ is half of the maximum value achievable, and depends on several characteristic parameters of the system, including the triplet diffusivity, by

$$I_{\mathrm{th}} (D_{\mathrm{tot}}, D_{\mathrm{T}}) = \frac{{k_{\mathrm{T}}}^2}{\alpha \gamma_{\mathrm{TT}} (D_{\mathrm{T}}) \phi_{\mathrm{ET}} (D_{\mathrm{tot}})}. \tag{4.9}$$

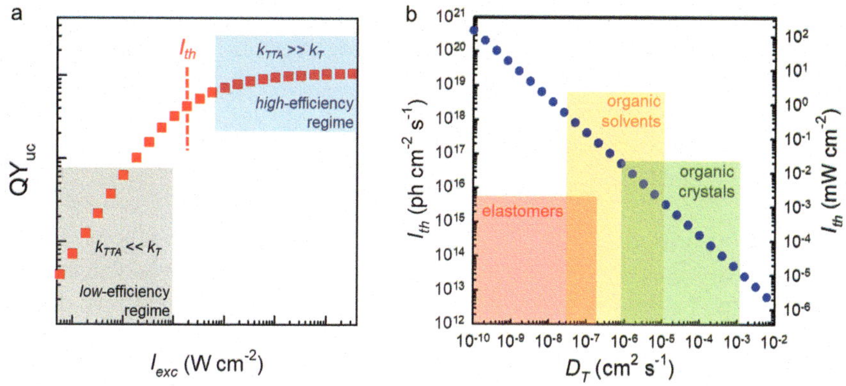

Fig. 4.1 (**a**) sTTA-UC quantum yield (QY_{uc}) as a function of the excitation intensity I_{exc}. (**b**) Excitation intensity threshold calculated using Eq. 4.9 as a function of the triplet diffusivity D_T considering $k_T = 1$ kHz, $\alpha = 10$ cm^{-1}, and an energy transfer yield $\phi_{ET} = 1$. The threshold values in W cm^{-2} are calculated considering an excitation beam of green photons at 532 nm (2.33 eV)

The key effect of the energy diffusivity on the threshold is pointed out in Fig. 4.1b, where I_{th} is calculated by considering typical average values for all the parameters in Eq. 4.9 except that for the triplet diffusivity D_T. Moving from low-viscosity solids like rubber amorphous polymers where molecule can still partially diffuse ($D_T \sim 10^{-10}$–10^{-9} cm^2 s^{-1}) [14] to crystalline systems ideal for triplet exciton diffusion ($D_T \sim 10^{-6}$–10^{-3} cm^2 s^{-1}) [15, 16], the energy diffusivity changes indeed by about four orders of magnitudes, thus demonstrating the potential of *s*TTA-UC solid-state systems for ultralow-power applications such as solar technologies and in vivo light-driven photochemistry, therapy, and optogenetics.

The rapid diffusion limit is therefore a regime where the efficiency of bimolecular phenomena is maximized. Once reached, a further increase of L_T does not have any effect on the process efficiency that is only limited by the distance of closest approach between the involved molecules. However, the realization of efficient solid-state sTTA-UC materials is not an easy task. The rapid diffusion condition requires indeed a large chromophore/exciton diffusivity, typical of low-viscosity materials, and/or a low decay rate of the involved excited states, a characteristic ensured by hosts with a large local rigidity in which the vibrational assisted quenching processes are suppressed. To date, the challenge to achieve solid-state upconverters suitable to fabricate solar devices for industrial processing is still open [17], since there are several difficulties especially with regard to the organization and accessibility of the two complementary components, i.e., the sensitizer and the annihilator/emitter, which should interact efficiently. The use of solid matrices implies further critical issues, such as the management of high emitter concentrations, control over quenching interactions, as well as molecular dynamics, which can affect negatively the other parameters in Eq. 4.9 and therefore vanish the beneficial effect of a high-diffusivity environment. Additionally, the solid-state arrangements of dyes must avoid phase separation and performance instability with time, in order to develop technologically appealing devices.

4.2 Solution-Mimicking sTTA-UC Materials

A straightforward strategy to fabricate solid *s*TTA-UC systems is to employ materials that allow a residual translational molecular mobility to the included dyes and thus enable inter-chromophore collisions [18]. A typical example of these systems is rubber polymers, such as polyacrylates and polyurethanes, which have been shown to be ideal matrices for the *s*TTA-UC with the best trade-off between their macroscopic and nanoscopic viscoelastic properties [19]. As illustrated in Fig. 4.2, macroscopic and nanoscopic properties of the polymer host play an important role in the *s*TTA-UC because they affect the annihilation rate k_{TTA} and the triplet spontaneous recombination rate k_{T}, respectively.

The residual molecular mobility that sets the triplet D_{T} in low-viscosity polymers is generally lower than the one in low-viscosity solvents. On the other side, in these soft hosts we can observe a decrease of k_{T} (i.e., an increment of the triplet lifetime) with respect to common organic solvents. This is due to a reduced vibration-assisted quenching in the rubber polymer matrix, which is, effectively, a more "rigid" environment for the metastable triplet state involved in the *s*TTA-UC. The recombination rate of a triplet T_1 can be written as

$$k_{\mathrm{T}} = k_{\mathrm{T}}^{\mathrm{rad}} + k_{\mathrm{T}}^{\mathrm{nr}}, \tag{4.10}$$

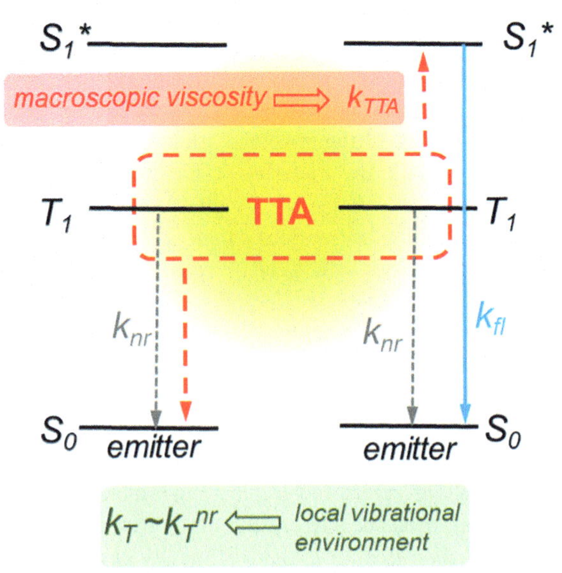

Fig. 4.2 Sketch of emitter electronic levels involved in *s*TTA-UC. A singlet fluorescent state S_1^* is generated upon annihilation of two emitter triplets T_1 (red dashed line). Dashed lines indicate non-radiative transitions. The *i*-transition is characterized by a rate k_i

which is the sum of radiative and non-radiative recombination probabilities to the ground state S_0. Considering that the triplet state is an optically dark state due to the spin conservation selection rules k_T^{rad}, its decay rate is determined by non-radiative recombination mechanisms that include the collision with inert moieties, such as solvent molecules, and the coupling with the vibrational bath of the environment. The local interactions with the surrounding polymer, even if not particularly efficient, offer in this case an additional decay pathway competing with the spontaneous decay of the long-living triplets. Therefore, also small variations of the matrix nanoscopic rigidity [20] give rise to large reduction of k_T and, according to Eq. 4.9, to an increase of the overall upconversion efficiency at low powers.

In rubber polymers, the optimized balance between residual molecular diffusivity and triplet recombination allows to achieve the rapid diffusion limit at excitation intensities close to the solar irradiance, thus supporting the use of these materials for the production of devices. Moreover, this class of polymers allows to fabricate high optical quality bulk materials with radical reaction under mild conditions. The reaction can be conducted directly from liquid solutions of dye-doped monomer, without using solvents and without subsequent mixing steps. In those conditions, branching and self-cross-linking reactions can also take place, ensuring the elastomeric behavior of the resulting materials. The good performance of rubber *s*TTA-UC films is demonstrated by their application in photovoltaic devices as shown in Fig. 4.3 [21, 22]. In perspective, the viscoelastic properties of rubber polymers can be tuned further by copolymerization or control of the cross-linking degree. In particular, in order to obtain ultralow-power upconverters Eq. 4.9 suggests that the better method is to design a sufficiently rigid host where the triplet state is almost unperturbed with respect to an organic solution, since the quadratic

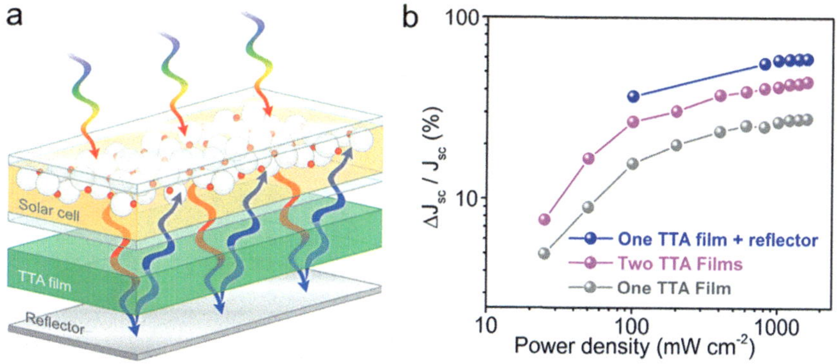

Fig. 4.3 (**a**) Schematic illustration of *s*TTA-UC film-assisted absorption of photons with energies below the HOMO−LUMO gap of the chromophore sensitizer in a dye-sensitized photoelectrochemical cell (DSPC). (**b**) Current density J_{sc} enhancement as a function of the number of *s*TTA-UC films under illumination at various intensities in DSPCs with a 10-μm-thick TiO_2 layer. Reproduced from [21], with the permission of ACS Publishing. Copyright © 2016, American Chemical Society

dependency of the intensity threshold on k_T could allow to compensate the loss in diffusivity that slows down the molecular collisions.

4.3 sTTA in the Solid State

In a molecular semiconductor even in a crystalline lattice the excitons are Frenkel type, and thus strongly localized in the generation volume to the contrary of inorganic semiconductors where excitons are widely delocalized in the crystalline periodic structure. Therefore, they can move only by hopping between resonant energy centers within the material along any allowed crystalline direction $n = x$, y, z with

$$D_n = \left(k_{\text{hop}} R\right)_n^2 \tag{4.11}$$

where the hopping rate k_{hop} between two molecules is separated by a center-to-center distance R along the n direction [23, 24]. In this weak coupling regime, the rate of exciton hopping between two molecules i and f can be expressed from Fermi's golden rule:

$$k_{\text{hop}} = k_{\text{if}} = \frac{2\pi}{\hbar} | J_{\text{if}}|^2 S_{\text{if}}, \tag{4.12}$$

where J_{if} is the excitonic coupling term and S_{if} is the overlap integral of the normalized density of states of vibronic states between the initial and the final state, i.e., the resonance required by energy conservation law. By considering triplet diffusion, the coupling term between two molecules can be written as [25]

$$J_{\text{TT}} \propto \frac{2\beta_{\text{et}}\beta_{\text{ht}}}{A} - Z \propto Ae^{-[2\theta R]}, \tag{4.13}$$

where β_{et} is the electron transfer matrix element between donor (d) and acceptor (a), and β_{ht} is the corresponding hole transfer matrix element. These are calculated from the HOMO-LUMO orbital wave functions of T_1 and S_0 states between donor and acceptor of the energy. A is the energy gap between the charge-separated (d^+a^-) and locally excited ($d^* a$) configurations, and Z is the two-electron exchange integral. Each of the β_{et} and β_{ht} are exponentially attenuated with distance; thus overall J_{TT} diminishes exponentially with the $d - a$ separation, and can often be modelled as a function of R as the center-to-center separation of the overlapping orbitals. This interaction is highly sensitive to molecular shape, size, and orientation, as evinced by the dependence on the atomic orbital coefficients and interactions determining J_{TT}. The parameter θ takes account of these factors. Similar considerations are made for Dexter-type ET, which can be treated as a hetero-molecular hopping mechanism. The nature of the physical mechanisms at the base of both ET and

TTA pushed the natural development of *s*TTA-UC systems in which triplet excitons effectively diffuse in densely organized molecular assemblies without the assistance of translational molecular diffusion [26–28]. Among various assembly systems, molecular crystals with ordered chromophore arrangements should be the potential best candidate to host *s*TTA-UC at low powers. Indeed, the close molecular packing in crystal ($R < 1$ nm) would result in a fast diffusion environment with D_T values that can outperform the ones of free molecules in low-viscosity organic solvents. Moreover, crystals have the potential ability to tune and drive energy diffusion by exploiting their intrinsic anisotropic nature. Indeed, because the molecular distances and orientations are different and fixed in any crystalline direction, there are definite $d - a$ pairs in which ET and TTA are fast or slow. This implies also that a proper alignment of TTA and ET pathways owing to their molecular packing is crucial for achieving higher upconversion efficiency in co-crystals or sensitizer-doped emitter crystals. Thus, the techniques applied for the fine-tuning of molecular packing could be tested, for example by introducing chemical groups into a DPA derivative as shown in Fig. 4.4, in order to synergistically design the ET and TTA pathways [29].

Several examples of sensitizer/emitter co-crystals have been developed and modelled, but usually they show two main drawbacks. First, co-crystals and doped crystals suffer from the aggregation of sensitizer molecules and their segregation in

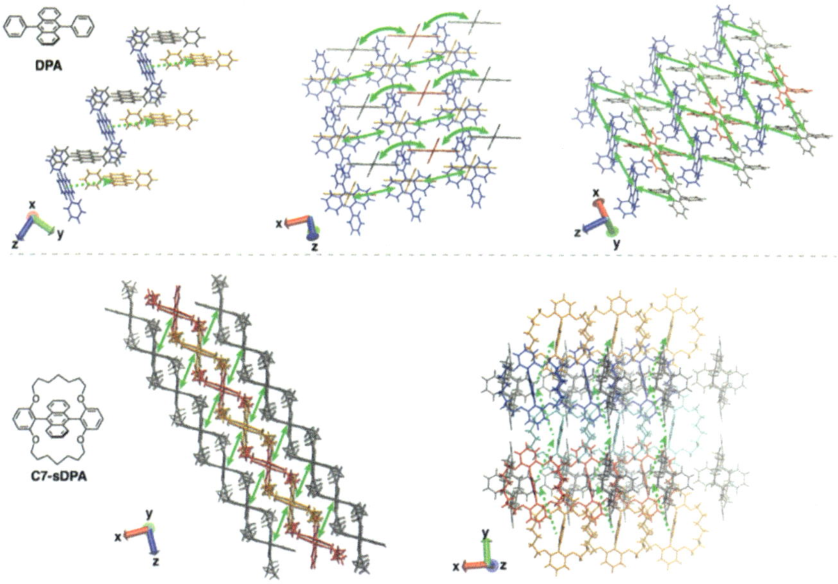

Fig. 4.4 Easy axes of triplet hopping processes of DPA and C7-sDPA drawn with their crystal structures. The green arrow means that hopping occurs between two monomers, while the green-dot arrow means that hopping does not occur due to the unfavorable orientation and distance in the *donor-acceptor* pair. Reproduced from [29], with the permission of RSC Publishing. Copyright © 2017, Royal Society of Chemistry

domains separated from the emitter crystals, especially at high sensitizer concentrations required to ensure an adequate optical density to the material. The segregation dramatically affects the efficiency of the ET by reducing the interface between sensitizer and emitter moieties, and, moreover, it represents a critical loss channel due to the formation of undesired aggregate species that can harvest energy but do not participate in the *s*TTA-UC process [30]. Second, organic crystals often show a large triplet spontaneous decay rate k_T with respect to the monomers in diluted solutions. In general, exciton recombination rates in crystals are strongly affected by the presence of defect states, which act as non-radiative recombination centers. In particular, this aspect can be especially true for triplet excitons, which have intrinsic long lifetimes and therefore high probability to encounter trap states, despite their low density, during their diffusion [4, 10]. Excluding trivial contamination by atmospheric moisture or chemical impurities, these defect-related states can result from aggregate formation, vacancies, domain interface states in polycrystalline system, or surface defects. In other words, the high diffusivity of triplet excitons determines not only the TTA rate, but also the average k_T that sets the threshold value. Moreover, the presence of defects can play a role also in determining the crystal fluorescence quantum yield QY_{fl}. The same considerations can be indeed applied to singlet emitting states, which can encounter trap states and therefore be quenching before radiative recombination resulting in a crystal PL efficiency usually lower than that for the building block monomers in dilute solution. To fabricate high-purity and defect-free organic single crystals is not an easy task. A possible partial solution of this problem has been found using metal-organic frameworks (MOFs) as annihilators/emitters [31], as shown in Fig. 4.5.

A MOF consists of an ordered network of emitter molecules interconnected by metal ions, obtained by means of self-assembly reaction [32]. These structures show high purity level and, importantly, preserve both the properties of the constituent fluorophores, which do not interact with each other, and the high exciton mobility of crystalline molecular aggregates. The MOF platform offers several advantages with respect to the single-component crystalline matrix. In MOF nanocrystals, the *s*TTA-UC is indeed successfully achieved thanks to the high triplet exciton diffusivity and to the high purity of the materials employed that limits the triplets k_T. Specifically, in MOF nanocrystal employed for sTTA-UC it has been demonstrated that a relevant role on the exciton recombination dynamic is played by surface defects that can be passivated avoiding energy losses [33]. Moreover, MOF are a platform that can realize a cooperative molecular level function due to their highly accessible and spatially discrete linkers in the framework, as shown in Fig. 4.5. With an *in situ* secondary functionality incorporation, a sensitizer can be included into a water-stable zirconium-based MOF nanocrystal, constructed from an annihilator linker. Alternatively, MOF nanocrystals can be employed as simple but efficient triplet harvesters and annihilators, which collect triplets upon interaction with sensitizers at the surfaces [34].

It is worth pointing out that order and crystallinity are not essential to achieve the *s*TTA-UC in the solid state. Besides MOFs, which can suffer from poor stability and low QY_{fl}, other disordered nanoporous architectures endowed with

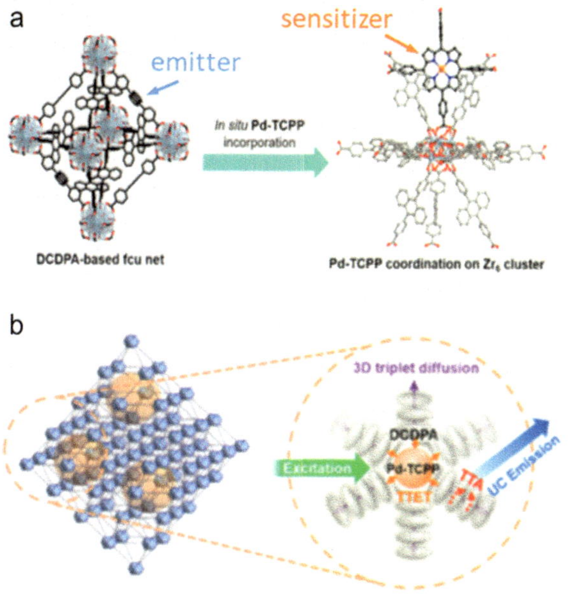

Fig. 4.5 Structure and characterization of *s*TTA-UC MOF nanocrystals. (**a**) *In situ* incorporation of Pd-TCPP (sensitizer) into DCDPA (emitter) MOF matrix through defective coordination on Zr_6 cluster. (**b**) 3D representation of *s*TTA-UC MOF nanocrystal structure with tunable Pd-TCPP-to-DCDPA ratio and schematic of *s*TTA-UC process in the system. Defective sites (orange spheres) created by missing linkers, allowing the incorporation of Pd-TCPP. Reproduced from [31], with the permission of ACS Publishing. Copyright © 2018, American Chemical Society

high surface area can be indeed engineered with active sites for targeted functions and with condensed optically active elements as building blocks. Also in certain cases, a wise choice of the struts and the constructive principles can preserve the individual electronic properties of optically active moieties, although inserted in a 3D framework. Additionally, the material porosity can be exploited to load in a controlled way the sensitizers, thus attaining full control on the absorption properties pivotal for upconversion. This is the case of nanoparticles of robust and emitting porous aromatic frameworks (*e*PAFs) with sensitizer-accessible pores [35], which were proved extremely effective as annihilators/emitters for sTTA upconversion, as shown in Fig. 4.6.

In *e*PAFs, the use of fully organic covalent bonds is a privileged route to chemically stable and highly emissive systems, since the absence of metal nodes and crystalline structure defects reduce the probability of energy trap generation (QY_{fl} is measured as high as ~70%) that quenches both singlet and triplet excitons. Indeed, particles of emitter-based *e*PAFs with included metalloporphyrin sensitizers enable green-to-blue photon upconversion with the record conversion yield of 15% in condensed-phase annihilators. Moreover, functionalized *e*PAF was engineered with permanent incorporation of the sensitizer as stably tethered through covalent bonds,

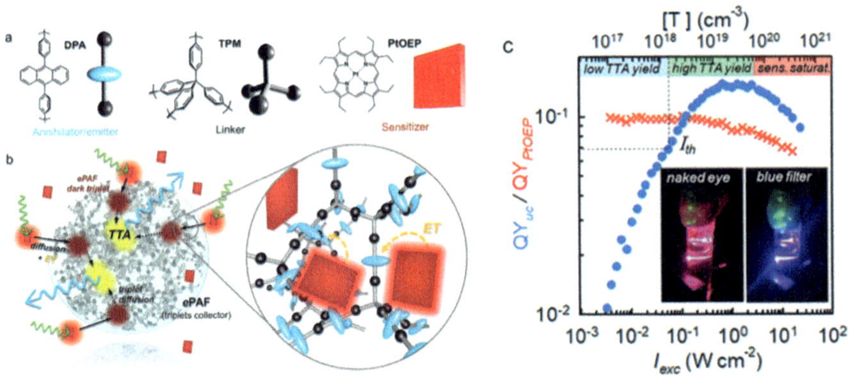

Fig. 4.6 (**a**) *s*TTA-UC in emitting *e*PAF nanoparticles. (**a**) Molecular structure of the *e*PAF building blocks; the fluorescent ligand DPA, the pore-generating building unit TPM, and the triplet sensitizer PtOEP. (**b**) Sketch of the sTTA-UC in each *e*PAF interacting with free sensitizers in dispersion. (**b**) Upconversion quantum yield QY_{uc} (dots) and residual phosphorescence yield QY_{PtOEP} (crosses) as function of the excitation intensity I_{exc}. Inset: Digital pictures of the sample under laser excitation at 532 nm with and without a blue optical filter. Reproduced from [35] with the permission of Wiley Publishing. Copyright Wiley-VCH Verlag GmbH & Co. KGaA

yielding an intimate bicomponent material at the molecular level with permanent performances.

Besides the porous case, many amorphous upconverters have also been investigated during the last decade, and upconverting films made of emitters and doped with sensitizers have been fabricated [36] using also alternative light harvester/sensitizers such as semiconductor quantum dots [37, 38]. Moreover, the magnetic field-dependent upconversion experiments demonstrated that the TTA is a process favored with respect to triplet dissociation in amorphous environments with respect to crystalline frameworks. Indeed, since the value of the dissociation rate constant is proportional to k_{hop}, and considering that disorder effects are known to reduce the hopping rate, smaller values of dissociation rates are therefore expected for films [6]. However, the sTTA-UC efficiency in rigid hosts is usually rather low because neither the molecules nor the excitons can diffuse quickly.

Due to limited diffusivity values, it is generally difficult to achieve the rapid diffusion limit in rigid amorphous systems at low powers. Nevertheless, this kind of material offers several possibilities to play with the other parameter that sets the exciton diffusion length, i.e., its spontaneous recombination rate k_T (or lifetime $\tau_T = (k_T)^{-1}$). As discussed above triplet states are extremely sensitive to the local environment, which, in typical emitter molecules employed for *s*TTA-UC as annihilator emitters where spin-orbit coupling effects are negligible, determines k_T. If the molecule is wrapped in a sufficiently rigid and dense environment also intramolecular vibrational modes can be switched off, thus eliminating internal quenching mechanisms. Therefore, the rapid diffusion limit can be obtained in ultra-rigid environments where vibrational assisted quenching is completely suppressed strongly reducing k_T and therefore compensating the low triplet exciton diffusivity. Figure 4.7 reports the measurement of k_T decrease in dual dye-doped, hyper-cross-linked polymeric polystyrene-based nanoparticles [39]. The nanoparticles have been loaded with a dye amount that enables to have intermolecular distances of ~2 nm (panel a), thus allowing both ET and hopping-mediated diffusion, even if slow. Remarkably, the high QY_{uc} observed has been achieved thanks to the extremely low value of k_T in the rigid environment, which decreases from values at around 1 kHz in organic solvents to 4 Hz in hyper-cross-linked nanoparticle. Unlike other solid-state systems in which good conversion outputs are obtained thanks to the residual mobility of the embedded dyes, here the possibility to generate high-energy singlets upon TTA is therefore granted by the extension of their natural lifetime up to a fraction of a second. Therefore, it is quite easy to achieve the *high*-efficiency regime ($k_{TTA} \gg k_T$). This finding paves the way for the realization of new classes of TTA-based upconverting solid materials that are no longer characterized by a low viscosity but, on the contrary, are extremely rigid. The demonstration of these *s*TTA-

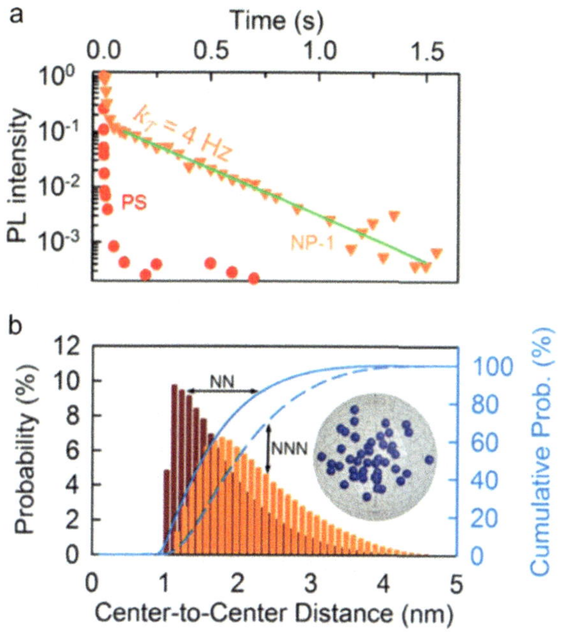

Fig. 4.7 (**a**) UC-PL decay dynamic in hyper-reticulated nanoparticles (NP-1, triangles) doped with PtOEP (sensitizer) and DPA (emitter) compared with the one in a polystyrene film (PS, circles) measured with the pump and probe method described in Sect. 4.4. (**b**) Histogram of the intermolecular stochastic nearest-neighbor (NN) and next-nearest-neighbor (NNN) distance distribution of an ensemble of 50 DPA molecules embedded in a spherical NP of 7 nm radius. Continuous and dashed lines are the corresponding cumulative probabilities. The inset depicts one example of the 10^5 randomly generated distributions. Reproduced from [39], with the permission of ACS Publishing. Copyright © 2016, American Chemical Society

UC dynamics in nanosized systems adds further degrees of freedom to the material design because upconverting particles can be used as is, for instance, in bioimaging applications, or can be included in suitable plastic matrixes for realizing optically active bulk materials. It is worth noting that, once properly optimized, these systems may be effective at exceptionally low photon fluxes for the very reason that the ultimate excitation power threshold is determined just by the triplet natural lifetime.

4.4 Kinetics of Confined sTTA-UC in Nanostructured Solids and Nanomaterials

Among *s*TTA-UC systems, many examples of nanosized/nanostructured materials have been developed so far [40–43]. Conceptually, the design of these systems is aimed to obtain easy-to-handle, individual upconverters, or to achieve closely packed emitter ensemble where triplet exciton can diffuse and experience TTA. It is worth noting that, depending on the triplet diffusion length L_T in the host material, these *s*TTA-UC systems can show quite different dynamics in the generation of the upconverted light, which require special care to be investigated in order to get insight on the intrinsic system properties necessary to optimize the design of efficient upconverters. Figure 4.8 is an illustrative sketch that shows the TTA dynamics in classical *s*TTA-UC systems under pulsed excitation.

Here triplet excitons are created and move in a bulk volume without boundaries, i.e., in a standard solution. After the excitation pulse, triplets immediately start to annihilate generating upconverted light, which intensity progressively decreases with the time until the density of triplets $[T]$ is reduced to the point that $[T] \ll R$, where R is the average intermolecular distance between two triplets, thus avoiding their encounter. This time-dependent TTA efficiency can be analytically treated, and the evolution in time of the upconverted PL intensity I_{uc} can be described by [44]

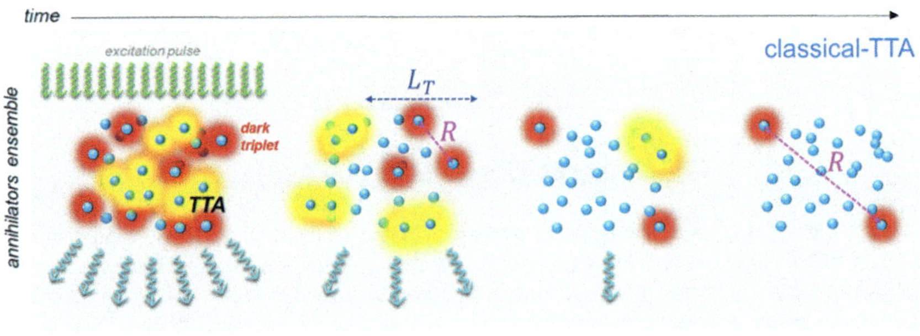

Fig. 4.8 Sketch of the TTA mechanism upon pulsed excitation in a classical bulk material, where the average distance between triplet pairs R changes in time. The diffusion length of the triplet exciton is marked as L_T. Reproduced from [34], with the permission of ACS Publishing. Copyright © 2019, American Chemical Society

$$I_{uc}(t) \propto \left(\frac{1 - \phi_{TTA}}{e^{k_T t} - \phi_{TTA}} \right)^2, \tag{4.14}$$

where ϕ_{TTA} is the TTA yield right after the excitation pulse. It is worth noting that at very long times the triplet density will become so low that the TTA rate is negligible with respect to k_T. In this condition, the ϕ_{TTA} is close to zero and therefore Eq. 4.10 is dominated by the denominator exponential term, thus resulting in

$$I_{uc}(t) \propto e^{-2k_T t}. \tag{4.15}$$

According to Eqs. 4.10 and 4.11, the investigation of upconverted PL recombination dynamics by means of time-resolved PL experiments is therefore paramount to fully model the *s*TTA-UC and to obtain key parameters such as k_T which cannot be obtained easily with other techniques.

Strikingly different UC-PL recombination dynamics can be observed in nano-sized/nanostructured systems where triplet excitons are generated in separated and noninteracting domains. Diffusion lengths of triplets range from hundreds of nanometers to several micrometers [45]. Therefore it is not unrealistic that, in a nanostructure of size *d*, the relationship $L_T \gg d$ is verified. This condition defines the regime of *confined* *s*TTA-UC, i.e., when triplets are physically confined in a limited volume smaller than the one potentially explored by random diffusion. This implies that, since the maximum intermolecular distance between two triplets created in the same structure is shorter than L_T, all the triplet pairs generated on a single triplet vessel annihilate before spontaneous recombination avoiding energy dissipation, thus achieving a $\phi_{TTA} = 100\%$. Consequently, each nanoparticle/nanocrystal/nanostructure can be considered as an individual, thresholdless annihilator under the only condition that at least two triplets are created simultaneously; otherwise, sTTA-UC is not allowed and isolated triplets decay spontaneously. This peculiar annihilation dynamic can be evidenced by means of time-resolved photoluminescence spectroscopy experiments. As sketched in Fig. 4.9, under pulsed

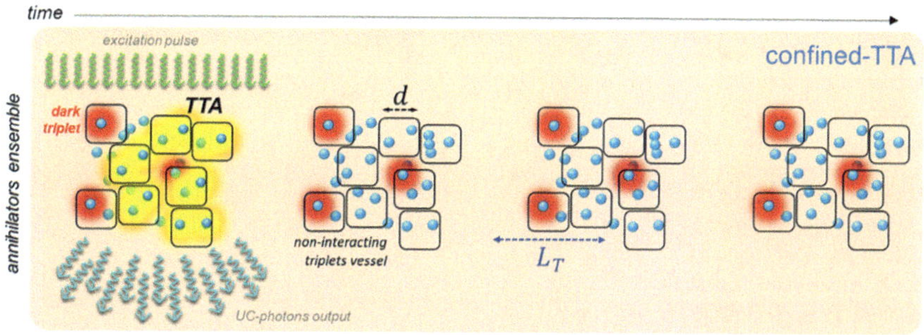

Fig. 4.9 Sketch of the TTA mechanism upon pulsed excitation in a confined system, where triplet pairs are collected in noninteracting vessels of size *d* shorter than their diffusion length L_T forcing their annihilation. Reproduced from [34], with the permission of ACS Publishing. Copyright © 2019, American Chemical Society

excitation the upconverted light is immediately generated in nanostructures where at least two triplets are created, while no emission can be detected at longer times from vessels where a single triplet survives.

This behavior is radically different from classical *s*TTA-UC in bulk materials and, importantly, features different experimental evidences. Because in a set of isolated objects, such as nanoparticles or nanocrystals where the *confined* TTA regime is reached, the optically detectable upconverted emission can be generated only when the same particle contains at least two excitons (Fig. 4.10a), the forced TTA prevents obtaining k_T from the standard time-resolved PL measurements.

Standard measurements with pulsed excitation allow therefore to estimate only k_{TTA} from the decay of upconverted PL with time, which does not show the usual dependency on the excitation intensity [46], with no clues on the intrinsic properties of the triplet state involved. In order to bypass this problem, the materials must be studied by means of a pump-and-probe experiment in which the system is first excited with a strong light pulse, that pushes the systems in steady-state condition, and then probed with additional low-intensity short pulses at different delays Δt. An example of the pulse sequence is reported in Fig. 4.10b [39]. With this method, we activate at different times the nanostructures where only one emitter triplet is present, that is, before its spontaneous recombination, without altering significantly the overall density of excited states, thus obtaining a direct measurement k_T since

$$I_{uc}^{confined}(t) \propto e^{-k_T t}. \tag{4.16}$$

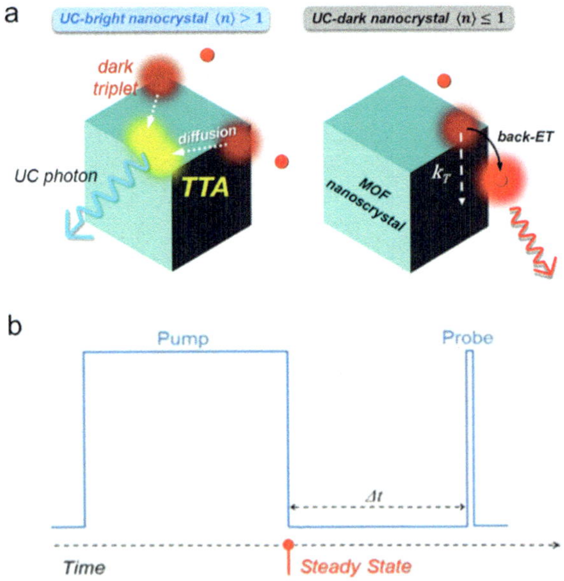

Fig. 4.10 (**a**) Sketch of the TTA mechanism upon pulsed excitation in a confined system, where triplet pairs are collected in noninteracting vessels of size *d* shorter than their diffusion length L_T forcing their annihilation. (**b**) Outline of the pump and probe laser excitation sequence employed to measure the spontaneous decay rate of triplet excitons in confined systems. Reproduced from [34, 39], with the permission of ACS Publishing. Copyright © 2016, American Chemical Society. Copyright © 2019, American Chemical Society

Remarkably, *confined* TTA can occur in materials with different structural and diffusivity properties because the key parameter to consider, L_T, depends on both the diffusivity and lifetime of triplet state that can be hardly quantified a priori. This peculiar *s*TTA-UC regime has been hinted firstly in hyper-reticulated nanoparticles loaded with both sensitizer and emitter moieties [39]. The triplet diffusion in this case is very slow, driven by random hopping of the exciton between the few emitters embedded in the nanoparticle. Nevertheless, the ultra-rigid environment wrapping the dyes increases the triplet lifetime up to second timescale, thus enabling much larger than the host nanoparticle size and thus reaching the confined regime condition. Organic crystals show the highest triplet diffusivity values observed and, it is not surprising that confined TTA has been observed in metal-organic framework nanocrystals with sizes of few tens of nanometers [34]. Conversely, in other porous but amorphous systems such as *e*PAFs, the triplet diffusivity by Dexter-mediated hopping is not fast enough to compensate the short triplet lifetime; thus the material behaves as a classical bulk upconverter [35].

In addition to different recombination dynamics, also the dependency of upconversion quantum efficiency QY_{uc} on the excitation intensity I_{exc} is different between classical and confined regimes. Figure 4.11 shows a comparison of QY_{uc} vs. I_{exc}

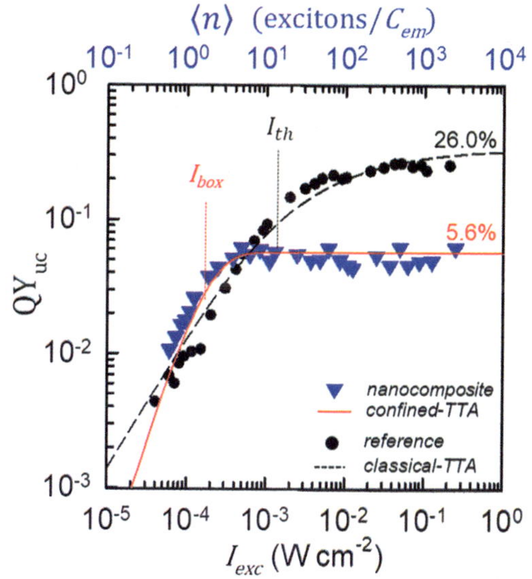

Fig. 4.11 Comparison of the upconversion quantum yield QY_{uc} measured as a function of the excitation intensity I_{exc} for the sTTA-UC nanocomposite (triangles) and for a standard DPA/PtOEP (10^{-2} M:10^{-4} M) solution in tetrahydrofuran (dots). The conversion efficiency for the nanocomposite is reported also as a function of the average number of excitons per nanocrystal n calculated for each I_{exc}. Dashed and solid lines are the theoretical QY_{uc} predicted, respectively, for the classical reference solution (classical-TTA) and for the thresholdless upconverting MOF nanocrystals (confined TTA). Reproduced from [34], with the permission of ACS Publishing. Copyright © 2016, American Chemical Society

for a *s*TTA-UC nanocomposite exploiting MOF nanocrystals as annihilator emitters (triangles) and for a standard dye solution (dots). The reference shows the typical behavior of bulk systems where triplet excitons are free to move in an infinite volume. Until saturation at high powers, the upconversion QY_{uc} rises as follows:

$$QY_{uc}(I_{exc}) = \frac{1}{2} f\ QY_{fl} \Phi_{ET} \Phi_{TTA}(I_{exc}) = \frac{1}{2} f\ QY_{fl} \Phi_{ET} \left[\frac{k_{TTA}(I_{exc})}{k_{TTA}(I_{exc}) + k_T} \right] \tag{4.17}$$

where f is the statistical probability to have a singlet upon TTA, QY_{fl} is the emitter fluorescence quantum yield, and Φ_{ET} is ET efficiency. The nanocomposite behaves differently, with a sharper transition from the low-efficiency regime to the saturation that derives from the different annihilation dynamics in a confined environment. As shown in Fig. 4.10a two scenarios are possible. If two excitons coexist in a MOF nanocrystal, they annihilate producing a fluorescent singlet, thus giving rise to a UC-bright nanocrystal with $\Phi_{TTA} = 1$ that is a thresholdless annihilator. Conversely, if only one exciton is present, the nanocrystal remains UC-dark. In this case, the triplet dissipates its energy by back-ET or spontaneous recombination. Therefore, the performance of the nanocomposite must be described by considering the statistical distribution of the exciton population in the nanocrystal ensemble as a function of the excitation intensity, instead of in terms of the power- and diffusion-dependent annihilation rates. In this case, the conversion yield is

$$QY_{uc}^{confined}(I_{exc}) = \frac{1}{2} f\ QY_{fl} \Phi_{ET} P_{n \geq 2}(I_{exc}), \tag{4.18}$$

where P_n is the cumulative probability to have at least n excitons in the same nMOF in the absence of annihilation that can be calculated by means of the Poisson distribution function [34]. $P_{n \geq 2}$ quantifies the fraction of UC-bright nanocrystals, i.e., the upconversion active population, and depends only on I_{exc}, which sets the average number of excitons per nanocrystals $\langle n \rangle$ that determines the center of the Poisson distribution.

4.5 General Remarks

Regardless of the application, *s*TTA-UC is a diffusion-limited process which efficiency depends on several parameters, such as energy diffusivity, excited-state lifetime, and density of available excited states. Any kind of solid-state material must be therefore designed in order to find the best trade-off between diffusivity and recombination kinetics in order to achieve the *high*-efficiency regime where the TTA rate k_{TTA} is much larger than the spontaneous decay rate of triplets k_T, thus completely avoiding energy losses, at the lowest excitation intensity possible, according to the application demands. Considering the intrinsic nature of the processes involved in the *s*TTA-UC several guidelines can be outlined for the design of solid-state upconverters:

1. The system must contain large amounts of annihilator/emitter moiety. This condition is required not only to maximize the ET, but also to guarantee to triplets the possibility to move by hopping-mediated diffusion within the emitter ensemble (both in ordered and disordered environments).
2. The triplet recombination rate plays a crucial role k_T. Given the quadratic dependency of the excitation intensity its control by tuning both the emitter molecular structure and its local environment appears as the most effective strategy to boost the upconversion yield at low powers.
3. The role of light harvester/sensitizer must be carefully considered. The absorption cross section and concentration of sensitizer ultimately set the density of triplets that are injected in the systems and participate in the TTA process. Consequently, a solid upconverter must contain a large amount of sensitizers in order to have a suitable optical density. In classical bulk upconverters, the coexistence of intimately mixed sensitizers and emitter populations at high concentrations can be made difficult by several problems related to solubility and phase segregation. To the contrary, several nanoparticles and nanostructured materials showed better versatility in the control of the dye loading level and composition properties. Therefore, a potential strategy to fabricate efficient bulk upconverters is to develop bulk host with embedded nanosized upconverters or directly fabricate nanostructured system materials as shown in Fig. 4.12. This

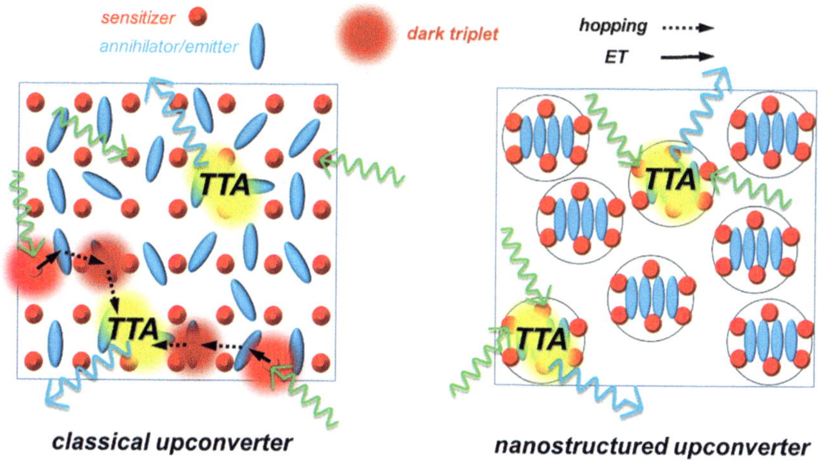

Fig. 4.12 Sketch of a classical TTA-based upconverter where dyes are homogenously dispersed in the available volume compared to a nanostructured material where the same amount of chromophores are confined in small domains to maximize *s*TTA-UC, thanks to the locally enhanced triplet density

configuration can present a fundamental advantage with respect to a classical bulk. For a given fixed dye concentration, the confinement of the active species in a fraction of the available volume, i.e., within the nanostructure, results in an increased local density of dyes with respect to the corresponding homogeneously loaded bulk. This incremented dye density translates to a locally enhanced absorptance, and therefore to a locally increased triplet density that boosts the TTA rate and yield with respect to the corresponding classical configuration.

References

1. S. Baluschev, T. Miteva, V. Yakutkin, G. Nelles, A. Yasuda, G. Wegner, Up-conversion fluorescence: noncoherent excitation by sunlight. Phys. Rev. Lett. **97**(14), 143903 (2006). https://doi.org/10.1103/PhysRevLett.97.143903
2. D.L. Dexter, A theory of sensitized luminescence in solids. J. Chem. Phys. **21**(5), 836–850 (1953). https://doi.org/10.1063/1.1699044
3. M. Inokuti, F. Hirayama, Influence of energy transfer by the exchange mechanism on donor luminescence. J. Chem. Phys. **43**(6), 1978–1989 (1965). https://doi.org/10.1063/1.1697063
4. A. Köhler, H. Bässler, What controls triplet exciton transfer in organic semiconductors? J. Mater. Chem. **21**(12), 4003–4011 (2011). https://doi.org/10.1039/C0JM02886J
5. R.C. Johnson, R.E. Merrifield, Effects of magnetic fields on the mutual annihilation of triplet excitons in anthracene crystals. Phys. Rev. B **1**(2), 896–902 (1970). https://doi.org/10.1103/PhysRevB.1.896
6. J. Mezyk, R. Tubino, A. Monguzzi, A. Mech, F. Meinardi, Effect of an external magnetic field on the up-conversion photoluminescence of organic films: the role of disorder in triplet-triplet annihilation. Phys. Rev. Lett. **102**(8), 087404 (2009). https://doi.org/10.1103/PhysRevLett.102.087404
7. F. Perrin, Law governing the diminution of fluorescent power as a function of concentration. Compt. Rend. **178**, 1978–1980 (1924)
8. A. Monguzzi, R. Tubino, F. Meinardi, Upconversion-induced delayed fluorescence in multicomponent organic systems: role of Dexter energy transfer. Phys. Rev. B **77**(15), 155122 (2008). https://doi.org/10.1103/PhysRevB.77.155122
9. O.V. Mikhnenko, P.W.M. Blom, T.-Q. Nguyen, Exciton diffusion in organic semiconductors. Energy Environ. Sci. **8**(7), 1867–1888 (2015). https://doi.org/10.1039/C5EE00925A
10. M. Pope, C.E. Swenberg, *Electronic Processes in Organic Crystals and Polymers* (Oxford University Press on Demand, New York, 1999)
11. L. Stryer, D.D. Thomas, C.F. Meares, Diffusion-enhanced fluorescence energy transfer. Annu. Rev. Biophys. Bioeng. **11**(1), 203–222 (1982)
12. M.A. Islam, Einstein–Smoluchowski diffusion equation: a discussion. Phys. Scr. **70**(2–3), 120–125 (2004). https://doi.org/10.1088/0031-8949/70/2-3/008
13. A. Monguzzi, J. Mezyk, F. Scotognella, R. Tubino, F. Meinardi, Upconversion-induced fluorescence in multicomponent systems: steady-state excitation power threshold. Phys. Rev. B **78**(19), 195112 (2008). https://doi.org/10.1103/PhysRevB.78.195112
14. A. Monguzzi, F. Bianchi, A. Bianchi, M. Mauri, R. Simonutti, R. Ruffo, R. Tubino, F. Meinardi, High efficiency up-converting single phase elastomers for photon managing applications. Adv. Energy Mater. **3**(5), 680–686 (2013). https://doi.org/10.1002/aenm.201200897
15. V. Ern, Anisotropy of triplet exciton diffusion in anthracene. Phys. Rev. Lett. **22**(8), 343–345 (1969). https://doi.org/10.1103/PhysRevLett.22.343
16. L. Grisanti, Y. Olivier, L. Wang, S. Athanasopoulos, J. Cornil, D. Beljonne, Roles of local and nonlocal electron-phonon couplings in triplet exciton diffusion in the anthracene crystal. Phys. Rev. B **88**(3), 035450 (2013). https://doi.org/10.1103/PhysRevB.88.035450

17. V. Gray, K. Moth-Poulsen, B. Albinsson, M. Abrahamsson, Towards efficient solid-state triplet–triplet annihilation based photon upconversion: supramolecular, macromolecular and self-assembled systems. Coord. Chem. Rev. **362**, 54–71 (2018). https://doi.org/10.1016/j.ccr.2018.02.011
18. Y.C. Simon, C. Weder, Low-power photon upconversion through triplet–triplet annihilation in polymers. J. Mater. Chem. **22**(39), 20817–20830 (2012)
19. A. Monguzzi, M. Mauri, A. Bianchi, M.K. Dibbanti, R. Simonutti, F. Meinardi, Solid-state sensitized upconversion in polyacrylate elastomers. J. Phys. Chem. C **120**(5), 2609–2614 (2016). https://doi.org/10.1021/acs.jpcc.6b00223
20. D.C. Thévenaz, A. Monguzzi, D. Vanhecke, R. Vadrucci, F. Meinardi, Y.C. Simon, C. Weder, Thermoresponsive low-power light upconverting polymer nanoparticles. Mater. Horiz. **3**(6), 602–607 (2016). https://doi.org/10.1039/C6MH00290K
21. C. Li, C. Koenigsmann, F. Deng, A. Hagstrom, C.A. Schmuttenmaer, J.-H. Kim, Photocurrent enhancement from solid-state triplet–triplet annihilation upconversion of low-intensity, low-energy photons. ACS Photonics **3**(5), 784–790 (2016). https://doi.org/10.1021/acsphotonics.5b00694
22. A. Monguzzi, A. Oertel, D. Braga, A. Riedinger, D.K. Kim, P.N. Knüsel, A. Bianchi, M. Mauri, R. Simonutti, D.J. Norris, F. Meinardi, Photocatalytic water-splitting enhancement by sub-bandgap photon harvesting. ACS Appl. Mater. Interfaces **9**(46), 40180–40186 (2017). https://doi.org/10.1021/acsami.7b10829
23. A. Einstein, Über die von der molekularkinetischen Theorie der Wärme geforderte Bewegung von in ruhenden Flüssigkeiten suspendierten Teilchen. Ann. Phys. **322**(8), 549–560 (1905)
24. M. Von Smoluchowski, Zur kinetischen theorie der brownschen molekularbewegung und der suspensionen. Ann. Phys. **326**(14), 756–780 (1906)
25. G.D. Scholes, Long-range resonance energy transfer in molecular systems. Annu. Rev. Phys. Chem. **54**(1), 57–87 (2003). https://doi.org/10.1146/annurev.physchem.54.011002.103746
26. S.H.C. Askes, V.C. Leeuwenburgh, W. Pomp, H. Arjmandi-Tash, S. Tanase, T. Schmidt, S. Bonnet, Water-dispersible silica-coated upconverting liposomes: can a thin silica layer protect TTA-UC against oxygen quenching? ACS Biomater Sci. Eng. **3**(3), 322–334 (2017). https://doi.org/10.1021/acsbiomaterials.6b00678
27. P. Duan, N. Yanai, H. Nagatomi, N. Kimizuka, Photon upconversion in supramolecular gel matrixes: spontaneous accumulation of light-harvesting donor–acceptor arrays in nanofibers and acquired air stability. J. Am. Chem. Soc. **137**(5), 1887–1894 (2015). https://doi.org/10.1021/ja511061h
28. T. Ogawa, N. Yanai, A. Monguzzi, N. Kimizuka, Highly efficient photon upconversion in self-assembled light-harvesting molecular systems. Sci. Rep. **5**(1), 10882 (2015). https://doi.org/10.1038/srep10882
29. K. Kamada, Y. Sakagami, T. Mizokuro, Y. Fujiwara, K. Kobayashi, K. Narushima, S. Hirata, M. Vacha, Efficient triplet–triplet annihilation upconversion in binary crystalline solids fabricated via solution casting and operated in air. Materials Horizons **4**(1), 83–87 (2017)
30. A. Monguzzi, R. Tubino, S. Hoseinkhani, M. Campione, F. Meinardi, Low power, non-coherent sensitized photon up-conversion: modelling and perspectives. Phys. Chem. Chem. Phys. **14**(13), 4322–4332 (2012)
31. J. Park, M. Xu, F. Li, H.-C. Zhou, 3D long-range triplet migration in a water-stable metal–organic framework for upconversion-based ultralow-power in vivo imaging. J. Am. Chem. Soc. **140**(16), 5493–5499 (2018). https://doi.org/10.1021/jacs.8b01613
32. H.-C. Zhou, J.R. Long, O.M. Yaghi, *Introduction to Metal–Organic Frameworks* (ACS Publications, 2012)
33. A. Monguzzi, M. Ballabio, N. Yanai, N. Kimizuka, D. Fazzi, M. Campione, F. Meinardi, Highly fluorescent metal–organic-framework nanocomposites for photonic applications. Nano Lett. **18**(1), 528–534 (2018). https://doi.org/10.1021/acs.nanolett.7b04536
34. F. Meinardi, M. Ballabio, N. Yanai, N. Kimizuka, A. Bianchi, M. Mauri, R. Simonutti, A. Ronchi, M. Campione, A. Monguzzi, Quasi-thresholdless photon upconversion in metal–organic framework nanocrystals. Nano Lett. **19**(3), 2169–2177 (2019). https://doi.org/10.1021/acs.nanolett.9b00543

35. J. Perego, J. Pedrini, C.X. Bezuidenhout, P.E. Sozzani, F. Meinardi, S. Bracco, A. Comotti, A. Monguzzi, Engineering porous emitting framework nanoparticles with integrated sensitizers for low-power photon upconversion by triplet fusion. Adv. Mater. **31**(40), 1903309 (2019)
36. R.R. Islangulov, J. Lott, C. Weder, F.N. Castellano, Noncoherent low-power upconversion in solid polymer films. J. Am. Chem. Soc. **129**(42), 12652–12653 (2007)
37. D. Beery, J.P. Wheeler, A. Arcidiacono, K. Hanson, CdSe quantum dot sensitized molecular photon upconversion solar cells. ACS Appl. Energy Mater. **3**(1), 29–37 (2020). https://doi.org/10.1021/acsaem.9b01765
38. M. Wu, D.N. Congreve, M.W. Wilson, J. Jean, N. Geva, M. Welborn, T. Van Voorhis, V. Bulović, M.G. Bawendi, M.A. Baldo, Solid-state infrared-to-visible upconversion sensitized by colloidal nanocrystals. Nat. Photonics **10**(1), 31–34 (2016)
39. A. Monguzzi, M. Mauri, M. Frigoli, J. Pedrini, R. Simonutti, C. Larpent, G. Vaccaro, M. Sassi, F. Meinardi, Unraveling triplet excitons photophysics in hyper-cross-linked polymeric nanoparticles: toward the next generation of solid-state upconverting materials. J. Phys. Chem. Lett. **7**(14), 2779–2785 (2016). https://doi.org/10.1021/acs.jpclett.6b01115
40. O.S. Kwon, J.-H. Kim, J.K. Cho, J.-H. Kim, Triplet–triplet annihilation upconversion in CdS-decorated SiO2 nanocapsules for sub-bandgap photocatalysis. ACS Appl. Mater. Interfaces **7**(1), 318–325 (2015)
41. S.N. Sanders, M.K. Gangishetty, M.Y. Sfeir, D.N. Congreve, Photon upconversion in aqueous nanodroplets. J. Am. Chem. Soc. **141**(23), 9180–9184 (2019)
42. R. Vadrucci, A. Monguzzi, F. Saenz, B.D. Wilts, Y.C. Simon, C. Weder, Nanodroplet-containing polymers for efficient low-power light upconversion. Adv. Mater. **29**(41), 1702992 (2017)
43. C. Wohnhaas, K. Friedemann, D. Busko, K. Landfester, S. Baluschev, D. Crespy, A. Turshatov, All organic nanofibers as ultralight versatile support for triplet–triplet annihilation upconversion. ACS Macro Lett. **2**(5), 446–450 (2013)
44. Y.Y. Cheng, B. Fückel, T. Khoury, R.G. Clady, M.J. Tayebjee, N. Ekins-Daukes, M.J. Crossley, T.W. Schmidt, Kinetic analysis of photochemical upconversion by triplet−triplet annihilation: beyond any spin statistical limit. J. Phys. Chem. Lett. **1**(12), 1795–1799 (2010)
45. H. Najafov, B. Lee, Q. Zhou, L.C. Feldman, V. Podzorov, Observation of long-range exciton diffusion in highly ordered organic semiconductors. Nat. Mater. **9**(11), 938–943 (2010)
46. S. Mattiello, A. Monguzzi, J. Pedrini, M. Sassi, C. Villa, Y. Torrente, R. Marotta, F. Meinardi, L. Beverina, Self-assembled dual dye-doped nanosized micelles for high-contrast up-conversion bioimaging. Adv. Funct. Mater. **26**(46), 8447–8454 (2016)

Chapter 5
Organic Triplet Photosensitizers for Triplet-Triplet Annihilation Upconversion

Zafar Mahmood, Shaomin Ji, Jianzhang Zhao, Mushraf Hussain, Farhan Sadiq, Noreen Rehmat, and Muhammad Imran

Abstract Triplet-triplet annihilation (TTA) upconversion, a photophysical phenomenon of conversion of low-energy photon into high-energy photon, is a promising approach due to its low excitation power requirement and high upconversion quantum yields. Most of the triplet photosensitizers (PSs) used for TTA upconversion are based on transition metal complexes and organic chromophores containing heavy atom. Recently TTA upconversion with heavy atom-free PSs has gained much attention due to its minimal detrimental effects. In this chapter we summarize organic triplet PSs with different intersystem crossing (ISC) mechanism used in TTA upconversion. We briefly describe different types of triplet PSs for TTA upconversion such as heavy atom-containing organic chromophores, BODIPY dimers involving exciton coupling, and electron spin convertor-bearing dyes. Recently new triplet PSs involving charge recombination (CR)-induced ISC and thermally activated delayed fluorescence (TADF) have been used in the TTA upconversion. We discuss the benefits of these PSs for TTA upconversion. This book chapter provides current progress in the field of TTA upconversion and designing new efficient triplet PSs.

Z. Mahmood
School of Chemical Engineering and Light Industry, Guangdong University of Technology, Guangzhou, P. R. China

State Key Laboratory of Fine Chemicals, School of Chemical Engineering, Dalian University of Technology, Dalian, P. R. China

S. Ji (✉)
School of Chemical Engineering and Light Industry, Guangdong University of Technology, Guangzhou, P. R. China
e-mail: smji@gdut.edu.cn

J. Zhao (✉) · M. Hussain · F. Sadiq · N. Rehmat · M. Imran
State Key Laboratory of Fine Chemicals, School of Chemical Engineering, Dalian University of Technology, Dalian, P. R. China
e-mail: zhaojzh@dlut.edu.cn

J. S. Lissau, M. Madsen (eds.), *Emerging Strategies to Reduce Transmission and Thermalization Losses in Solar Cells*, https://doi.org/10.1007/978-3-030-70358-5_5

Keywords Excited state · Intersystem crossing · Photochemistry · Triplet photosensitizer · Triplet-triplet annihilation upconversion

5.1 Introduction

Luminescent materials gained much attention due to their wide range of applications in the area of molecular sensors and optical devices. These applications are based on efficient luminescent materials including organic dyes [1], metal complexes [2–6], and fluorescent proteins [7]. However, these luminescent materials share a common drawback of emitting low-energy photons as compared to excitation light; that is, these materials are down-conversion materials [8].

Photon upconversion is the process of conversion of low-energy photon into higher energy emission [9, 10]. In other words, it is a phenomenon which involves absorption of low-energy photon and upconverted emission of high-energy photon at shorter wavelength (anti-Stokes shift) than the excitation light. Upconversion materials gained considerable attention due to its potential applications for artificial photosynthesis, photocatalysis [11–13], optics [9], nanomaterial for biomedicines [10, 14–16], and especially solar cells with improved efficiency [17–20]. Solar light comprises more than 40% of near-infrared radiations which is difficult to be utilized by dye-sensitized solar cell (DSCs; Fig. 5.1). The efficiency of DSCs can be improved by using the upconversion material that can convert radiations of longer wavelength into radiations of shorter wavelengths.

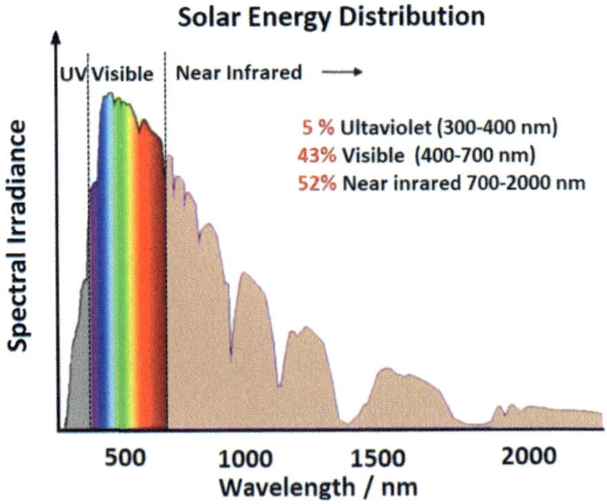

Fig. 5.1 Modified solar energy distribution diagram

Brief Introduction of Upconversion Mechanisms

The process of photon upconversion was first discovered by F. Auzel over 60 years ago [21]. Since then, it has been a focus of much interest, especially NIR-to-visible upconversion, to enhance the utilization of solar energy and improve the efficiency of solar cells. The normal fluorescence of a compound emits photons which are lower in energy than the excitation photons. The energy difference between the emitted photon and the excitation photon is termed Stokes shift. Photon upconversion is with anti-Stokes; that is, it emits high-energy photons after excitation with low-energy photons. In photon upconversion normally two or more photons of low energy are absorbed to emit one photon of higher energy. Upconversion process, especially for rare earth materials, is mainly divided into three broad categories: excited-state absorption (ESA), energy transfer upconversion (ETU), and photon avalanche (PA) [22]. All these upconversion processes involve sequential absorption of two or more photons and are different from simple multiphoton absorption process that involves simultaneous absorption of multiple photons. Currently different techniques are used for upconversion, including upconversion with inorganic crystals, upconversion using two photon-absorbing dyes (TPA), and upconversion with rare earth materials [10, 23–25]. However, all these techniques suffer from some drawbacks such as weak absorption of visible light, low upconversion quantum yield, and requirement of high-power coherent excitation source (i.e., laser, picosecond, or femtosecond pulsed laser) [26]. These drawbacks limit their application, and make them unsuitable to implement in applications such as solar cells which are based on low-power-density excitation source (solar light).

5.2 Triplet-Triplet Annihilation Upconversion

Recently a new promising approach via triplet-triplet annihilation (TTA) upconversion has drawn much attention because it requires only low-power-density non-coherent excitation source (e.g., sunlight is sufficient) [27–30]. About 60 years ago, Parker and Hatchard observed this phenomenon in the solution containing energy donor/acceptor, i.e., phenanthrene and anthracene [31, 32]. Upon selective excitation of donor, ultraviolet upconverted emission from the acceptor was observed. Although the phenomenon of TTA upconversion has a long history, it was sparingly investigated in the last decade. Recently, studies of TTA upconversion have experienced significant progress [26–28], because it could be used in solar cells to increase the efficiency, etc. Numerous PSs with efficient ISC have been devised to achieve high upconversion quantum yield in the visible range [33–36].

TTA upconversion has a few advantages over the abovementioned conventional upconversion techniques. The main advantages are that TTA upconversion requires very low excitation source, only few mW cm^{-2} power density is sufficient to excite the upconversion process, and it is not necessary to be coherent [37]. Another major

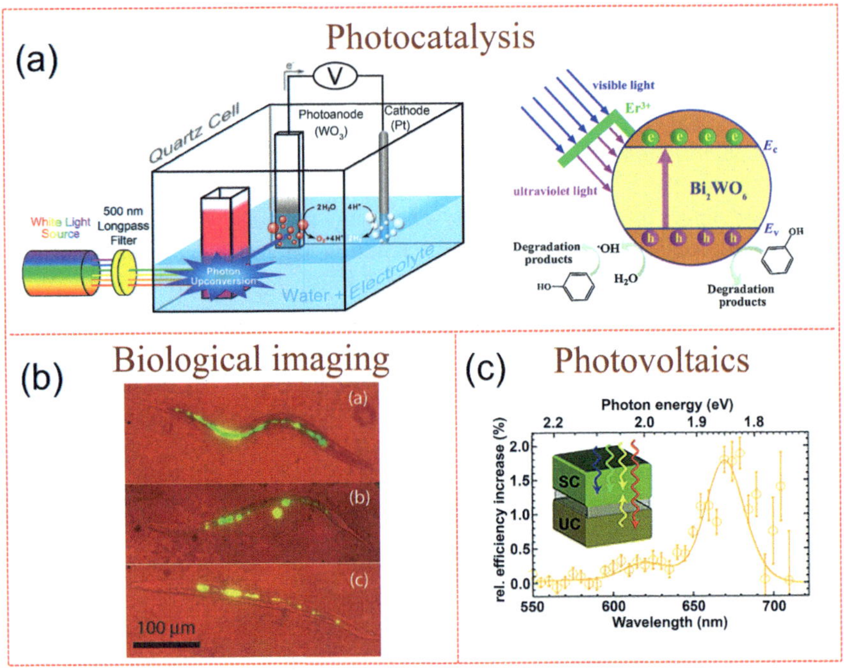

Fig. 5.2 TTA upconversion applications; (**a**) photocatalysis, reproduced from [11] with permission of Elsevier publisher and from [13] with permission of the Royal Chemical Society; (**b**) biological imaging, reproduced from [15] with permission of the American Chemical Society; (**c**) photovoltaics, from [17] with permission of the Royal Chemical Society

advantage of TTA upconversion is that the wavelength of upconverted emission can be tuned by careful selection of donor and acceptor units. In short words, TTA upconversion is a promising approach for the applications of photocatalysis, photovoltaics, and some other light-activated photophysical processes (Fig. 5.2).

Principle of TTA Upconversion

The mechanism of TTA upconversion involves a bimolecular system comprising a triplet photosensitizer (donor) and triplet acceptor (emitter), and can be explained by the following Jablonski diagram (Fig. 5.3).

(I) Initially the excitation of triplet PS (donor) using low-power coherent excitation source populates its singlet excited state ($^1S^*$) which subsequently populates its triplet excited state ($^3S^*$) through ISC (non-radiative relaxation pathway). (II) Then triplet state of acceptor ($^3A^*$) populates through triplet-triplet energy transfer (TTET) from triplet PS to acceptor. (III) Afterwards, the two triplet states of acceptor annihilate producing one high-energy singlet excited state of acceptor

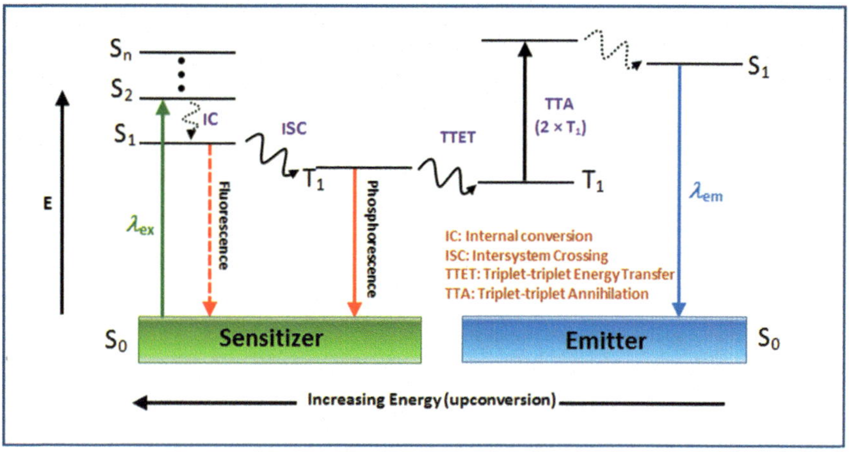

Fig. 5.3 Schematic demonstration of TTA-assisted upconversion process

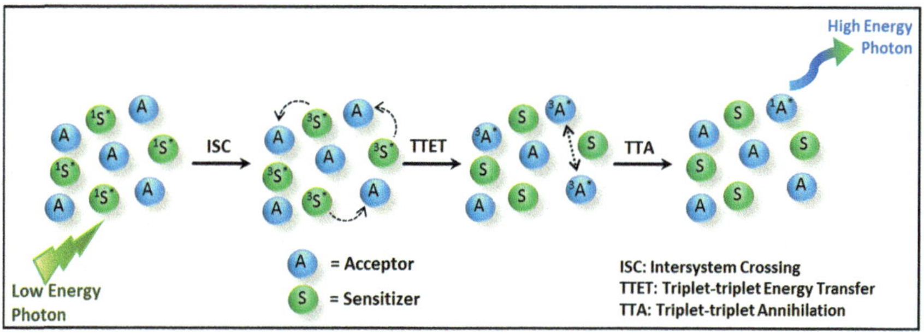

Fig. 5.4 Schematic demonstration of TTA upconversion, reproduced from [39] with permission of the Royal Chemical Society

(^{1}A*) which relaxes to ground state by emitting a high-energy photon (Fig. 5.4). It is worthy to note that the emission from the acceptor has shorter wavelength as compared to initial excitation of donor.

TTA upconversion process mainly involves the intermolecular energy transfer from triplet-PS molecules to triplet-energy acceptor molecules; therefore, its efficiency is based upon a number of factors [38]. (1) Firstly triplet PS should have high ISC efficiency, so that triplet excited state will be efficiently populated. TTA upconversion greatly depends upon ISC ability of triplet photosensitizer. On the other hand, the triplet state acceptor/annihilator should be with high fluorescence efficiency. (2) TTA upconversion process also depends upon the triplet energy level of photosensitizer and acceptor. To promote the energy transfer process, the triplet energy level of acceptor should be lower than the triplet energy level of photosensitizer. Large gap between the triplet energy levels of photosensitizer and acceptor favors TTET process, because the driving force for energy transfer is large.

(3) Another major requirement of TTA process is that the total energy of two triplet excited states from two triplet acceptors should be greater or equal to the energy of singlet excited state of the acceptor. In simple words, the triplet excited-state (T_1) energy level and singlet excited-state energy level of acceptor should fulfill the condition $2 \times E_{T1} > E_{S1}$, where E_{T1} and E_{S1} are the triplet excited-state energy level and singlet excited-state energy level of acceptor, respectively. (4) The light-harvesting ability of triplet PS is also a major factor which affects the upconversion quantum yield. The molar absorption coefficient of triplet PSs should be high so that it can be excited with low-power excitation source. If the absorption capability of photosensitizer is good, large number of molecules will populate the triplet excited state, and hence triplet concentration will be high which will enhance the TTA process significantly. (5) The fluorescent quantum yield of triplet acceptor should be high, so that the upconversion quantum yield will be high. (6) At last, the triplet excited-state lifetime of triplet photosensitizer should be long. If decay of triplet excited state is slow, there will be large chances of triplet-state energy transfer from PS to acceptor through diffusion and TTET process will be efficient.

5.3 Organic Triplet Photosensitizers for TTA Upconversion

TTA upconversion process involves a bimolecular system, a triplet PS to populate triplet excited state after absorbing incident excitation light, and an acceptor (triplet emitter) molecule undergoing triplet-triplet annihilation and emitting process. Triplet PS and acceptor both are indispensable components of TTA upconversion process.

Triplet PSs are compounds which have efficient ISC ability to populate triplet excited state [40]. These versatile compounds have potential applications in the field of photocatalysis [41, 42], photovoltaics [43], photodynamic therapy [44–47], and catalytic hydrogen (H_2) production [48]. To date, a large number of triplet photosensitizers based on metal complexes and organic dyes have been devised. The triplet photosensitizers based on metal complexes include Pd(II), Pt(II) [26, 27], Ru(II) [49, 50], Re(I) [51], and Ir(III) complexes [52, 53]. In the past, triplet PSs were limited to Pt(II)/Pd(II) porphyrin complexes for TTA upconversion but the upconversion yield was moderate [27]. Among the porphyrin complexes, PtOEP (**5.1**) and PdOEP (**5.2**) are known efficient sensitizers in TTA upconversion (Fig. 5.5). Compound **5.1** exhibits a high phosphorescence quantum yield of 50% and upconversion quantum yield of 23% [54].

In 2005, Castellano et al. used the Ru(II)-bipyridine complex $[Ru(dmb)_3][PF_6]_2$ (**5.4**) in TTA upconversion [55]. The mixed solution of photosensitizer **5.4** and 9,10-diphenylanthracene (DPA, **5.11** in Fig. 5.6) was excited with green laser and blue upconverted emission was observed. A covalently linked Ru(II) complex with anthracene was also studied for intramolecular TTA upconversion [56]. Upon excitation at longer wavelength (450 nm) an anti-Stokes delayed emission was observed through bimolecular TTA process. Moreover, this study also demonstrated

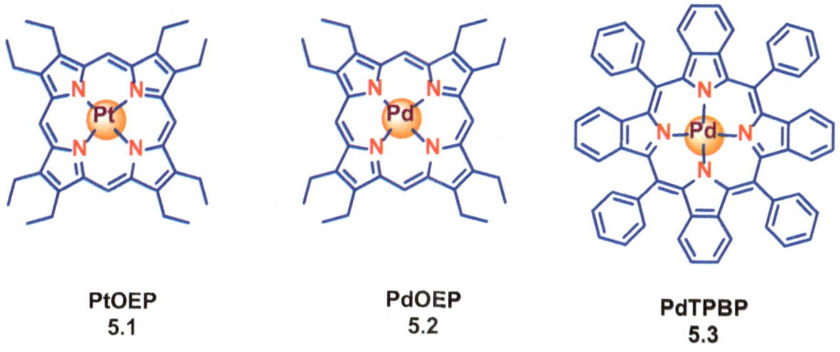

Fig. 5.5 Molecular structures of porphyrin-based triplet photosensitizers used in TTA upconversion process [26, 27]

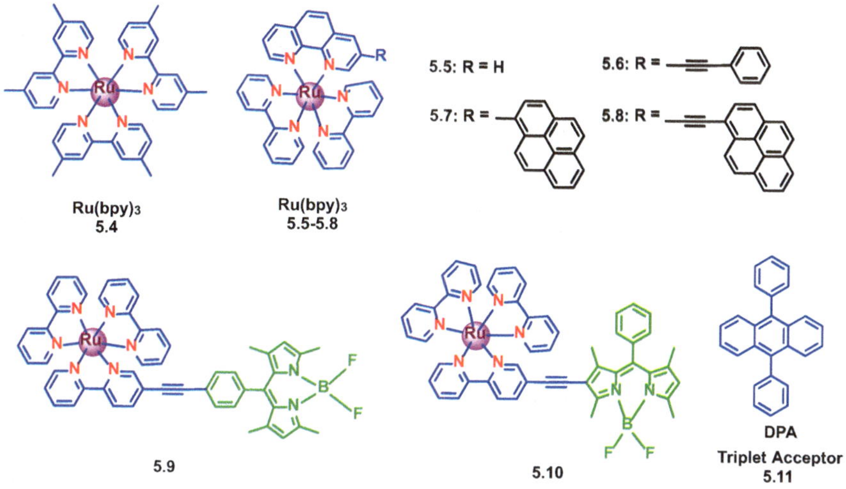

Fig. 5.6 Molecular structures of some conventional ruthenium Ru(II) complexes and triplet photosensitizers for TTA upconversion [49, 58]

that upconverted emission enhanced by ~2.9-fold for intermolecular TTA system as compared to covalently linked system due to inhibition of the undesired backward of Förster resonance energy transfer to the triplet PS molecules [56]. Due to weak absorption in the visible region and short triplet-state lifetimes, the Ru(II) bipyridine complex exhibited low upconversion quantum yield.

To overcome these drawbacks, Zhao and coworkers synthesized some Ru(II) complexes using different ligands to extend the triplet-state lifetimes [30, 49, 50, 57, 58]. Ru(II) complex (**5.8**) with intraligand triplet excited state (^{3}IL, localized on the ligand) exhibiting long-lived triplet lifetime (108 μs) was devised and used in TTA upconversion utilizing DPA (**5.11**) as triplet acceptor emitter, and a significant upconversion quantum yield (9.8%) was observed (Fig. 5.6) [49].

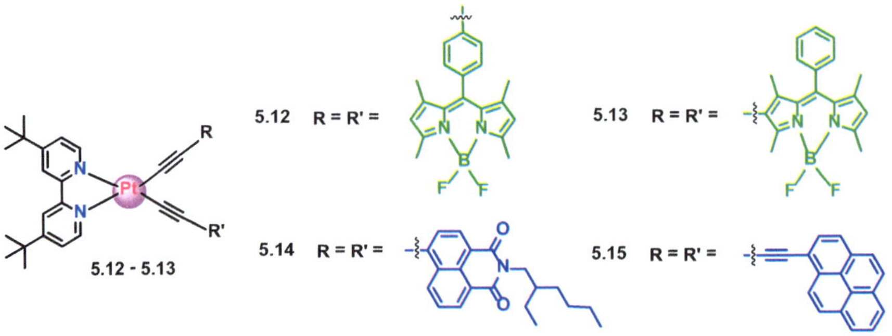

Fig. 5.7 Molecular structure of Pt(II) bisacetylide complexes for TTA upconversion [59, 60]

Zhao and coworkers devised Pt(II) acetylide complexes (**5.12–51.5**) for TTA upconversion (Fig. 5.7) [59, 60]. The Pt(II) acetylide complexes have the following advantages: high phosphorescence quantum yield and their photophysical properties can feasibly be tuned by using different acetylide ligands [61–64]. Transition metal complexes have good ISC ability, as the heavy atom effect of metals enhances the spin-orbit coupling. However, these complexes are synthetically demanding and are expensive. Moreover, the conventional transition metal complexes have weak absorption in visible spectral region and short triplet excited-state lifetime [40]. Due to these drawbacks the ordinary transition metal complexes are not ideal for the application in TTA upconversion.

Heavy Atom-Based Triplet Photosensitizers (Metal Atom-Free)

The ordinary transition metal complexes are usually not ideal for TTA upconversion due to low molar absorption coefficient in visible region and short-lived triplet excited state. In order to avoid these drawbacks, pure organic triplet PSs have been developed. To induce the spin-orbital coupling (SOC) effect, heavy atoms such as iodine and bromine were incorporated which enhances the ISC in these compounds.

Iodo-Substituted Triplet Photosensitizers

Halogens, especially iodine, enhance the ISC efficiency of organic compounds due to the SOC. In 2005, Nagano and coworkers reported a BODIPY-based triplet PS (**5.17**) by introducing the iodine atom at the 2 and 6 positions of BODIPY core [65]. Iodo-substituted BODIPY chromophore (**5.17**) shows the redshifted absorption ($\lambda_{max} = 535$ nm) and low fluorescence quantum yield ($\Phi_F = 2\%$) as compared to the unsubstituted BODIPY derivative **5.16** ($\lambda_{max} = 505$ nm), which is highly fluorescent ($\Phi_F = 70\%$). The reduction in fluorescence quantum yield in iodo-substituted BODIPY is due to the efficient ISC ability, which was confirmed by its high singlet oxygen quantum yield ($\Phi_{\Delta} = 89\%$) [65].

Due to efficient ISC ability of iodo-substituted BODIPY, Zhao and coworkers used it in the application of TTA upconversion. In order to study the TTA upconversion in iodo-substituted organic compounds, a small library of iodo-substituted organic triplet photosensitizers (**5.18–5.25**) was synthesized (Fig. 5.8) [66]. These iodo-substituted derivatives exhibit high molar absorption coefficient (up to 2×10^5 M^{-1} cm^{-1}) in visible region and efficient ISC. The absorption of these derivatives was bathochromic shifted by increasing the π-conjugation system as compared to simple iodo-substituted derivatives to attain the large anti-Stokes shift in TTA upconversion. These triplet PSs show intense absorption in the range of 510–629 nm. Nanosecond transient absorption spectral study confirmed the long-lived triplet excited state of these photosensitizers (τ_T up to 66.3 μs for **5.18**) upon excitation. The energy level of triplet excited state was estimated through DFT calculations. TTA upconversion was performed in solution and also in thin films by using these PSs. Noteworthy upconversion quantum yield (Φ_{UC}) up to 6.1% was observed for **5.21** by using perylene (**5.26**) as triplet acceptors upon excitation with 532 nm laser, while **5.24** exhibited only 1.2% upconversion quantum yield using ICBPEA (**5.23**) as triplet acceptor [66].

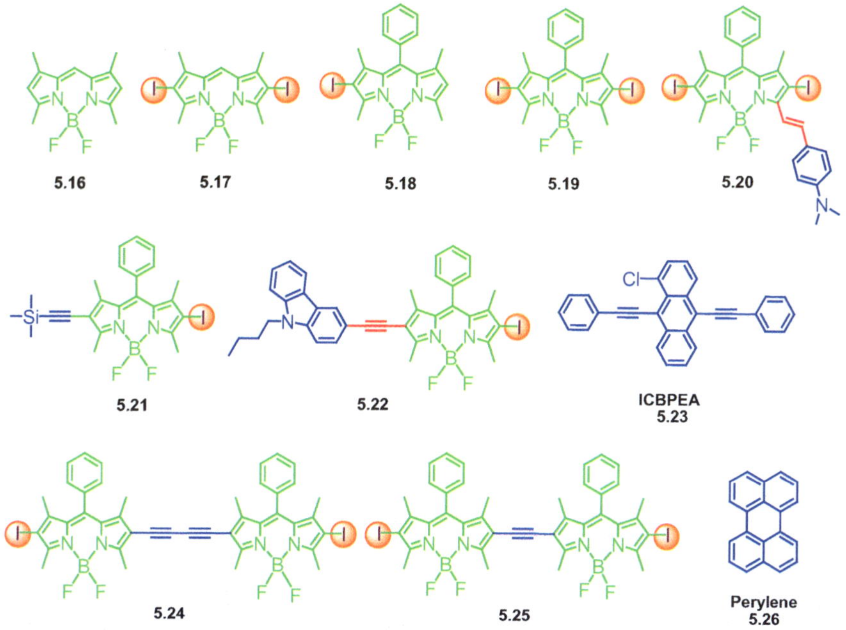

Fig. 5.8 Iodo-substituted BODIPY derivatives for TTA upconversion [66]. Triplet acceptors **5.23** and **5.26** are also presented

In 2014, Zhao and coworkers found that the upconversion quantum yield also largely depends upon the energy level of triplet acceptor and can be improved by the selection of triplet acceptor with suitable T_1 energy level and high fluorescence quantum yield. It was observed that the upconversion quantum yield of compound **5.24** can be improved up to 17.8% using perylene as triplet acceptor instead of ICBPEA [67]. The optimization of energy levels of S_1 and T_1 states of acceptor is also crucial to satisfy the requirement of TTA, i.e., ($2E_{T1} > E_{S1}$), and to enhance the efficiency of upconversion. Following this line, different perylene derivatives were devised as triplet acceptors (**5.27–5.31**) for TTA upconversion, by attaching aryl groups or BODIPY moiety to perylene which showed decreased S_1-state energy level as compared to perylene (Fig. 5.9) [67]. TTA was performed by using **5.24** as triplet PS and TTA efficiency of different triplet acceptors was compared. The results showed that the new perylene derivatives are with high TTA efficiency than perylene. The high efficiency of TTA can be attributed to the lower S_1 energy of perylene derivatives. The acceptor **5.31** gives the highest TTA efficiency among other derivatives due to enhanced TTET since the triplet energy level of perylene is lower than BODIPY unit, whereas BODIPY unit is with lower S_1-state energy level than that of perylene; thus the TTA is with larger driving force than the perylene as triplet acceptor alone [67].

In 2017, Zhou et al. observed that solvent polarity has a great impact on the upconversion quantum yield [68]. For this study, the conventional upconversion system based on PS **5.19** and perylene (**5.26**) as triplet acceptor was selected and upconversion was performed in five inert solvents with different polarities (*n*-hexane, heptane, toluene, 1,4-dioxane, and dimethyl sulfoxide). Femtosecond time-resolved transient difference absorption spectroscopy showed that decay of singlet excited state (S_1) becomes faster and intersystem crossing (ISC) from S_1 to T_1 state increases with increase in solvent polarity (from hexane to DMSO)

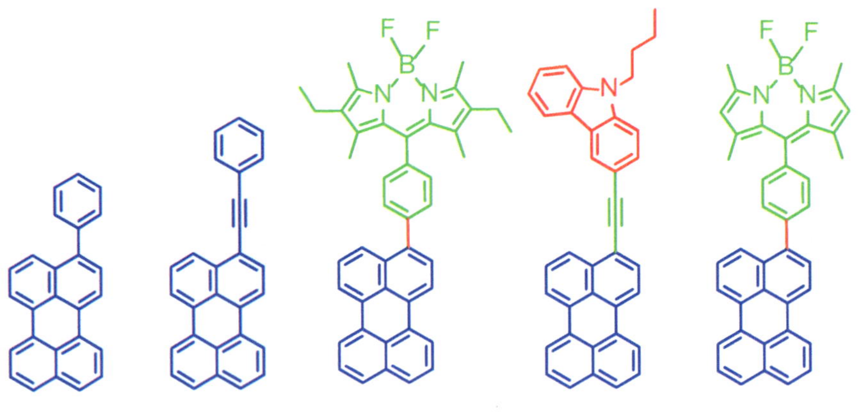

Fig. 5.9 Molecular structures of perylene-based triplet emitters used in TTA upconversion [67]

which is in agreement with the calculated energy gap between S_1 and T_1 states. Nanosecond transient difference absorption spectra showed that the triplet lifetime (τ_T) of **5.19** increases (13.0 μs–95.8 μs) with increasing solvent polarity from hexane to DMSO. The rate constants of bimolecular quenching (k_q) obtained in the five solvents decreased with increasing solvent viscosity. Furthermore, all of the k_q values were close to the diffusion rate $k_{diffuse}$ which indicates that the TTET process is a diffusion-controlled process. TTA upconversion quantum yield was derived from the upconversion emission spectra in five different solvents. The highest upconversion quantum yield (Φ_{UC} = 19.16%) was observed in dioxane while hexane (Φ_{UC} = 5.77%), heptane (Φ_{UC} = 5.59%), and toluene (Φ_{UC} = 8.75%) show modest quantum yields. Surprisingly, very low upconversion quantum yield (Φ_{UC} = 1.51%) was observed in DMSO [68].

Recently Zhao et al. studied intramolecular rotation (IMR) effect on the singlet and triplet excited states of BODIPY chromophore (compounds **5.32** and **5.33**; Fig. 5.10) [69]. The steady state and time-resolved transient absorption/emission spectroscopy studies revealed that IMR quenched the singlet excited state of BODIPY but not the triplet excited state. The unquenched triplet excited state of the iodinated BODIPY (**5.33**) was used for TTA upconversion and upconversion quantum yield of Φ_{UC} = 6.3% was observed [69].

A triplet PS based on 2,4,5,7-tetraiodo-6-hydroxy-3-fluorone (TIHF; compound **5.34** in Fig. 5.11) was implemented in TTA upconversion using DPA as triplet acceptor emitter [70]. Upon photoexcitation with 532 nm laser, upconverted emission of DPA (**5.11**) was observed, but the upconversion quantum yield was very low (0.6%). The transient absorption spectroscopy showed that the compound **5.34** has short triplet excited-state lifetime (25 μs) due to which the upconversion was not efficient. In order to improve the upconversion efficiency and anti-Stokes shift, Zhao and coworkers designed some BODIPY-based broadband triplet photosensitizers with long-lived triplet excited state. Two new thienyl iodo-BODIPY compounds (compounds **5.35** and **5.36**) were prepared and used as triplet PS for TTA upconversion process (Fig. 5.11) [71]. The main purpose to attach thienyl moiety at BODIPY core was to attain bathochromic shift in absorbance through extending

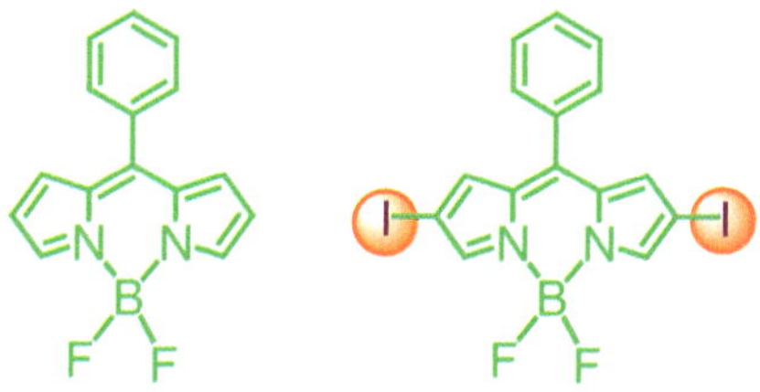

5.32 **5.33**

Fig. 5.10 Molecular structure of BODIPY rotor **5.33** used in TTA upconversion [69]. Note that the triplet state was not quenched in **5.33**, but the singlet state was quenched in reference compound **5.32** (both compounds share similar molecular structures, except the iodine substitution)

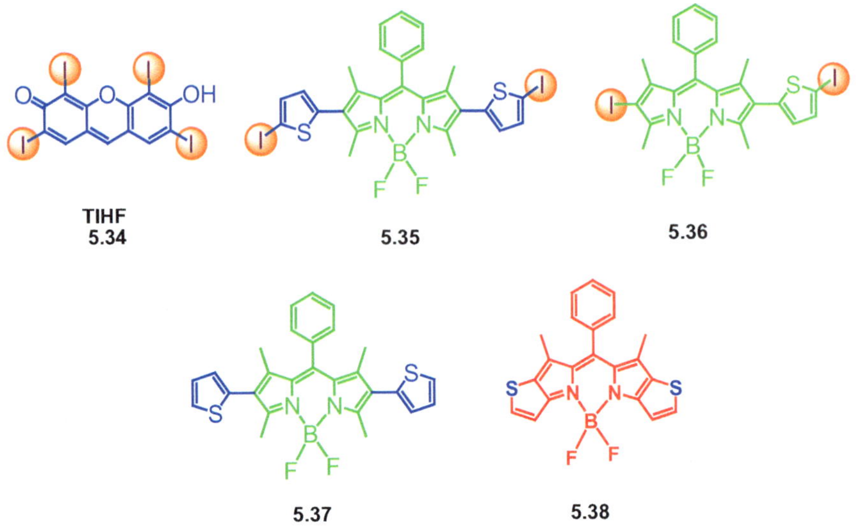

Fig. 5.11 BODIPY-derived triplet photosensitizers (**5.35**–**5.38**) for TTA upconversion [70–72]

π-conjugation system. The iodine atom was introduced to enhance ISC ability. The devised triplet PS exhibited redshifted absorption-emission wavelength as compared to unsubstituted BODIPY (absorption bands of **5.35** and **5.36** are centered as 526 nm and 532 nm, respectively) but the T_1 energy level remains unchanged. The compounds show large Stokes shift (up to 86 nm) and DFT calculations suggest that geometry relaxation may be accountable for large Stokes shift. Compounds **5.35** and **5.36** exhibited long triplet lifetime of $\tau_T = 94.7$ μs and $\tau_T = 95.2$ μs, respectively. The triplet PSs were used for TTA upconversion with perylene (**5.26**) as triplet acceptor/emitter and high quantum yields $\Phi_{UC} = 16.5\%$ for compound **5.35** and $\Phi_{UC} = 10.3\%$ for compound **5.36** were observed [71]. To study the effect of mild heavy atom effect of sulfur on the ISC, two thienyl-substituted derivatives (**5.37** and **5.38**) were prepared and their photophysical properties were investigated [72]. Compound **5.37**, in which the two thienyl moieties were attached at the 2 and 6 positions of BODIPY core, exhibited a low triplet quantum yield of 6.1%. Compound **5.38**, in which two thienyl moieties were fused with BODIPY core to extend the π-conjugation, exhibited efficient ISC and high triplet yield of 63.7% was observed. Through DFT computations, it was found that in compound **5.38** the sulfur atom participated at the lowest-lying excited state, due to which largest SOC value (3.2 cm^{-1}) and smallest singlet-triplet energy gap were observed for compound **5.38** as compared to compound **5.37** in which 0.2 cm^{-1} SOC value was observed, which is the reason for efficient ISC in compound **5.38** [72].

To enhance the visible light absorption capability of triplet PSs, Zhao and coworkers devised some dyads (**5.39**–**5.43**) involving the intramolecular resonance energy transfer (RET) from donor (BODIPY) to acceptor part (iodo-BODIPY; Fig. 5.12) [73]. The singlet energy transfer was confirmed through femtosecond transient

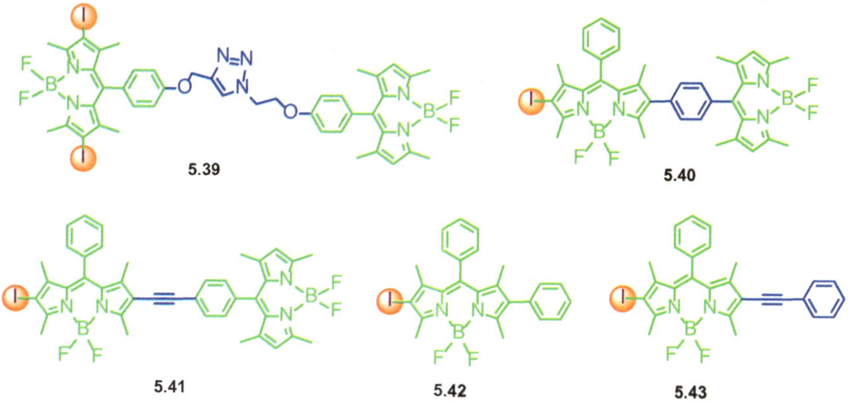

Fig. 5.12 Broadband light-absorbing RET-based triplet PSs **5.39–5.41** and reference compounds **5.42** and **5.43** for TTA upconversion [73]

Fig. 5.13 Ping-pong energy transfer in RET-based dyads of **5.39–5.41**. FRET stands for Förster resonance energy transfer, TTET stands for triplet-triplet energy transfer

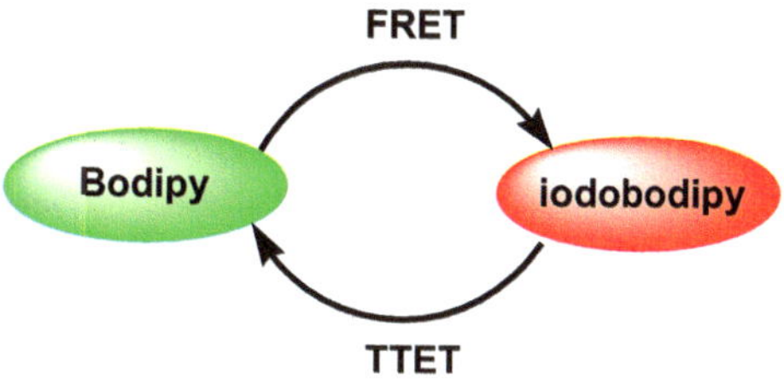

absorption spectroscopy and fluorescence quenching study. The dyads exhibited enhanced absorption as compared to triplet photosensitizers consisting of single chromophore (i.e., iodo-BODIPY). Nanosecond transient absorption study revealed that triplet excited state was localized on both the energy donor and acceptor moiety (it is dependent on the chromophores used in the dyad) [73].

On the basis of steady-state and time-resolved spectroscopic studies, ping-pong energy transfer was proposed for the dyads presented in Fig. 5.12; first *singlet* energy transfer from the energy donor (BODIPY) to the iodo-BODIPY (energy acceptor) occurs, followed by ISC of energy acceptor (iodo-BODIPY) moiety, and in turn the backward triplet energy transfer from iodo-BODIPY to the BODIPY part (an equilibrium of the triplet states may be established; Fig. 5.13). This kind of "ping-pong" energy transfer was rare for organic RET molecular arrays. The dyads involving intramolecular energy transfer were used as triplet PSs in TTA upconversion and perylene was used as triplet acceptor/emitter. Improved TTA upconversion yield was observed as compared to monochromophore-based triplet PSs. The triplet PSs **5.39**, **5.40**, and **5.41** exhibited upconversion quantum yield of Φ_{UC} = 8.1%, 8.0%, and 4.9%, respectively, which are higher than monochromophore compounds, 7.8% for **5.42** and 4.5% for **5.43** [73].

In 2018, a broadband absorbing triplet PS based on BODIPY donor and naphthalenediimide (NDI) acceptor was reported (Fig. 5.14) [74]. In these

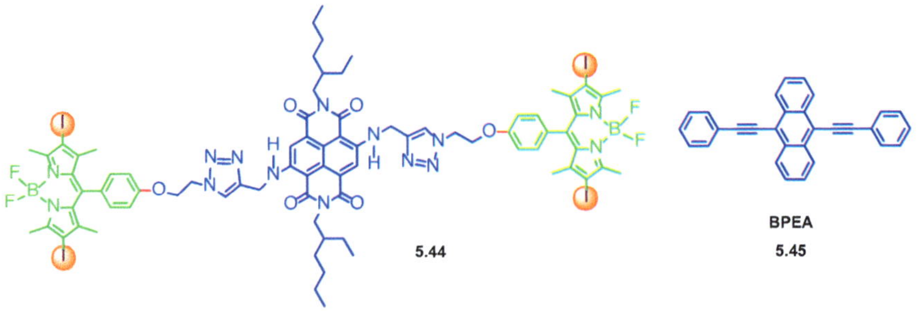

Fig. 5.14 Broadband excitable BODIPY-naphthalenediimide-based triplet PS **5.44** and the triplet acceptor **5.45** [74]

donor/acceptor triads, no triplet generation was observed through charge recombination process, and heavy atom was introduced at the BODIPY core to access the triplet excited state. Nanosecond transient absorption study indicated that the triplet excited state is localized on the NDI chromophore. On selective photoexcitation of iodo-BODIPY part, the ISC competes with the FRET between iodo-BODIPY and NDI parts. Femtosecond transient absorption spectroscopy measurements revealed that the rate of FRET ($k_{ISC} = 2.5 \times 10^{11}\ s^{-1}$) is faster than the ISC rate ($k_{ISC} = 3.3 \times 10^{9}\ s^{-1}$) which hinders the triplet-state formation in the triad via the direct ISC of the iodo-BODIPY moiety. However, the triplet state populates efficiently on selective excitation of the NDI part in the triad. This formation of triplet state was attributed to intramolecular heavy atom effect. The triad (**5.44**) was used as a broadband excitable (the absorption of triplet PS is in the range of 500–620 nm) triplet PS and moderate upconversion quantum yield ($\Phi_{UC} = 2.8\%$) was observed by using BPEA (**5.45**, Fig. 5.14) as triplet acceptor [74].

A new pyrrole-BF_2-based bis(difluoro-boron)1,2-bis((pyrrol-2-yl)methylene) hydrazine (BOPHY) was prepared (**5.46**) and used as triplet photosensitizer for TTA upconversion process (Fig. 5.15) [75]. Long triplet excited-state lifetime (177.2 μs) was observed with nanosecond time-resolved transient absorption spectroscopy. The triplet PS **5.46** exhibited high singlet oxygen quantum yield ($\Phi_{\Delta} = 58\%$) and DFT/TDDFT calculation shows high triplet-state energy level ($E_{T1} = 2.0$ eV). It was utilized as triplet PS in TTA upconversion with DPA (**5.11**) as triplet acceptor and an upconversion quantum yield of $\Phi_{UC} = 2.8\%$ was observed [75].

Broadband RET-based triplet PSs (compounds **5.47–5.49**) were prepared in which the triplet excited-state properties were controlled (Fig. 5.15) [76]. The dyad consisted of two units: BODIPY chromophore (energy donor) and acid-responsive iodo-styryl BODIPY (energy acceptor). The triplet excited-state properties of dyad were regulated through the protonation/deprotonation of amino group of iodo-styryl BODIPY. Noteworthily it was the first attempt to control the triplet excited-state properties of dyad using this approach. The compound **5.47** lacks ISC ability due to PET process but after addition of acid, the amino group gets pronated and PET

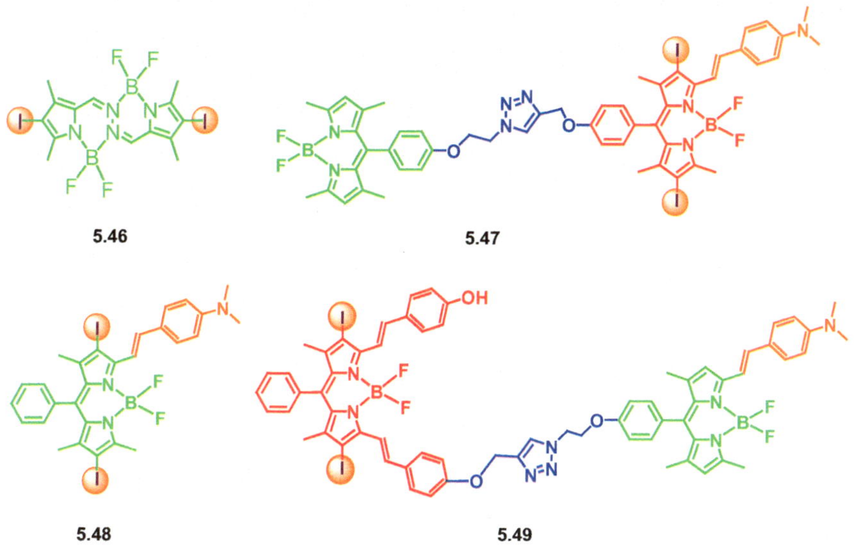

Fig. 5.15 Molecular structures of BOPHY-derived triplet PS **5.46** and acid-responsive triplet photosensitizers **5.47**–**5.49** [75, 76]

process is stopped which results in the restoration of long-lived triplet excited state ($\tau_T = 3.1\ \mu s$) of dyad [76].

To bring more functionality in TTA upconversion, Zhao and coworkers devised some external stimulus-switchable TTA upconversion systems. A reversible photo-controllable TTA upconversion was investigated for the first time using diiodo-BODIPY (compound **5.19**) as triplet photosensitizer, perylene (**5.26**) as triplet acceptor, and dithienylethene (DTE) as photochromic switch [77]. DTE shows reversible changes from closed form to open form and vice versa upon photoexcitation with visible and UV-light irradiation, respectively. The singlet-state S_1 energy change accompanying the photochromic process is responsible for the color change. The T_1-state energy will also change, but it was less investigated. It was observed that the TTA upconversion system remained unaffected with open form of DTE while it was switched off upon irradiation of the mixture with 254 nm light due to formation of closed form of DTE, which is with lower triplet-state energy level than its open form [77]. The switching process is reversible.

To attain deep insight into the photophysical process involved in photo-switching of TTA upconversion, a covalently linked triplet photosensitizer (compound **5.50**) with DTE was devised and used in TTA upconversion with perylene (**5.26**) as triplet acceptor (Fig. 5.16) [78]. The compound **5.50** has two isomers (**5.50-o** and **5.50-c**) due to open and closed DTE structure; both forms can change into each other upon selective irradiation with UV-visible light, respectively. The energy transfer (RET) in **5.50-c** (DTE-c as intramolecular energy acceptor and 2,6-diiodo-BODIPY as energy donor) results in reduction of the apparent triplet-state lifetime of BODIPY

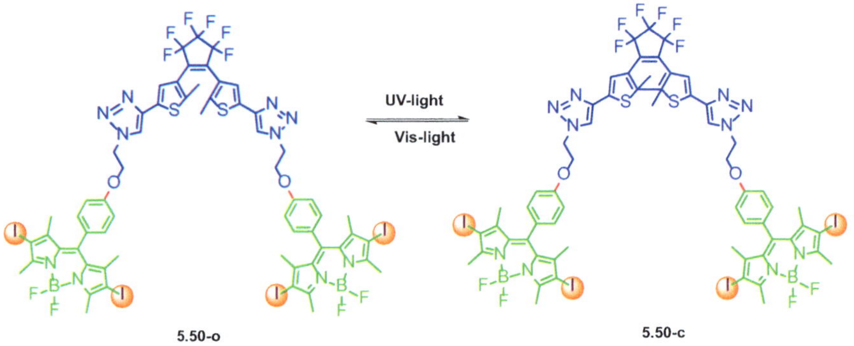

Fig. 5.16 DTE-linked BODIPY derivative for controlled TTA upconversion [78]

from 105.1 to 40.9 μs. In addition, the fluorescence of BODIPY was quenched and the singlet oxygen quantum yield was reduced from 87.6 to 36.9%, as a result of photo-switching of the DTE from opened form to closed form.

The compound **5.50** was utilized to develop controlled TTA upconversion system [78]. TTA upconversion was observed upon exciting the mixture of compound **5.50** and perylene (**5.26**) with 532 nm CW laser, and upconversion quantum yield of 11.6% was observed. The upconverted emission intensity decreased upon excitation of the mixture with 254 nm UV light and the Φ_{UC} reduced to 0.67% due to conversion of compound **5.50** from open form to closed form. The decrease in the efficiency of upconversion is presumably due to decreased triplet yield and shorter triplet lifetime of **5.50-c** than **5.50-o** [78].

In 2015, Zhao et al. developed thiol-activatable 2,4-dinitrobenzene-sulfonyl (DNBS)-caged 2,6-diiodo-BODIPY-based triplet PSs (compounds **5.51**–**5.53**) and thiol stimulus-responsive TTA upconversion with perylene (**5.26**) as triplet acceptor/emitter (Fig. 5.17) [79]. A shorter triplet-state lifetime ($\tau_T = 24.7$ μs, $\Phi_\Delta = 74\%$) was observed for DNBS-caged 2,6-diiodo-BODIPY (**5.51**-caged) as compared to uncaged PS ($\tau_T = 86$ μs, $\Phi_\Delta = 88\%$, **5.51**-uncaged, Fig. 5.17) formed after reaction with thiol and photoinduced electron transfer (PET) process was proposed for this quenching effect. Moreover, the TTA upconversion was more efficient ($\Phi_{UC} = 0.5\%$) for DNBS-uncaged triplet photosensitizer (after cleavage of DNBS moiety by thiol) than DNBS-caged ($\Phi_{UC} = 0.2\%$) [79]. This was a good attempt to develop thiol-responsive TTA upconversion system but caging effect of DNBS on triplet excited-state lifetime was unsatisfactory. The lifetime of triplet excited-state was not much altered before and after the addition of thiol which results in the poor switching of upconversion. Moreover, upconversion quantum yield was very low due to formation of some side products during reaction with thiol, which was demonstrated to be able to quench the upconversion.

Considering this effect, Zhao and coworkers modified the structure of triplet acceptor/emitter instead of triplet PS. An efficient thiol-responsive TTA upconversion system was devised by using perylene-maleimide (**5.54**, Fig. 5.18) as

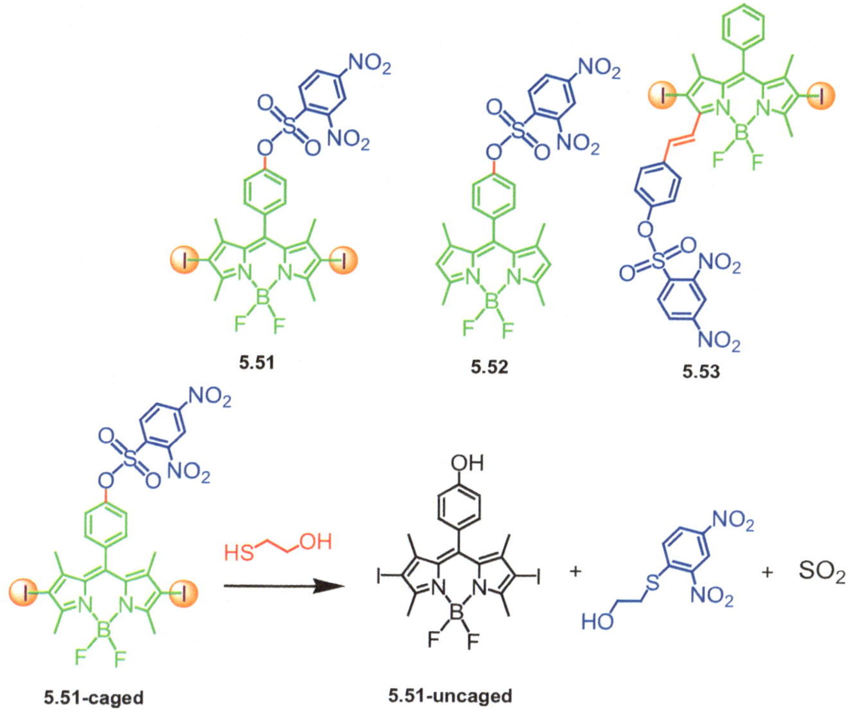

Fig. 5.17 Chemical structure of thiol-activated BODIPY derivatives along with possible reaction demonstration [79]

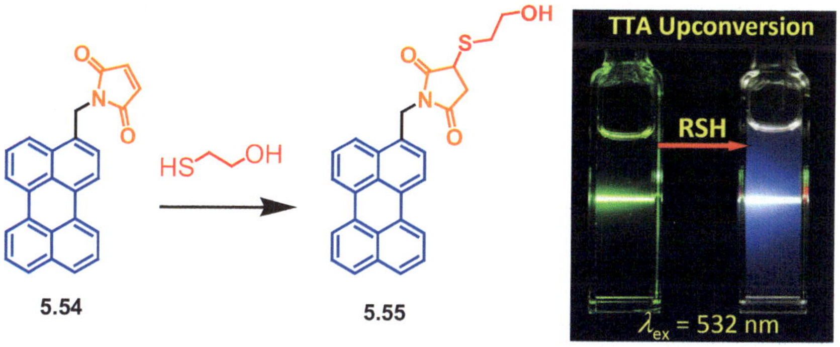

Fig. 5.18 Molecular structure of perylene-maleimide triplet acceptor/emitter and photograph demonstrating thiol-activated TTA upconversion, reproduced from [80] with permission of the American Chemical Society

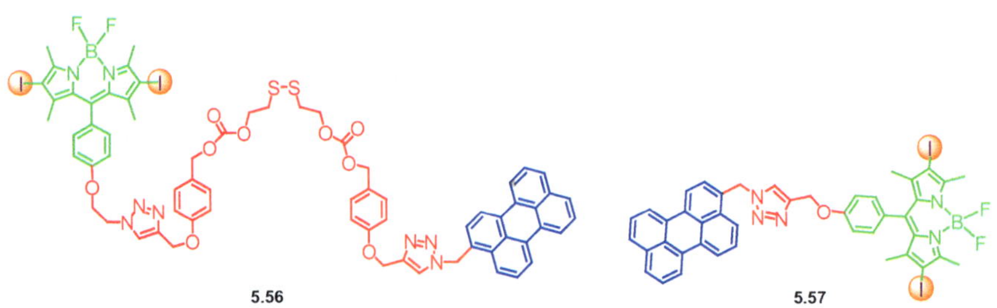

Fig. 5.19 Molecular structure of thiol-activatable triplet PS **5.56** and reference compound **5.57** (un-cleavable) used in TTA upconversion [82]

triplet acceptor/emitter and compound **5.19** as triplet PS [80]. The fluorescence of perylene was caged ($\Phi_F = 0.08\%$) through PET process in **5.54**, with maleimide moiety as the electron acceptor. No upconversion was observed in the absence of thiol. However, after addition of thiol, the maleimide moiety underwent adduction reaction, the PET process was inhibited, the fluorescence of perylene moiety was restored ($\Phi_F = 98\%$), and efficient TTA upconversion was observed ($\Phi_{UC} = 5.9\%$) [80]. In this study, it was concluded that the PET has different quenching effects on the singlet excited state and triplet excited state of chromophore, which was further verified by devising zinc (II)-responsive fluorophore based on BODIPY moiety [80, 81].

In 2017, a thiol-cleavable compound (**5.56**) was synthesized by linking triplet-PS unit (diiodo-BODIPY) and triplet emitter (perylene) through disulfide bridge [82]. This disulfide bond can be selectively cleaved by thiols (Fig. 5.19). A reference (**5.57**) was also prepared by using 1,2,3-triazole linker between the two components. In both of these dyads ping-pong energy transfer, i.e., forward singlet energy transfer and backward triplet energy transfer, was observed. The steady-state and time-resolved transient spectroscopies indicate that the rate for forward intramolecular Förster resonance energy transfer (k_{FRET}) from perylene to diiodo-BODIPY was faster than backward triplet-triplet energy transfer (TTET) process from diiodo-BODIPY to perylene (for compound **5.56**, $k_{FRET} = 1.9 \times 10^8$ s^{-1} and $k_{TTET} = 3.7 \times 10^7$ s^{-1}, while for compound **5.57**, $k_{FRET} = 3.1 \times 10^8$ s^{-1} and $k_{TTET} = 8.4 \times 10^7$ s^{-1}). It was also revealed that the rate constant of FRET is more dependent on the linker's length than the TTET process. The cleavage of disulfide bond by thiol effectively shuts down the intramolecular FRET channel and thiol-activatable TTA upconversion was observed due to *intermolecular* TTET process. The Φ_{UC} efficiency of **5.56** was improved from 0.03 to 0.5% in the presence of thiols [82].

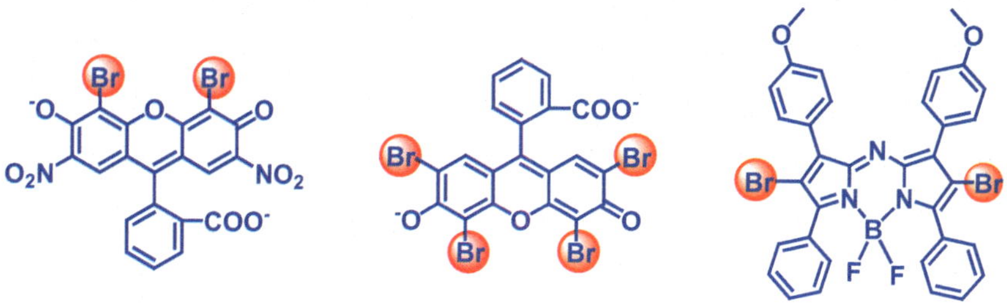

Fig. 5.20 Molecular structures of some bromo-substituted triplet photosensitizers [83, 84]

Bromo-Substituted Triplet Photosensitizers

Besides the transition metal and iodine atoms, bromine atom was also often used to enhance the SOC in organic compounds due to their large atomic number. Eosin Y (compound **5.59**, Fig. 5.20) is known as triplet PS which can generate singlet oxygen due to heavy atom effect of bromine and it was used for photocatalytic aerobic cross-dehydrogenation coupling reaction [83]. However, the application of this PS in TTA upconversion was not explored. In 2004, O'Shea et al. reported brominated aza-BODIPY derivative which showed absorption in the NIR spectral region [84]. Low fluorescence quantum yield (1–10%) and moderate singlet oxygen generation were observed with compound **5.60**. Due to NIR absorption and efficient ISC ability, compound **5.60** was used in the photodynamic therapy (PDT) [84].

In last few years, naphthalenediimide (NDI) was extensively used in molecular assemblies and fluorescent molecular probes. To study the triplet-state properties of NDI derivatives several derivatives were synthesized by introducing bromine atom at the core (Fig. 5.21) [85]. Native NDI has absorption at short wavelength, which is not suitable for some applications such as TTA upconversion and PDT. To attain bathochromic shifted absorption, several amino moieties were introduced at the bay position of NDI. Nanosecond transient absorption study confirmed formation of long-lived triplet excited state in these derivatives. These compounds were used as triplet PSs for TTA upconversion with perylene (**5.26**) as triplet acceptor. The upconversion quantum yield is up to 18.5% for **5.63** [85].

Heavy Atom-Free Triplet Photosensitizers

Heavy atom effect can ensure the SOC in organic compounds; however, these heavy atom-based triplet PSs suffer from some drawbacks, such as dark toxicity and short triplet excited-state lifetime. Therefore, it is desired to develop heavy atom-free triplet PSs. A few strategies have been developed to devise heavy atom-free triplet

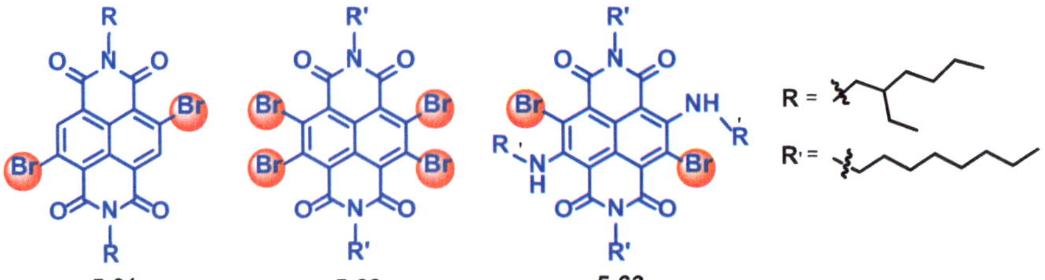

Fig. 5.21 Bromo-naphthalenediimide triplet PSs used in TTA upconversion [85]

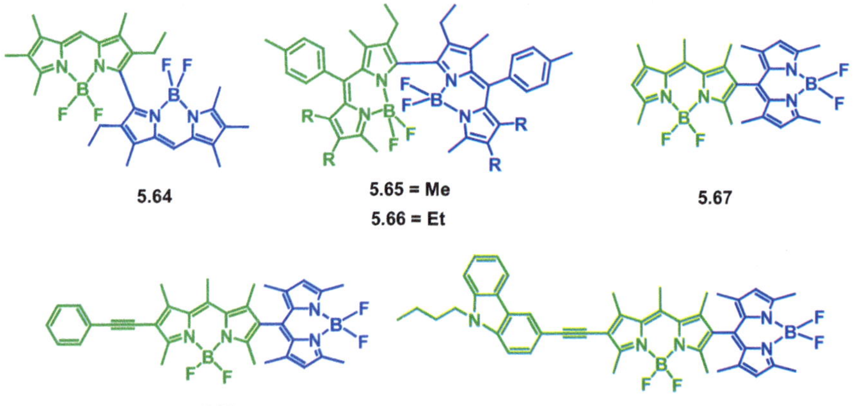

Fig. 5.22 BODIPY dimer-based triplet photosensitizers [86, 87]

PSs. Some of these PSs were utilized in TTA upconversion and efficient TTA was observed.

Exciton Coupling-Induced ISC

Two directly connected identical chromophores, without π-conjugation, but with specific geometry, may exhibit the exciton-coupling effect [86, 87]. A few examples based on BODIPY dimers were reported and successfully used in the application of TTA upconversion. BODIPY dimers (compounds **5.64**, **5.65**, and **5.66**) were devised (Fig. 5.22), which demonstrated the exciton-coupling effect as confirmed by the UV-Vis absorption which splits into two bands as compared to the monomer unit [86, 87]. These compounds show high fluorescence quantum yield in the range of 67–71% and low singlet oxygen quantum yield in the range of 10–50%.

Zhao and coworkers modified the molecular structures of BODIPY dimers to modulate triplet yield. A series of compounds (**5.67–5.69**) were devised by varying the π-conjugation system and it was observed that triplet quantum yield decreased with increased π-conjugation (Fig. 5.22) [88]. Compounds **5.67** and **5.68** exhibited 60% and 42% single oxygen quantum yield, respectively, while compound **5.69** showed lack of ISC ability. These compounds were used in TTA upconversion using perylene (**5.26**) as triplet emitter and upconversion quantum yield of 3.7% was observed for compound **5.67** [88].

Use of Electron Spin Converter to Attain Efficient ISC: Application in TTA Upconversion

Fullerene C_{60} itself is not a good triplet photosensitizer because it shows very weak absorption in the visible range, although its ISC efficiency is about 100%. On the other hand, C_{60} is an ideal electron spin converter due to its low-lying singlet excited state (1.72 eV) and efficient ISC. Previously, C_{60} has been widely used as an electron acceptor in the study of photoinduced charge separation, because of its small reorganization energy [89, 90]. The absorption wavelength of C_{60}-organic chromophore dyads can be tuned by linking suitable energy donor with spin convertor C_{60}. Singlet energy transfer from the organic chromophore to C_{60} in the dyads will populate the singlet excited state (S_1) of C_{60}, and then via ISC of the C_{60} unit, the triplet state of C_{60} unit will be produced. Note that in some cases the charge recombination-induced ISC was also observed.

A dyad (**5.70**, Fig. 5.23), based on BODIPY (energy and electron donor) and C_{60} (acceptor), was reported and charge transfer dynamics as well as triplet-state population mechanism in different solvents were investigated [91]. In DCM, charge transfer (CT) state was observed (lifetime: 160 ps); the charge recombination produces $^3C_{60}{}^*$ state, not the ^{3}BODIPY* state. This initial study provided a way for further investigation by using C_{60} to attain efficient ISC. Following this approach, Zhao et al. proposed the concept of using C_{60} as electron spin converter to prepare heavy atom-free triplet PSs, and C_{60}-PBI dyad (**5.71**) was prepared (Fig. 5.23) [92]. The fluorescence of PBI-C_{60} was significantly quenched, indicating the efficient intramolecular energy transfer from PBI unit to C_{60} unit. Upon nanosecond pulsed

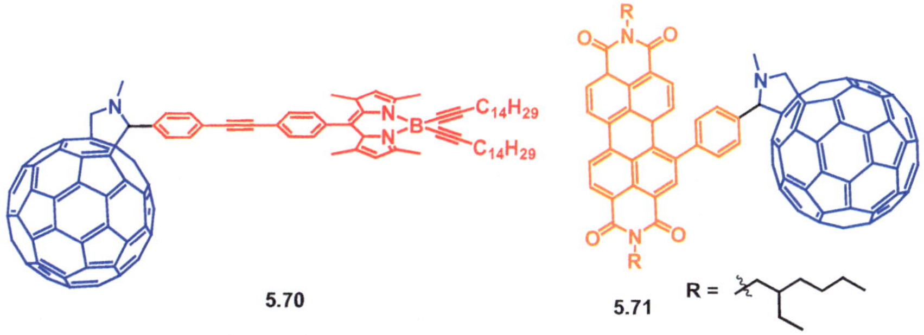

Fig. 5.23 Visible light-harvesting C_{60}-BODIPY and C_{60}-PBI dyads [91, 92]

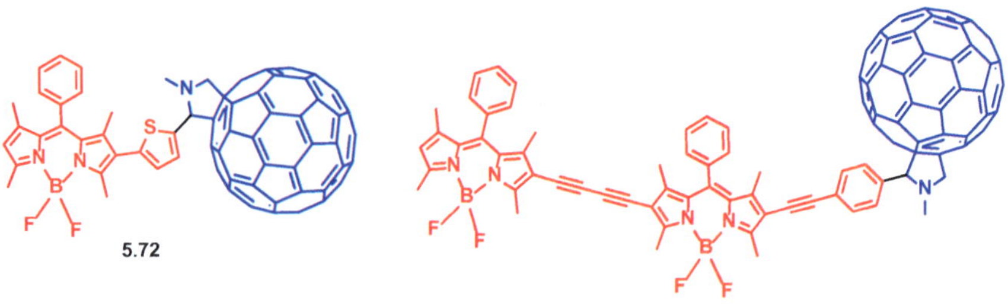

Fig. 5.24 Visible light-harvesting C_{60}-BODIPY triplet photosensitizers for TTA upconversion [93]

laser excitation ^{3}PBI*-localized triplet state was observed. In conclusion, a "ping-pong" energy transfer process was observed (singlet energy transfer from PBI to C_{60} and then the backward triplet excited-state transfer from C_{60} to PBI). The approach of using C_{60}-organic chromophore dyads to attain strong absorption in visible region and long-lived triplet states is useful for the design of new triplet PSs suitable for PDT and photocatalysis, as well as for photo-upconversion.

Zhao et al. prepared some visible light-absorbing broadband BODIPY-C_{60} dyads (Fig. 5.24); the fluorescence of BODIPY units in these dyads was strongly quenched, indicating the efficient energy transfer or electron transfer from BODIPY unit to C_{60} unit [93]. Notably, the singlet-state energy of C_{60} is 1.72 eV, which is lower than the singlet state of BODIPY unit (2.40 eV); thus it is possible for singlet energy transfer from BODIPY to C_{60} unit to occur. By nanosecond transient absorption spectroscopy, efficient formation of C_{60}-localized triplet production was observed for these dyads, with 33.3 μs triplet-state lifetime. This long-lived triplet state and strong absorption in visible spectral region made these triplet PSs suitable to be used in TTA upconversion. With perylene as triplet energy acceptor and BODIPY-C_{60} dyads as triplet PS, upconversion quantum yields of 2.9% and 7.0% were observed for these dyads **5.72** and **5.73** (Fig. 5.24), respectively [93].

Charge Transfer-Induced ISC and the Related Triplet Photosensitizer for TTA Upconversion

There are two types of charge transfer-induced ISC: (1) spin-orbit charge transfer ISC (SOCT-ISC) and (2) radical pair ISC (RP-ISC). In RP-ISC, the electron donor and accepter moieties are usually connected through a lengthy spacer group and the dyad shows weak coupling and lower *J* value for the CT state. RP-ISC-based triplet PS was not reported for TTA upconversion. SOCT-ISC is observed in compact electron donor and accepter systems where two moieties are directly connected through an orthogonal geometry. For this system, strong coupling is observed and *J* value is expected to be larger than that required for RP-ISC. Recently, SOCT-ISC-based triplet PSs gained attention in the area of TTA upconversion and PDT.

Zhao and coworkers reported BODIPY-phenothiazine dyad (compound **5.74**), where BODIPY was directly connected to phenothiazine moiety, via the *meso*-position of BODIPY moiety (Fig. 5.25) [35]. The photophysical properties of dyad were investigated in detail by nanosecond and femtosecond transient absorption spectroscopies and ISC was confirmed to be due to charge recombination, based on solvent polarity-dependent ISC efficiency and observation of the CT state and CR process. This SOCT-ISC-based compound was used as a triplet PS in TTA upconversion with perylene (**5.26**) as triplet emitter. By excitation of the triplet PS at 532 nm, upconversion quantum yield of 3.2% was observed.

A new donor/acceptor system was devised by replacing phenothiazine with phenoxazine to study the effect of the redox potential of the electron donors on triplet-state formation (Fig. 5.25) [94]. Directly connected BODIPY-phenoxazine dyads (compound **5.75**) were reported to undergo SOCT-ISC. The compound was used as a PS in TTA upconversion using 510 nm excitation and perylene as an acceptor; very high upconversion quantum yield (12.3%) was observed which can be attributed to longer triplet lifetime of **5.75** (539 μs) as compared to BODIPY-phenothiazine dyad **5.74** (116 μs) [35, 94]. BODIPY-phenoxazine dyads (**5.76** and

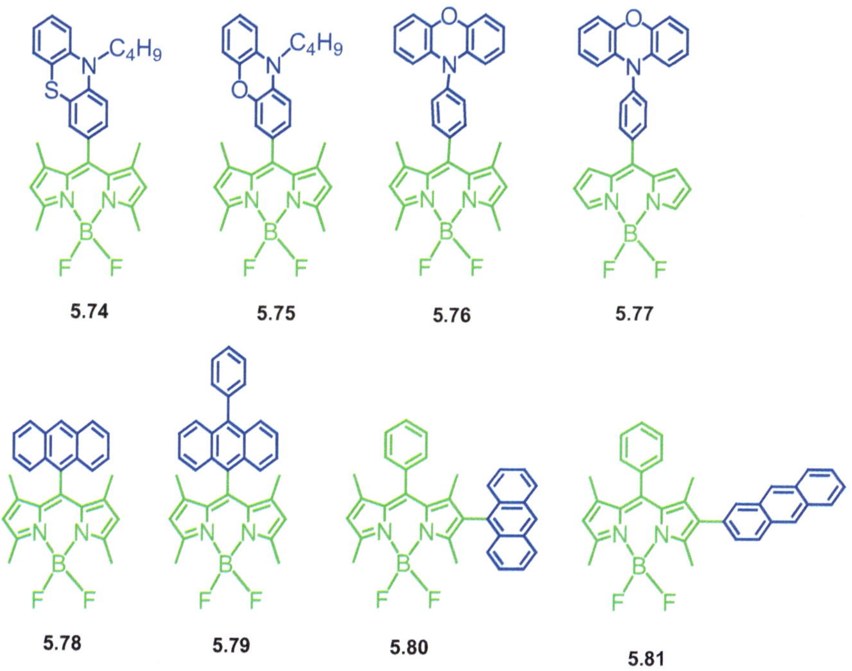

Fig. 5.25 BODIPY-based donor/acceptor dyads exhibiting SOCT-ISC for TTA upconversion [34, 35, 94]

5.77) having a phenyl spacer were also reported for TTA upconversion but lower upconversion quantum yields were observed (3.7% and 1.3%, respectively, for **5.76** and **5.77**) due to their lower triplet quantum yields (27% and 25%, respectively) [94].

The SOCT-ISC of BODIPY-anthracene-based donor/acceptor dyads was studied, as well as the application in TTA upconversion. A series of BODIPY-anthracene dyads (**5.78–5.81**, Fig. 5.25) were prepared by varying chromophore orientation between the two moieties and it was observed that SOCT-ISC not just depends on orthogonal geometry but is also significantly affected by mutual orientation of the chromophore (Fig. 5.25) [34]. These derivatives were used as triplet PSs in TTA upconversion with perylene (**5.26**) as acceptor and high upconversion quantum yield up to 15.8% was observed for compound **5.79**.

To Increase the Anti-Stokes Shift of TTA Upconversion with Triplet PSs Showing CT Absorption Band and TADF Property

In TTA upconversion, larger anti-Stokes shifts are desired. However, the energy loss due to the ISC of the triplet PSs is a bottleneck, because for normal aromatic chromophores, the electron exchange energy for the two electrons in frontier molecular orbitals is large; thus the energy gap between the S_1 and the T_1 state ($2J$) is large. As a result, the energy loss during ISC will usually lead to small anti-Stokes shift for TTA upconversion. In order to tackle this drawback, triplet PSs with small S_1/T_1-state energy gap are desired.

The molecules based on electron donor accepter moieties have a distinct feature in the UV-Vis absorbance spectra. With appropriate electronic coupling magnitude between the electron donor and acceptor units, such an electron donor/accepter system results in a new absorbance band at longer wavelength as compared to the respective components, i.e., a broad absorption band which is not found in parent donor or accepter molecule. This band is very useful for attaining large anti-Stokes shifts in TTA upconversion, due to two reasons; firstly, its redshifted absorbance range could be used for excitation and hence an increased anti-Stokes shift could be observed in upconversion. Secondly, in such a system although a CT absorption band at longer wavelength is observed, the triplet energy level is not compromised as compared to the components of the triplet PS. Therefore, it could result in decreased energy difference between singlet state and triplet state (Scheme 5.1a), which is beneficial for attaining smaller energy loss during ISC, and larger anti-Stokes shift is expected for the TTA upconversion with these compounds as triplet PS.

Recently Zhao et al. reported a BODIPY-perylene electron donor accepter system (compound **5.82**, Fig. 5.26) [95]. BODIPY shows an absorption band centered at ca. 500 nm and perylene moiety shows absorption band in the range of 400–450 nm; however, in BODIPY-perylene dyad a new absorption band was observed ranging up

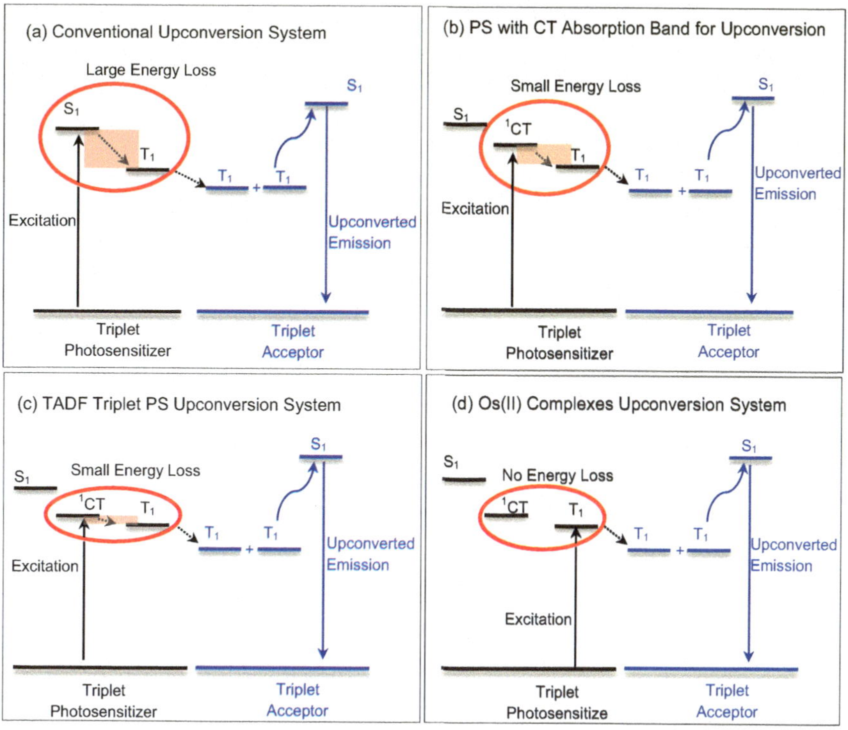

Scheme 5.1 Diagram demonstrating TTA upconversion in (**a**) conventional system with more energy loss, (**b**) triplet PSs showing CT absorption bands with less energy loss, (**c**) TADF substance with minimal energy loss, and (**d**) compounds with direct triplet excitation such as Os(II) complexes

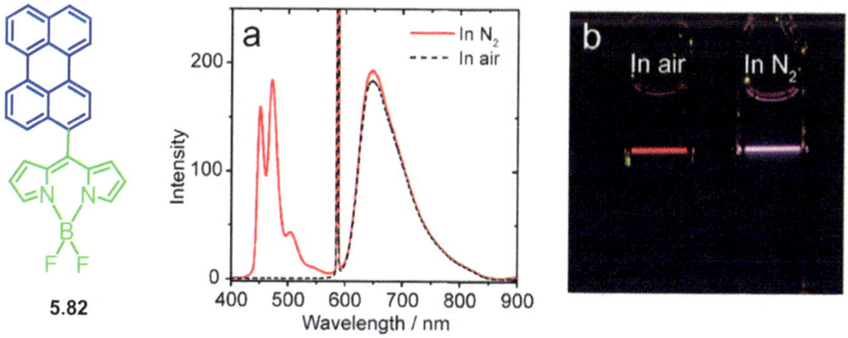

Fig. 5.26 Molecular structure of BODIPY-perylene donor/acceptor dyad (**5.82**) and (**a**) upconversion with dyad as the triplet PS ($c = 1 \times 10^{-5}$ M) using perylene ($c = 2.8 \times 10^{-4}$ M) as acceptor excited with 589 nm CW laser; (**b**) photograph of upconversion, reproduced from [95] with permission of the Royal Chemical Society

to 635 nm. This redshifted absorption band is assigned as CT absorption ($S_0 \rightarrow {}^1CT$ transition). The dyad was used for TTA upconversion with perylene by excitation at CT band (589 nm) and high upconversion quantum yield (5.7%) was observed (Fig. 5.26). The anti-Stokes shift for excitation at CT absorption band is 0.65 eV which is larger than that obtained by excitation at LE absorption band (0.37 eV) [95].

When a compound shows a much-prolonged emission fluorescence lifetime in N_2 environment as compared to that in aerated environment, it indicates possible involvement of triplet state in the emission and it may involve the reverse intersystem crossing (rISC) [96]. The emission is termed as delayed fluorescence. There are two types of delayed fluorescence, namely (1) triplet-triplet annihilation (TTA) and (2) thermally activated delayed fluorescence (TADF) [97–101]. In TTA, sum of two triplets is similar or higher in energy than the S_1 state. Hence rISC takes place when two molecules are in T_1 state with sufficient energy to produce a S_1 state [97, 102]. As compared to TTA, in TADF, which is usually based on an electron donor/accepter system, a ${}^1CT$ state is formed upon photoexcitation. This ${}^1CT$ state has energy very similar to the LE triplet state, i.e., usually less than 150 meV, and hence thermally activated rISC takes place [101]. By excitation at CT band of donor-accepter molecules some of the energy loss is prevented. TADF molecules are superior than normal electron donor accepter systems because in TADF compounds not only CT state exists but also the energy difference between ${}^1CT$ and triplet is minimum (Scheme 5.1c). This property has been exploited in upconversion.

Baldo reported **5.83**, a TADF molecule based on carbazole and dicyanobenzene moieties (Fig. 5.27) [103]. In this compound the singlet-triplet-state energy difference was found to be less than 100 meV [104]. This compound was used in TTA upconversion with DPA (**5.11**) as acceptor/emitter, in a solid-state double-layer film.

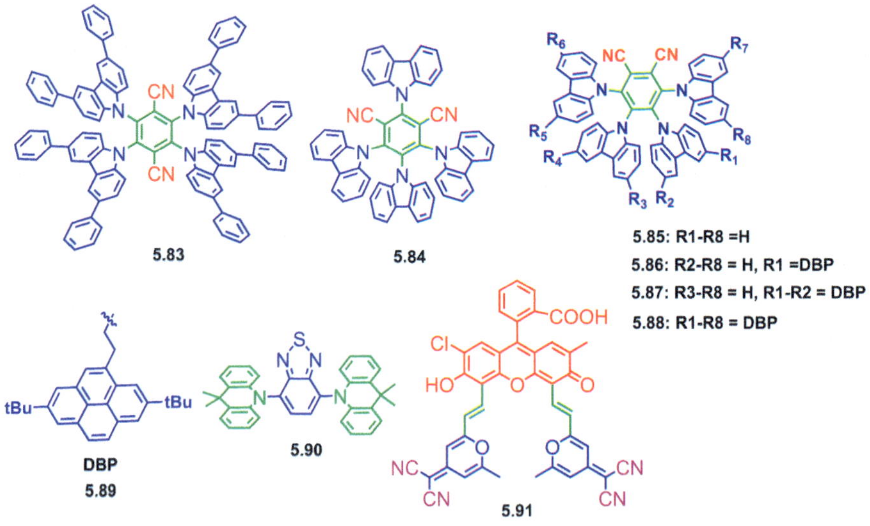

Fig. 5.27 Molecular structure of some TADF molecules used in TTA upconversion

They achieved 1.1% upconversion quantum yield by exciting the system with a 532 nm laser.

Kimizuka et al. reported **5.84** showing TADF property for upconversion in order to achieve large anti-Stokes shift (Fig. 5.27) [105]. The singlet-triplet energy gap in this compound is 83 meV. This compound was used for visible-to-UV upconversion with *p*-terphenyl and *p*-quaterphenyl as triplet accepter. The compound **5.84** with *p*-quaterphenyl as acceptor gives anti-Stokes shift of 0.73 eV and upconversion quantum yield of 3.9%. The upconversion quantum yield with *p*-terphenyl was observed to be 2.8%. A large anti-Stokes shift of 0.83 eV was observed for the upconversion [105].

In 2016, Zhao and coworkers reported a series of carbazole and dicyanobenzene-based photosensitizers (**5.85–5.88**) along with **5.84** for TTA upconversion [106]. These TADF-based molecular scaffolds were utilized as a photosensitizer in the upconversion with 2,7-di-*tert*-butylpyrene (DBP) as an acceptor (compound **5.89**, Fig. 5.27). Hence TTA was enhanced as triplet energy was transferred to the triplet accepter. They achieved upconversion quantum yield up to 4.4% in solution and also implemented it into polymer film [106].

In 2017, a TADF molecule based on dimethylacridine and thiadiazole (compound **5.90**) was reported (Fig. 5.27) [107]. Its triplet quantum yield was 51.2%. This TADF compound was used as a triplet photosensitizer with DPA (**5.11**) as triplet accepter and upconversion was observed with anti-Stokes shift of 0.52 eV. The upconversion quantum yield was reported as 1.9%. The luminescence lifetime of compound **5.89** was found to be 4.05 μs in deaerated atmosphere, and it was greatly reduced in aerated solution (0.42 μs). This result indicated that the triplet state is involved in the luminescence, and upconversion was also observed in air for this system [107].

In 2019, Peng et al. reported compound **5.91**, a TADF molecule for upconversion with perylene (**5.26**) and DPA (**5.11**) as triplet acceptor/emitter [108]. This compound can absorb red light and hence can be excited with a 635 nm laser. The singlet/triplet energy gap in compound **5.91** is found to be very small, i.e., 28.36 meV. The upconversion with DPA results in a large anti-Stokes shift (0.94 eV) and a high upconversion quantum yield (11.2%). Compound **5.91** was also used as PS with perylene (**5.26**) as accepter and in this case anti-Stokes shift was found to be 0.80 eV and upconversion quantum yield was observed to be 7.0% [108].

Triplet PSs Showing $S_0 \rightarrow T_1$ Absorption Band for Upconversion

In TADF molecules due to small energy gap between singlet and triplet states, energy loss is minimized and large anti-Stokes shift was observed in TTA upconversion. An alternative approach will be the direction excitation of the $S_0 \rightarrow T_1$ absorption band of a triplet PS, although this kind of absorption band is extremely weak for normal triplet PSs. Although direct population of T_1 state by excitation of S_0 state is forbidden, due to strong spin-orbit coupling in some metal complexes, a non-negligible $S_0 \rightarrow T_1$ absorption band can be resulted. For example, in some Os(II) complexes strong $S_0 \rightarrow T_1$ absorption bands were observed. Thus, using of the $S_0 \rightarrow T_1$ absorption band of Os(II) complex for excitation can bypass this energy loss during ISC of the triplet PS.

Kimizuka and coworkers reported an Os(II) complex (**5.92**) which shows $S_0 \rightarrow {}^1$MLCT absorption band in NIR region (938 nm) [109]. Compound **5.92** can be directly excited by $S_0 \rightarrow T_1$ absorption band at 938 nm. It was used as PS with rubrene as an acceptor and upconverted emission was observed at 570 nm upon photoexcitation at 938 nm. The devised PS gives NIR-to-Vis upconversion with anti-Stokes shift of 0.86 eV. Donor/acceptor nano-hybrid polymer was also devised with compound **5.92** as triplet PS and rubrene as acceptor/emitter. PVA film exhibited very high upconversion quantum yield (3.1%) even in the air [109]. An Os complex (compound **5.93**) was synthesized and used as a triplet photosensitizer from NIR to blue TTA upconversion upon direct excitation at MLCT band (Fig. 5.28) [110]. Using 2,5,8,11-tetra-*tert*-butylperylene as acceptor/emitter, **5.93** exhibited high upconversion quantum yield of 2.7% and anti-Stokes shift of 0.97 eV. It was further found that when the photosensitizer and acceptor were mechanically mixed and encapsulated in PVA matrix to form a solid film, the solid film manifested TTA upconversion which is independent of air [110].

$Os(bpy)_3$ complex has a very short triplet lifetime (56 ns). However, the TTET process, which is an important step during TTA upconversion, is highly dependent on the triplet-state lifetime of the triplet PS molecule. A longer triplet lifetime is desired which provides sufficient TTET. In an effort to prolong the triplet lifetime of $Os(bpy)_3$, Zhao and coworkers have attached a BODIPY unit in a $Os(bpy)_3$ complex (**5.94**) (Fig. 5.28) [111]. The triplet lifetime of the compound increased from 56 ns to 1.73 μs due to equilibrium between coordination center-localized and BODIPY ligand-localized triplet state. The complex was used in TTA upconversion and due to increased lifetime, the complex showed fourfold increase in TTA upconversion quantum yield (1.2%) as compared to the reference Os complex **5.95** (0.3%) lacking BODIPY ligand [111].

5.4 General Summary

Triplet-triplet annihilation (TTA) upconversion is an important photophysical process which gained significant attention of the researchers due to its several advantages over the conventional upconversion approaches and excellent outcomes in the various photochemical applications; it has been summarized in this book chapter along with its basic principles and requirements. As has been demonstrated, a number of triplet photosensitizers (PSs) based on different ISC mechanisms have been developed and used in TTA upconversion to achieve high upconversion quantum yield. The conventional triplet PSs based on heavy atom effect or metal

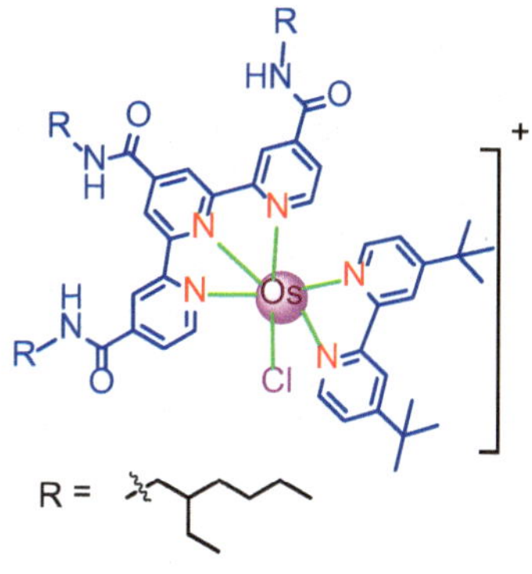

5.92

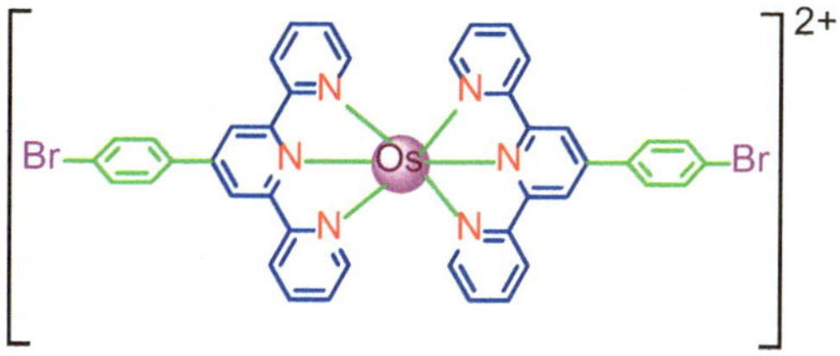

5.93

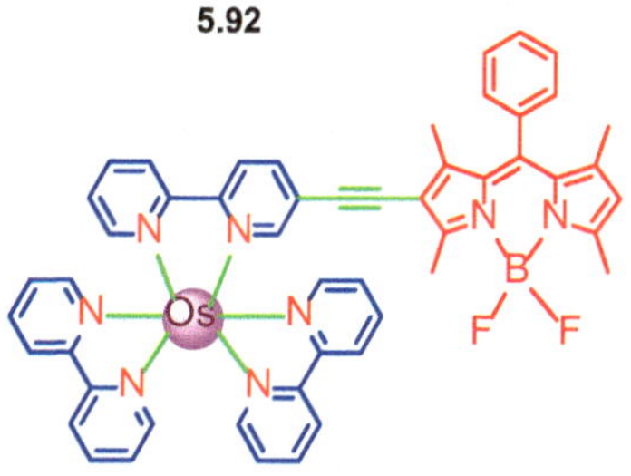

5.94

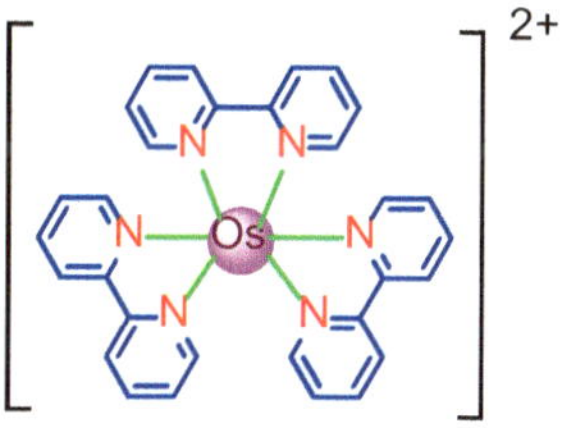

5.95

Fig. 5.28 Os (II) complexes with $S_0 \longrightarrow T_1$ absorption for TTA upconversion [109–111]. The anions are omitted for clarity

complexes are extensively investigated for TTA upconversion but such PSs suffer from some drawbacks which limit their practical applications. Thus, attention was paid to devise heavy atom-free triplet PSs for TTA upconversion adopting different approaches which include BODIPY dimers based on exciton-coupling effect and use of spin convertor and donor/acceptor system based on charge recombination (CR)-induced ISC. The book chapter also focused on and demonstrated the methods to minimize energy loss in TTA upconversion by utilizing the triplet photosensitizers based on, e.g., electron donor/acceptor system based on CR-induced ISC and with charge transfer absorption character and osmium-based metal complexes with direct $S_0 \rightarrow T_1$ excitation feature. This study is expected to have an impact on the field of photochemistry for the development of new efficient triplet PSs that will be useful not only for the TTA upconversion but also in other photochemical applications.

Acknowledgment J. Z. thanks the NSFC (U2001222, 21673031, 21761142005, and 21911530095), the Fundamental Research Funds for the Central Universities (DUT19TD28), and State Key Laboratory of Fine Chemicals (ZYTS201901) for financial support.

References

1. H. Kobayashi, M. Ogawa, R. Alford, P.L. Choyke, Y. Urano, New strategies for fluorescent probe design in medical diagnostic imaging. Chem. Rev. **110**, 2620–2640 (2010)
2. C.P. Montgomery, B.S. Murray, E.J. New, R. Pal, D. Parker, Cell-Penetrating Metal complex optical probes: targeted and responsive systems based on Lanthanide luminescence. Acc. Chem. Res. **42**, 925–937 (2009)
3. V. Fernández-Moreira, F.L. Thorp-Greenwood, M.P. Coogan, Application of d_6 transition metal complexes in fluorescence cell imaging. Chem. Commun. **46**, 186–202 (2010)
4. J.-C.G. Bünzli, Lanthanide luminescence for biomedical analyses and imaging. Chem. Rev. **110**, 2729–2755 (2010)
5. Q. Zhao, F. Li, C. Huang, Phosphorescent chemosensors based on heavy-metal complexes. Chem. Soc. Rev. **39**, 3007–3030 (2010)
6. Q. Zhao, C. Huang, F. Li, Phosphorescent heavy-metal complexes for bioimaging. Chem. Soc. Rev. **40**, 2508–2524 (2011)
7. R.Y. Tsien, THE green fluorescent protein. Annu. Rev. Biochem. **67**, 509–544 (1998)
8. J. Zhou, Q. Liu, W. Feng, Y. Sun, F. Li, Upconversion luminescent materials: advances and applications. Chem. Rev. **115**, 395–465 (2015)
9. F. Auzel, Upconversion and Anti-Stokes processes with f and d Ions in solids. Chem. Rev. **104**, 139–174 (2004)
10. M. Haase, H. Schäfer, Upconverting nanoparticles. Angew. Chem. Int. Ed. **50**, 5808–5829 (2011)
11. Z. Zhang, W. Wang, W. Yin, M. Shang, L. Wang, S. Sun, Inducing photocatalysis by visible light beyond the absorption edge: effect of upconversion agent on the photocatalytic activity of Bi_2WO_6. Appl. Catal. B Environ. **101**, 68–73 (2010)
12. W. Qin, D. Zhang, D. Zhao, L. Wang, K. Zheng, Near-infrared photocatalysis based on YF_3 : $Yb_3{}^+$,$Tm_3{}^+$/TiO_2 core/shell nanoparticles. Chem. Commun. **46**, 2304–2306 (2010)
13. R.S. Khnayzer, J. Blumhoff, J.A. Harrington, A. Haefele, F. Deng, F.N. Castellano, Upconversion-powered photoelectrochemistry. Chem. Commun. **48**, 209–211 (2012)
14. Q. Liu, T. Yang, W. Feng, F. Li, Blue-emissive upconversion nanoparticles for low-power-excited bioimaging in Vivo. J. Am. Chem. Soc. **134**, 5390–5397 (2012)
15. S.F. Lim, R. Riehn, W.S. Ryu, N. Khanarian, C.-K. Tung, D. Tank, R.H. Austin, In Vivo and Scanning Electron Microscopy Imaging of upconverting nanophosphors in Caenorhabditis elegans. Nano Lett. **6**, 169–174 (2006)
16. R.S. Niedbala, H. Feindt, K. Kardos, T. Vail, J. Burton, B. Bielska, S. Li, D. Milunic, P. Bourdelle, R. Vallejo, Detection of analytes by immunoassay using up-converting phosphor technology. Anal. Biochem. **293**, 22–30 (2001)
17. Y.Y. Cheng, B. Fückel, R.W. MacQueen, T. Khoury, R.G.C.R. Clady, T.F. Schulze, N.J. Ekins-Daukes, M.J. Crossley, B. Stannowski, K. Lips, T.W. Schmidt, Improving the light-harvesting of amorphous silicon solar cells with photochemical upconversion. Energy Environ. Sci. **5**, 6953–6959 (2012)
18. W. Shockley, H.J. Queisser, Detailed balance limit of efficiency of p-n Junction solar cells. J. Appl. Phys. **32**, 510–519 (1961)
19. T. Trupke, M.A. Green, P. Würfel, Improving solar cell efficiencies by up-conversion of sub-band-gap light. J. Appl. Phys. **92**, 4117–4122 (2002)
20. A. Shalav, B.S. Richards, M.A. Green, Luminescent layers for enhanced silicon solar cell performance: up-conversion. Sol. Energy Mater. Sol. Cells **91**, 829–842 (2007)
21. F. Auzel, Up-conversions in re-doped solids, in *Spectroscopic Properties of Rare Earths in Optical Materials*, ed. by R. Hull, J. Parisi, B. Jacquier, (Springer, Berlin, Heidelberg, 2005), pp. 266–319
22. F. Wang, X. Liu, Recent advances in the chemistry of lanthanide-doped upconversion nanocrystals. Chem. Soc. Rev. **38**, 976–989 (2009)

23. C. Reinhard, R. Valiente, H.U. Güdel, Exchange-induced upconversion in Rb_2MnCl_4:Yb_3^+. J. Phys. Chem. B **106**, 10051–10057 (2002)
24. I. Etchart, M. Bérard, M. Laroche, A. Huignard, I. Hernández, W.P. Gillin, R.J. Curry, A.K. Cheetham, Efficient white light emission by upconversion in Yb_3^+-, Er_3^+- and Tm_3^+-doped Y_2BaZnO_5. Chem. Commun. **47**, 6263–6265 (2011)
25. W. Niu, S. Wu, S. Zhang, L. Li, Synthesis of colour tunable lanthanide-ion doped $NaYF_4$ upconversion nanoparticles by controlling temperature. Chem. Commun. **46**, 3908–3910 (2010)
26. J. Zhao, S. Ji, H. Guo, Triplet-triplet annihilation based upconversion: from triplet sensitizers and triplet acceptors to upconversion quantum yields. RSC Adv. **1**, 937–950 (2011)
27. T.N. Singh-Rachford, F.N. Castellano, Photon upconversion based on sensitized triplet–triplet annihilation. Coord. Chem. Rev. **254**, 2560–2573 (2010)
28. A. Monguzzi, R. Tubino, S. Hoseinkhani, M. Campione, F. Meinardi, Low power, non-coherent sensitized photon up-conversion: modelling and perspectives. Phys. Chem. Chem. Phys. **14**, 4322–4332 (2012)
29. R.R. Islangulov, F.N. Castellano, Photochemical upconversion: anthracene dimerization sensitized to visible light by a Ru(II) Chromophore. Angew. Chem. Int. Ed. **45**, 5957–5959 (2006)
30. S. Ji, H. Guo, W. Wu, W. Wu, J. Zhao, Ruthenium(II) polyimine–coumarin dyad with non-emissive ^{3}IL excited state as sensitizer for triplet–triplet annihilation based upconversion. Angew. Chem. Int. Ed. **50**, 8283–8286 (2011)
31. C.A. Parker, C.G. Hatchard, Delayed fluorescence from solutions of anthracene and phenanthrene. Proc. R. Soc. Lond. A **269**, 574–584 (1962)
32. C.A. Parker, C.G. Hatchard, T.A. Joyce, Selective and mutual sensitization of delayed fluorescence. Nature **205**, 1282–1284 (1965)
33. Z. Wang, J. Zhao, A. Barbon, A. Toffoletti, Y. Liu, Y. An, L. Xu, A. Karatay, H.G. Yaglioglu, E.A. Yildiz, M. Hayvali, Radical-enhanced intersystem crossing in new Bodipy derivatives and application for efficient triplet–triplet annihilation upconversion. J. Am. Chem. Soc. **139**, 7831–7842 (2017)
34. Z. Wang, J. Zhao, Bodipy–Anthracene dyads as triplet photosensitizers: effect of chromophore orientation on triplet-state formation efficiency and application in triplet–triplet annihilation upconversion. Org. Lett. **19**, 4492–4495 (2017)
35. K. Chen, W. Yang, Z. Wang, A. Iagatti, L. Bussotti, P. Foggi, W. Ji, J. Zhao, M. Di Donato, Triplet excited state of Bodipy accessed by charge recombination and its application in triplet–triplet annihilation upconversion. J. Phys. Chem. A **121**, 7550–7564 (2017)
36. A. Nattestad, Y.Y. Cheng, R.W. MacQueen, T.F. Schulze, F.W. Thompson, A.J. Mozer, B. Fückel, T. Khoury, M.J. Crossley, K. Lips, G.G. Wallace, T.W. Schmidt, Dye-sensitized solar cell with integrated triplet–triplet annihilation upconversion system. J. Phys. Chem. Lett. **4**, 2073–2078 (2013)
37. A. Monguzzi, J. Mezyk, F. Scotognella, R. Tubino, F. Meinardi, Upconversion-induced fluorescence in multicomponent systems: steady-state excitation power threshold. Phys. Rev. B **78**, 195112 (2008)
38. T. Miteva, V. Yakutkin, G. Nelles, S. Baluschev, Annihilation assisted upconversion: all-organic, flexible and transparent multicolour display. New J. Phys. **10**, 103002 (2008)
39. F.N. Castellano, C.E. McCusker, MLCT sensitizers in photochemical upconversion: past, present, and potential future directions. Dalton Trans. **44**, 17906–17910 (2015)
40. J. Zhao, W. Wu, J. Sun, S. Guo, Triplet photosensitizers: from molecular design to applications. Chem. Soc. Rev. **42**, 5323–5351 (2013)
41. D.P. Hari, B. Konig, Synthetic applications of eosin Y in photoredox catalysis. Chem. Commun. **50**, 6688–6699 (2014)
42. S. Fukuzumi, K. Ohkubo, Selective photocatalytic reactions with organic photocatalysts. Chem. Sci. **4**, 561–574 (2013)
43. Y. Shao, Y. Yang, Efficient organic heterojunction photovoltaic cells based on triplet materials. Adv. Mater. **17**, 2841–2844 (2005)

44. P. Majumdar, R. Nomula, J. Zhao, Activatable triplet photosensitizers: magic bullets for targeted photodynamic therapy. J. Mater. Chem. C **2**, 5982–5997 (2014)
45. S.G. Awuah, Y. You, Boron dipyrromethene (BODIPY)-based photosensitizers for photodynamic therapy. RSC Adv. **2**, 11169–11183 (2012)
46. O.J. Stacey, S.J.A. Pope, New avenues in the design and potential application of metal complexes for photodynamic therapy. RSC Adv. **3**, 25550–25564 (2013)
47. A. Kamkaew, S.H. Lim, H.B. Lee, L.V. Kiew, L.Y. Chung, K. Burgess, BODIPY dyes in photodynamic therapy. Chem. Soc. Rev. **42**, 77–88 (2013)
48. M. Kobayashi, S. Masaoka, K. Sakai, Photoinduced hydrogen evolution from water by a simple Platinum(II) terpyridine derivative: a Z-Scheme photosynthesis. Angew. Chem. Int. Ed. **51**, 7431–7434 (2012)
49. S. Ji, W. Wu, W. Wu, H. Guo, J. Zhao, Ruthenium(II) polyimine complexes with a long-lived ^{3}IL excited state or a ^{3}MLCT/^{3}IL equilibrium: efficient triplet sensitizers for low-power upconversion. Angew. Chem. Int. Ed. **50**, 1626–1629 (2011)
50. W. Wu, S. Ji, W. Wu, J. Shao, H. Guo, T.D. James, J. Zhao, Ruthenium(II)–polyimine–coumarin light-harvesting molecular arrays: design rationale and application for triplet–triplet-annihilation-based upconversion. Chem. Eur. J. **18**, 4953–4964 (2012)
51. X. Yi, J. Zhao, W. Wu, D. Huang, S. Ji, J. Sun, Rhenium(I) tricarbonyl polypyridine complexes showing strong absorption of visible light and long-lived triplet excited states as a triplet photosensitizer for triplet-triplet annihilation upconversion. Dalton Trans. **41**, 8931–8940 (2012)
52. J. Sun, W. Wu, J. Zhao, Long-lived room temperature deep-red-emissive intraligand triplet excited state of Naphthalimide in cyclometalated Ir(III) complexes and its application in triplet-triplet annihilation based upconversion. Chem. Eur. J. **18**, 8100–8112 (2012)
53. J. Sun, W. Wu, H. Guo, J. Zhao, Visible-light harvesting with cyclometalated Iridium(III) complexes having long-lived ^{3}IL excited states and their application in triplet–triplet annihilation based upconversion. Eur. J. Inorg. Chem. **2011**, 3165–3173 (2011)
54. A.K. Bansal, W. Holzer, A. Penzkofer, T. Tsuboi, Absorption and emission spectroscopic characterization of platinum-octaethyl-porphyrin (PtOEP). Chem. Phys. **330**, 118–129 (2006)
55. R.R. Islangulov, D.V. Kozlov, F.N. Castellano, Low power upconversion using MLCT sensitizers. Chem. Commun. **30**, 3776–3778 (2005)
56. D.V. Kozlov, F.N. Castellano, Anti-Stokes delayed fluorescence from metal–organic bichromophores. Chem. Commun., 2860–2861 (2004)
57. W. Wu, J. Sun, X. Cui, J. Zhao, Observation of the room temperature phosphorescence of Bodipy in visible light-harvesting Ru(II) polyimine complexes and application as triplet photosensitizers for triplet-triplet-annihilation upconversion and photocatalytic oxidation. J. Mater. Chem. C **1**, 4577–4589 (2013)
58. S. Ji, W. Wu, W. Wu, P. Song, K. Han, Z. Wang, S. Liu, H. Guo, J. Zhao, Tuning the luminescence lifetimes of ruthenium(II) polypyridine complexes and its application in luminescent oxygen sensing. J. Mater. Chem. **20**, 1953–1963 (2010)
59. W. Wu, L. Liu, X. Cui, C. Zhang, J. Zhao, Red-light-absorbing diimine Pt(II) bisacetylide complexes showing near-IR phosphorescence and long-lived ^{3}IL excited state of Bodipy for application in triplet–triplet annihilation upconversion. Dalton Trans. **42**, 14374–14379 (2013)
60. S. Ji, W. Wu, J. Zhao, H. Guo, W. Wu, Efficient triplet–triplet annihilation upconversion with Platinum(II) bis(arylacetylide) complexes that show long-lived triplet excited states. Eur. J. Inorg. Chem. **2012**, 3183–3190 (2012)
61. Y. Liu, W. Wu, J. Zhao, X. Zhang, H. Guo, Accessing the long-lived near-IR-emissive triplet excited state in naphthalenediimide with light-harvesting diimine platinum(II) bisacetylide complex and its application for upconversion. Dalton Trans. **40**, 9085–9089 (2011)
62. H. Sun, H. Guo, W. Wu, X. Liu, J. Zhao, Coumarin phosphorescence observed with *N*/*N*Pt(II) bisacetylide complex and its applications for luminescent oxygen sensing and triplet-triplet-annihilation based upconversion. Dalton Trans. **40**, 7834–7841 (2011)

63. H. Guo, M.L. Muro-Small, S. Ji, J. Zhao, F.N. Castellano, Naphthalimide phosphorescence finally exposed in a Platinum(II) diimine complex. Inorg. Chem. **49**, 6802–6804 (2010)
64. I.E. Pomestchenko, C.R. Luman, M. Hissler, R. Ziessel, F.N. Castellano, Room temperature phosphorescence from a Platinum(II) diimine bis(pyrenylacetylide) complex. Inorg. Chem. **42**, 1394–1396 (2003)
65. T. Yogo, Y. Urano, Y. Ishitsuka, F. Maniwa, T. Nagano, Highly efficient and photostable photosensitizer based on BODIPY chromophore. J. Am. Chem. Soc. **127**, 12162–12163 (2005)
66. W. Wu, H. Guo, W. Wu, S. Ji, J. Zhao, Organic triplet sensitizer library derived from a single chromophore (BODIPY) with long-lived triplet excited state for triplet–triplet annihilation based upconversion. J. Org. Chem. **76**, 7056–7064 (2011)
67. X. Cui, A. Charaf-Eddin, J. Wang, B. Le Guennic, J. Zhao, D. Jacquemin, Perylene-derived triplet acceptors with optimized excited state energy levels for triplet–triplet annihilation assisted upconversion. J. Org. Chem. **79**, 2038–2048 (2014)
68. Q. Zhou, M. Zhou, Y. Wei, X. Zhou, S. Liu, S. Zhang, B. Zhang, Solvent effects on the triplet–triplet annihilation upconversion of diiodo-Bodipy and perylene. Phys. Chem. Chem. Phys. **19**, 1516–1525 (2017)
69. Z. Lou, Y. Hou, K. Chen, J. Zhao, S. Ji, F. Zhong, Y. Dede, B. Dick, Different quenching effect of intramolecular rotation on the singlet and triplet excited states of Bodipy. J. Phys. Chem. C **122**, 185–193 (2018)
70. H.-C. Chen, C.-Y. Hung, K.-H. Wang, H.-L. Chen, W.S. Fann, F.-C. Chien, P. Chen, T.J. Chow, C.-P. Hsu, S.-S. Sun, White-light emission from an upconverted emission with an organic triplet sensitizer. Chem. Commun. (27), 4064–4066 (2009)
71. Y. Chen, J. Zhao, L. Xie, H. Guo, Q. Li, Thienyl-substituted BODIPYs with strong visible light-absorption and long-lived triplet excited states as organic triplet sensitizers for triplet-triplet annihilation upconversion. RSC Adv. **2**, 3942–3953 (2012)
72. S. Ji, J. Ge, D. Escudero, Z. Wang, J. Zhao, D. Jacquemin, Molecular structure intersystem crossing relationship of heavy atom free BODIPY triplet photosensitizers. J. Org. Chem. **80**, 5958–5963 (2015)
73. C. Zhang, J. Zhao, S. Wu, Z. Wang, W. Wu, J. Ma, S. Guo, L. Huang, Intramolecular RET enhanced visible light-absorbing Bodipy organic triplet photosensitizers and application in photooxidation and triplet–triplet annihilation upconversion. J. Am. Chem. Soc. **135**, 10566–10578 (2013)
74. M. Hussain, A.M. El-Zohry, H.B. Gobeze, J. Zhao, F. D'Souza, Mohammed OF, Intramolecular energy and electron transfers in Bodipy Naphthalenediimide triads. J. Phys. Chem. A **122**, 6081–6088 (2018)
75. C. Zhang, J. Zhao, Triplet excited state of diiodo-BOPHY derivatives: preparation, study of photophysical properties and application in triplet–triplet annihilation upconversion. J. Mater. Chem. C **4**, 1623–1632 (2016)
76. L. Huang, W. Yang, J. Zhao, Switching of the triplet excited state of Styryl 2,6-diiodo-bodipy and its application in acid-activatable singlet oxygen photosensitizing. J. Org. Chem. **79**, 10240–10255 (2014)
77. X. Cui, J. Zhao, Y. Zhou, J. Ma, Y. Zhao, Reversible photoswitching of triplet–triplet annihilation upconversion using Dithienylethene photochromic switches. J. Am. Chem. Soc. **136**, 9256–9259 (2014)
78. J. Ma, X. Cui, F. Wang, X. Wu, J. Zhao, X. Li, Photoswitching of the triplet excited state of Diiodo-Bodipy-Dithienylethene triads and application in photo-controllable triplet–triplet annihilation upconversion. J. Org. Chem. **79**, 10855–10866 (2014)
79. C. Zhang, J. Zhao, X. Cui, X. Wu, Thiol-activated triplet–triplet annihilation upconversion: study of the different quenching effect of electron acceptor on the singlet and triplet excited states of Bodipy. J. Org. Chem. **80**, 5674–5686 (2015)
80. Z. Mahmood, J. Zhao, Thiol-activatable triplet–triplet annihilation upconversion with Maleimide-Perylene as the caged triplet acceptor/emitter. J. Org. Chem. **81**, 587–594 (2016)

81. Z. Mahmood, J. Zhao, The unquenched triplet excited state of the fluorescent OFF/ON Bodipy-derived molecular probe based on photo-induced electron transfer. Photochem. Photobiol. Sci. **15**, 1358–1365 (2016)
82. K. Xu, J. Zhao, E.G. Moore, Covalently bonded Perylene–Diiodo-Bodipy dyads for thiol-activatable triplet–triplet annihilation upconversion. J. Phys. Chem. C **121**, 22665–22679 (2017)
83. Q. Liu, Y.-N. Li, H.-H. Zhang, B. Chen, C.-H. Tung, L.-Z. Wu, Reactivity and mechanistic insight into visible-light-induced aerobic cross-dehydrogenative coupling reaction by organophotocatalysts. Chem. Eur. J. **18**, 620–627 (2012)
84. A. Gorman, J. Killoran, C. O'Shea, T. Kenna, W.M. Gallagher, D.F. O'Shea, In vitro demonstration of the heavy-atom effect for photodynamic therapy. J. Am. Chem. Soc. **126**, 10619–10631 (2004)
85. S. Guo, W. Wu, H. Guo, J. Zhao, Room-temperature long-lived triplet excited states of Naphthalenediimides and their applications as organic triplet photosensitizers for photooxidation and triplet–triplet annihilation upconversions. J. Org. Chem. **77**, 3933–3943 (2012)
86. B. Ventura, G. Marconi, M. Broring, R. Kruger, L. Flamigni, Bis(BF_2)-2,2′-bidipyrrins, a class of BODIPY dyes with new spectroscopic and photophysical properties. New J. Chem. **33**, 428–438 (2009)
87. M. Bröring, R. Krüger, S. Link, C. Kleeberg, S. Köhler, X. Xie, B. Ventura, L. Flamigni, Bis(BF_2)-2,2′-Bidipyrrins (BisBODIPYs): highly fluorescent BODIPY dimers with large stokes shifts. Chem. Eur. J. **14**, 2976–2983 (2008)
88. W. Wu, X. Cui, J. Zhao, Hetero Bodipy-dimers as heavy atom-free triplet photosensitizers showing a long-lived triplet excited state for triplet-triplet annihilation upconversion. Chem. Commun. **49**, 9009–9011 (2013)
89. T. Yamazaki, Y. Murata, K. Komatsu, K. Furukawa, M. Morita, N. Maruyama, T. Yamao, S. Fujita, Synthesis and electrolytic polymerization of the ethylenedioxy-substituted terthiophene−fullerene dyad. Org. Lett. **6**, 4865–4868 (2004)
90. T.L. Chen, Y. Zhang, P. Smith, A. Tamayo, Y. Liu, B. Ma, Diketopyrrolopyrrole-containing oligothiophene-fullerene triads and their use in organic solar cells. ACS Appl. Mater. Interfaces **3**, 2275–2280 (2011)
91. R. Ziessel, B.D. Allen, D.B. Rewinska, A. Harriman, Selective triplet-state formation during charge recombination in a Fullerene/Bodipy molecular dyad (Bodipy= Borondipyrromethene). Chem. Eur. J. **15**, 7382–7393 (2009)
92. Y. Liu, J. Zhao, Visible light-harvesting perylene bisimide-fullerene (C_{60}) dyads with bidirectional "ping-pong" energy transfer as triplet photosensitizers for photooxidation of 1,5-dihydroxynaphthalene. Chem. Commun. **48**, 3751–3753 (2012)
93. W. Wu, J. Zhao, J. Sun, S. Guo, Light-harvesting fullerene dyads as organic triplet photosensitizers for triplet–triplet annihilation upconversions. J. Org. Chem. **77**, 5305–5312 (2012)
94. Y. Dong, A.A. Sukhanov, J. Zhao, A. Elmali, X. Li, B. Dick, A. Karatay, V.K. Voronkova, Spin–Orbit charge transfer intersystem crossing (SOCT-ISC) in Bodipy-Phenoxazine dyads: effect of chromophore orientation and conformation restriction on the photophysical properties. J. Phys. Chem. C **123**, 22793–22811 (2019)
95. Z. Wang, J. Zhao, M. Di Donato, G. Mazzone, Increasing the anti-Stokes shift in TTA upconversion with photosensitizers showing red-shifted spin-allowed charge transfer absorption but a non-compromised triplet state energy level. Chem. Commun. **55**, 1510–1513 (2019)
96. F.B. Dias, J. Santos, D.R. Graves, P. Data, R.S. Nobuyasu, M.A. Fox, A.S. Batsanov, T. Palmeira, M.N. Berberan-Santos, M.R. Bryce, A.P. Monkman, The role of local triplet excited states and D-A relative orientation in thermally activated delayed fluorescence: photophysics and devices. Adv. Sci. **3**, 1600080 (2016)
97. T. Serevičius, R. Komskis, P. Adomėnas, O. Adomėnienė, G. Kreiza, V. Jankauskas, K. Kazlauskas, A. Miasojedovas, V. Jankus, A. Monkman, S. Juršėnas, Triplet–triplet annihilation in 9,10-Diphenylanthracene derivatives: the role of intersystem crossing and exciton diffusion. J. Phys. Chem. C **121**, 8515–8524 (2017)

98. Y. Tao, K. Yuan, T. Chen, P. Xu, H. Li, R. Chen, C. Zheng, L. Zhang, W. Huang, Thermally activated delayed fluorescence materials towards the breakthrough of organoelectronics. Adv. Mater. **26**, 7931–7958 (2014)
99. Y. Im, S.Y. Byun, J.H. Kim, D.R. Lee, C.S. Oh, K.S. Yook, J.Y. Lee, Recent progress in high-efficiency blue-light-emitting materials for organic light emitting diodes. Adv. Funct. Mater. **27**, 1603007 (2017)
100. H. Ye, D.H. Kim, X. Chen, A.S.D. Sandanayaka, J.U. Kim, E. Zaborova, G. Canard, Y. Tsuchiya, E.Y. Choi, J.W. Wu, F. Fages, J.-L. Bredas, A. D'Aléo, J.-C. Ribierre, C. Adachi, Near-Infrared electroluminescence and low threshold amplified spontaneous emission above 800 nm from a thermally activated delayed fluorescent emitter. Chem. Mater. **30**, 6702–6710 (2018)
101. F.B. Dias, T.J. Penfold, A.P. Monkman, Photophysics of thermally activated delayed fluorescence molecules. Methods Appl. Fluoresc. **5**, 012001 (2017)
102. C.-J. Chiang, A. Kimyonok, M.K. Etherington, G.C. Griffiths, V. Jankus, F. Turksoy, A.P. Monkman, Ultrahigh efficiency fluorescent single and bi-Layer organic light emitting diodes: the key role of triplet fusion. Adv. Funct. Mater. **23**, 739–746 (2013)
103. T.C. Wu, D.N. Congreve, M.A. Baldo, Solid state photon upconversion utilizing thermally activated delayed fluorescence molecules as triplet sensitizer. Appl. Phys. Lett. **107**, 031103 (2015)
104. H. Uoyama, K. Goushi, K. Shizu, H. Nomura, C. Adachi, Highly efficient organic light-emitting diodes from delayed fluorescence. Nature **492**, 234–238 (2012)
105. N. Yanai, M. Kozue, S. Amemori, R. Kabe, C. Adachi, N. Kimizuka, Increased vis-to-UV upconversion performance by energy level matching between a TADF donor and high triplet energy acceptors. J. Mater. Chem. C **4**, 6447–6451 (2016)
106. J. Peng, X. Guo, X. Jiang, D. Zhao, Y. Ma, Developing efficient heavy-atom-free photosensitizers applicable to TTA upconversion in polymer films. Chem. Sci. **7**, 1233–1237 (2016)
107. D. Wei, F. Ni, Z. Zhu, Y. Zou, C. Yang, A red thermally activated delayed fluorescence material as a triplet sensitizer for triplet–triplet annihilation up-conversion with high efficiency and low energy loss. J. Mater. Chem. C **5**, 12674–12677 (2017)
108. W. Chen, F. Song, S. Tang, G. Hong, Y. Wu, X. Peng, Red-to-blue photon up-conversion with high efficiency based on a TADF fluorescein derivative. Chem. Commun. **55**, 4375–4378 (2019)
109. S. Amemori, Y. Sasaki, N. Yanai, N. Kimizuka, Near-Infrared-to-visible photon upconversion sensitized by a metal complex with spin-forbidden yet strong S_0–T_1 absorption. J. Am. Chem. Soc. **138**, 8702–8705 (2016)
110. Y. Sasaki, S. Amemori, H. Kouno, N. Yanai, N. Kimizuka, Near infrared-to-blue photon upconversion by exploiting direct S_0–T_1 absorption of a molecular sensitizer. J. Mater. Chem. C **5**, 5063–5067 (2017)
111. D. Liu, Y. Zhao, Z. Wang, K. Xu, J. Zhao, Exploiting the benefit of $S_0 \rightarrow T_1$ excitation in triplet–triplet annihilation upconversion to attain large anti-stokes shifts: tuning the triplet state lifetime of a tris(2,2′-bipyridine) osmium(II) complex. Dalton Trans. **47**, 8619–8628 (2018)

Chapter 6
Plasmon-Enhanced Homogeneous and Heterogeneous Triplet-Triplet Annihilation

Emily Westbrook, Xian Cao, and Peng Zhang

Abstract Triplet-triplet annihilation (TTA) process includes two categories, a homogeneous TTA occurring between two triplet excited molecules of the same type such as the homogeneous TTA upconversion (TTA-UC) and a heterogeneous TTA occurring between two triplet excited molecules of different types such as the heterogeneous TTA-UC, or between a triplet excited state and a triplet ground state such as the sensitized singlet oxygen generation. To the other front, noble metal nanostructures are known to exhibit an extraordinary capability to manipulate light through the collective oscillations of their conduction-band electrons, the so-called localized surface plasmon resonances (LSPR). Plasmonic nanostructures have been shown to be able to dramatically enhance the performances of many optical systems. In this book chapter, we will use a few examples to demonstrate that LSPR of noble metal nanoparticles can enhance the efficiency of both categories of TTA, and to discuss the conditions where such plasmonic enhancement would occur. The results shed light onto ways to improve the overall TTA efficiency, which would be relevant to the broad applications involving TTA-UC or sensitized singlet oxygen generation.

Keywords Plasmonic enhancement · Triplet-triplet annihilation · Singlet oxygen generation · Photon upconversion

Noble metal nanostructures have the ability to manipulate light by the collective oscillations of their conduction-band electrons, the so-called localized surface plasmon resonances (LSPR). These plasmonic nanostructures lend this unique optical property to the performance enhancement of many optical systems in the nanometer scale [1]. Optical systems are typically composed of atoms, ions, molecules, and quantum dots, which have diameters between 0.1 and 10 nm. The plasmon is a strong, localized electric field that can help concentrate macroscopic visible radiation to the subwavelength scale needed by conventional optical systems,

E. Westbrook · X. Cao · P. Zhang (✉)
Department of Chemistry, University of Cincinnati, Cincinnati, OH, USA
e-mail: peng.zhang@uc.edu; zhangph@ucmail.uc.edu

J. S. Lissau, M. Madsen (eds.), *Emerging Strategies to Reduce Transmission and Thermalization Losses in Solar Cells*, https://doi.org/10.1007/978-3-030-70358-5_6

in order to overcome the typical diffraction limit [2]. The best known examples of plasmon-enhanced light-matter interactions include surface-enhanced Raman scattering and plasmon-enhanced photoluminescence [3–5]. This chapter focuses on the latter.

Since the 1980s, there have been many studies that used the LSPR of plasmonic nanostructures to enhance the photoluminescence intensity of chromophores. The earliest examples of plasmon-enhanced fluorescence used rough noble metal surfaces, with broad SPR due to their nonuniformity [6, 7]. The broad SPR of the rough noble metal surfaces did enhance the fluorescence, but it was not possible to quantify the average enhancement or the mode of the SPR that led to enhancement. Since then, focus has shifted to the use of plasmonic nanostructures with controlled size and shape in order to tune the LSPR.

The photoluminescence enhancement is realized in two ways. The first is by the increase in the rate of excitation due to the enhanced excitation and emission light intensities from the increased local electric field of the LSPR [1]. Secondly, the LSPR can manipulate the chromophore non-radiative and radiative decay rates. The enhanced or weakened decay rates depend on the size and shape of the plasmonic nanostructure, specifically its absorption and scattering cross sections at the chromophore emission wavelength [1].

Depending on the magnitude and polarization of the absorption and scattering cross sections, it is possible that the plasmonic nanostructures could absorb or scatter the photoluminescence emitted from other plasmonic nanostructure-chromophore couples within a solution, which alters the radiative and non-radiative decay rates. Nanostructure scattering efficiency increases with increasing particle size and the non-radiative decay rate is reduced when the plasmonic nanostructures and chromophores are used in the solid state [8, 9]. Furthermore, when chromophores are in direct contact with the plasmonic nanostructure surface, non-radiative decay via energy or charge transfer from the chromophore to the nanostructure dominates, resulting in fluorescence quenching [10, 11]. Fluorescence quenching is minimized when chromophores are at least 5 nm from the nanostructure surface [10, 11].

The role of the LSPR band energy and nanoparticle scattering efficiency in the enhancement of a weakly fluorescent dye was examined [4]. Gold nanospheres and gold-silica nanoshells with varying sizes (radii between 24 and 130 nm) were coated in human serum albumin (HSA). The HSA provided a spacer layer between the nanostructure surface and the chromophore to minimize fluorescence quenching. Additionally, it provided a source of conjugation to the nanostructure surface for the chromophore used, indocyanine green (ICG). ICG is a commonly used dye for biomedical imaging, but only has a quantum yield of 1.2%. It was discovered that an enhancement by more than a factor of 50 could be realized for ICG when the nanostructure had an LSPR band overlapping the fluorescence emission band of ICG and a large scattering cross section.

More recently, it was demonstrated that the SPR of thin silver films enhanced triplet-triplet annihilation upconversion (TTA-UC) [12]. However, once again the undefined band of the SPR made it difficult to quantify the enhancement of the

TTA-UC quantum yield. The enhancement of TTA-UC is further complicated by the fact that it is a bichromophoric process, which requires even more precise tuning of the LSPR band in order to avoid excited-state quenching. The TTA-UC mechanism provides additional decay pathways that are affected by the LSPR of plasmonic nanostructures. In addition to increased incident excitation light intensity and excitation rate of the sensitizer and enhanced fluorescence of the acceptor or phosphorescence of the sensitizer, the intersystem crossing (ISC), triplet-triplet energy transfer (ET), and triplet-triplet annihilation (TTA) mechanisms can also be affected by the LSPR.

The earliest example of simple silver nanoparticle plasmon-enhanced TTA-UC used a silver nanoplate embedded in a polymethyl methacrylate (PMMA) thin film with the benchmark palladium octaethylporphyrin (PdOEP)/diphenylanthracene (DPA) pair showing TTA-UC enhancement [13]. With a 20% w/w loading of silver nanoplates, the TTA-UC emission was enhanced by 8.5-fold. This enhancement was attributed to an increased number of triplet excited states due to increased absorption of the sensitizer, PdOEP, as the silver nanoplates' LSPR overlaps with the absorption wavelength of the PdOEP. A similar study was done using gold nanoparticles and platinum octaethylporphyrin (PtOEP) sensitizer/DPA acceptor embedded in PMMA thin films and similar results were found [14]. At 3% w/w loading of gold nanoparticles, a 1.8-fold enhancement of TTA-UC was reported. The final example of a PtOEP/DPA system in solid-state PMMA thin films showed that in the presence of silver nanowires, the TTA-UC intensity was enhanced by 1.1-fold [15]. Furthermore, when the chromophores were near the LSPR hot spot at the junction of two or more pieces of nanowire, the TTA-UC intensity was enhanced up to 15-fold and the PtOEP phosphorescence was enhanced by 4- to 5-fold.

We have investigated the plasmon-enhanced homogeneous and heterogeneous TTA processes by gold nanoparticles (AuNPs) [16]. Rose Bengal (RB) was covalently conjugated to the AuNPs, and served as a hybrid sensitizer (AuNP@RB) in both cases, with 1,3-diphenylisobenzofuran (DPBF) as the acceptor in homogeneous TTA-UC and oxygen as the heterogeneous TTA acceptor (Fig. 6.1).

The overall quantum yield of TTA-UC (Φ_{UC}) for the AuNP@RB-DPBF system can be described using Eq. (6.1):

$$\Phi_{UC} = \gamma_{ex} \cdot \Phi_{ISC} \cdot \Phi_{ET} \cdot \Phi_{TTA} \cdot \Phi_{F} \cdot \varepsilon_{coll} \quad (6.1)$$

where γ_{ex}, Φ_{ISC}, Φ_{ET}, Φ_{TTA}, Φ_{F}, and ε_{coll} are the excitation rate of the sensitizer, intersystem crossing efficiency of the sensitizer, energy transfer efficiency between sensitizer and acceptor, triplet-triplet annihilation efficiency of the acceptor, fluorescence quantum yield of the acceptor, and light collection efficiency of the instrument. The parameters γ_{ex}, Φ_{ISC}, and Φ_{ET} are associated with the sensitizer while Φ_{TTA} and Φ_{F} are associated with the acceptor. Thus, it is expected that plasmon enhancement would reflect in the first three terms, since the sensitizer is covalently bound to the AuNPs.

There needs to be significant overlap between the sensitizer absorption and the AuNP LSPR band in order to observe the plasmon enhancement of TTA-

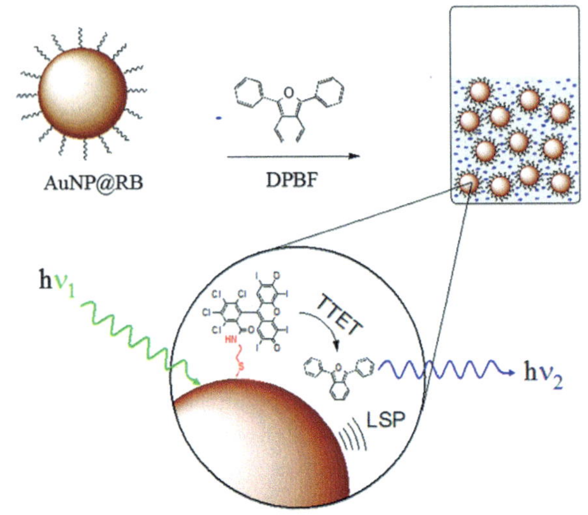

Fig. 6.1 Schematic illustration of plasmon-enhanced TTA-UC of AuNP@RB-DPBF system. Reprinted with permission from Fig. 1, ***Phys. Chem. Chem. Phys.***, 2015, **17**, 14479–14483

UC. In the AuNP@RB-DPBF system, the RB absorption peak wavelength was 560 nm, matching well with the LSPR band of the AuNPs, and plasmon-enhanced homogeneous TTA-UC was observed (Fig. 6.2a). As a comparison, we conjugated RB to silver nanoparticles (AgNPs). The AgNPs had an LSPR band centered at 420 nm, which overlapped with the absorption of DPBF. The AgNP@RB-DPBF system showed little difference from the corresponding RB-DPBF system. The AgNPs did not enhance TTA-UC, which supported the notion that plasmon enhancement of TTA-UC in the AuNP@RB-DPBF system was due to the LSPR-sensitizer absorption resonance coupling.

The three terms γ_{ex}, Φ_{ISC}, and Φ_{ET} were enhanced by the LSPR-sensitizer resonance coupling. As described previously, the AuNP LSPR enhanced the excitation rate of RB. The Φ_{ISC} was increased due to enhanced spin-orbit coupling between the singlet-triplet spin of the molecules. This was seen in the AuNP@RB-DPBF system as the observation of lower fluorescence (Fig. 6.2b) and shorter fluorescence lifetime from RB, as compared to that of RB-DPBF, which suggested an increase of Φ_{ISC}. The Φ_{ET} associated with energy transfer between the triplet states of sensitizer and acceptor, which is of Dexter type, was increased due to the enhanced localized electromagnetic field that promoted the electron exchange. A longer decay time of upconversion was observed from AuNP@RB-DPBF compared to RB-DPBF (Fig. 6.3), which indicated a longer sensitizer triplet lifetime. This promoted energy transfer between the triplets of sensitizer and acceptor. The AuNP@RB-DPBF system displayed an increase in the Φ_{UC} (Fig. 6.2c) and a 1.6-fold enhancement of TTA-UC over RB-DPBF (Fig. 6.2d).

We then examined the role of plasmon enhancement in the heterogeneous TTA between AuNP@RB as the sensitizer and ground-state oxygen as the acceptor, leading to the sensitized generation of singlet oxygen. A significant enhancement of the singlet oxygen generation was observed with AuNP@RB as compared to RB

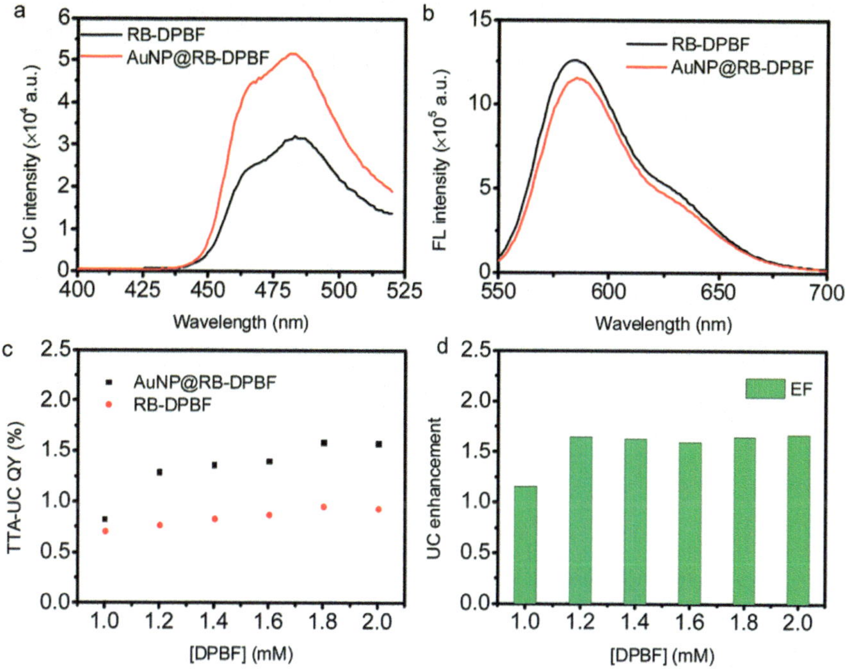

Fig. 6.2 (**a**) Upconversion spectrum of RB-DPBF (black) and AuNP@RB-DPBF (red), excited by a 532 nm laser. (**b**) Fluorescence spectrum of RB-DPBF (black) and AuNP@RB-DPBF (red), excited by a 532 nm laser. (**c**) Quantum yield of RB-DPBF (red) and AuNP@RB-DPBF (black) systems as a function of the [DPBF]. (**d**) Enhancement factor of the TTA-UC as a function of the [DPBF]. Reprinted with permission from Fig. 2, ***Phys. Chem. Chem. Phys.***, 2015, **17**, 14479–14483

Fig. 6.3 Normalized intensity decay curves of TTA-UC emission at 485 nm of AuNP@RB-DPBF (red) and RB-DPBF (blue) with equal molar ratios in deoxygenated *N,N*-dimethylformamide. Reprinted with permission from Fig. 4, ***Phys. Chem. Chem. Phys.***, 2015, **17**, 14479–14483

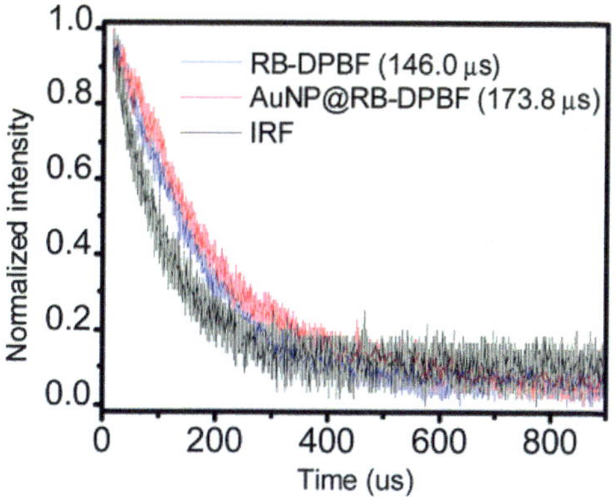

alone, especially with increasing concentration of AuNP@RB (Fig. 6.4a–c). The observed singlet oxygen emission intensity from the AuNP@RB sensitizer can be described using Eq. (6.2):

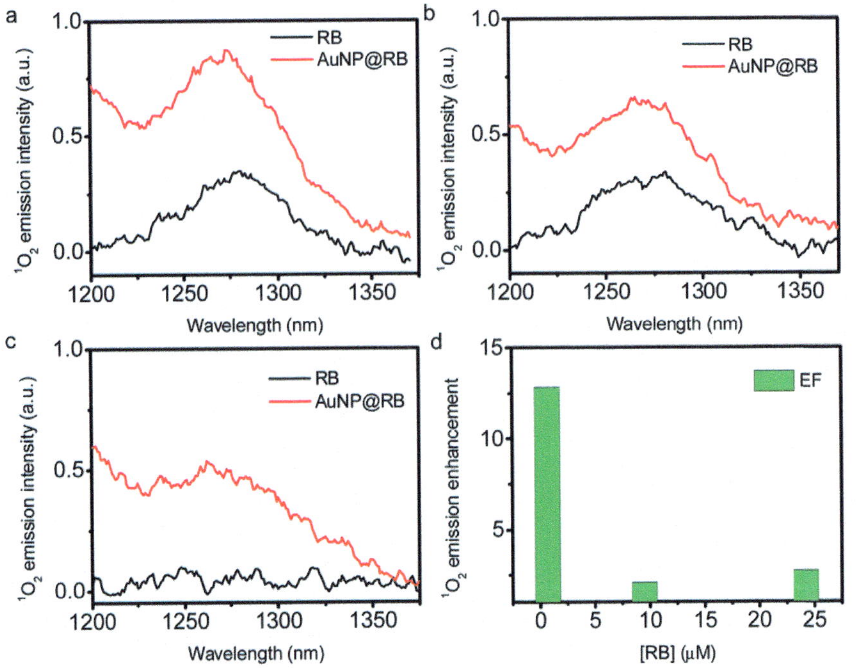

Fig. 6.4 (**a–c**) Singlet oxygen phosphorescence emission with varying [RB] from AuNP@RB (red) or RB by itself (black): (**a**) highest to (**c**) lowest [RB]. (**d**) Enhancement factor of singlet oxygen generation from AuNP@RB as a function of [RB]. Reprinted with permission from Fig. 5, ***Phys. Chem. Chem. Phys.***, 2015, **17**, 14479–14483

$$I = \gamma_{ex} \cdot \Phi_{ISC} \cdot \Phi_{ET} \cdot \Phi_{P} \cdot \varepsilon_{coll} \quad (6.2)$$

where γ_{ex}, Φ_{ISC}, Φ_{ET}, Φ_{P}, and ε_{coll} are the excitation rate of the sensitizer, intersystem crossing efficiency of the sensitizer, efficiency of energy transfer from triplet sensitizer to ground-state oxygen, singlet oxygen emission efficiency, and light collection efficiency of the instrument, respectively. The LSPR band of the AuNPs was 520 nm, far from the emission peak of singlet oxygen at around 1270 nm. There was very little resonance coupling between the LSPR of the AuNPs and the singlet oxygen emission. Therefore, only the terms γ_{ex}, Φ_{ISC}, and Φ_{ET} were enhanced by the plasmonic AuNPs. These enhanced mechanisms were similar to those in the homogeneous TTA-UC. A 13-fold enhancement of singlet oxygen generation sensitized with AuNP@RB was observed, as compared to free RB (Fig. 6.4d).

In both the homogeneous and heterogeneous TTA processes described, the plasmonic nanoparticles were tuned to interact with the sensitizer. The enhancement in both cases came from the strong overlap of the LSPR with the absorption wavelength of the sensitizer, with no effect on the acceptor. However, the efficiency of mechanisms involving the acceptor can be enhanced with the proper tuning of the LSPR band to the acceptor.

We investigated the plasmon-enhanced homogeneous TTA-UC of a polymeric acceptor covalently linked to AgNPs [17]. In this study, PtOEP was used as the sensitizer. Anthracene was covalently conjugated to a methacrylate copolymer backbone (polyACA), which was coated on the surface of AgNPs with stoichiometry set at one polymer chain per nanoparticle. These nanocomposites served as a hybrid acceptor (Ag@polyACA), as depicted in Fig. 6.5. Additionally, the percentage of anthracene per polymer varied between 5 and 52% in order to gauge the effect of the LSPR on the TTA process and concentration quenching of fluorescence.

The Φ_{UC} of the Ag@polyACA system could still be described by Eq. (6.1). The terms Φ_{TTA} and Φ_{F} are associated with the acceptor. Thus, it is expected that plasmon enhancement would reflect in these terms, since the polymeric acceptor is bound to the AgNPs. There should be significant overlap between the acceptor fluorescence emission and the AgNPs LSPR band in order to observe the plasmon enhancement of TTA-UC. In the Ag@polyACA system, the polyACA fluorescence emission peak wavelength was 460 nm and the LSPR band of the AgNPs was centered at 420 nm, which produced good overlap.

Fluorescence spectroscopy was first performed on Ag@polyACA, as compared to polyACA alone, to determine the fluorescence enhancement (Fig. 6.6). At equivalent concentrations of anthracene, the fluorescence intensities of the Ag@polyACA series were enhanced over polyACA except at moderate anthracene percentages (10 and 16%). The fluorescence intensity of Ag@polyACA trended down at first with increasing anthracene percentage, similar to the polyACA series alone. This suggested that concentration quenching of the fluorescence was dominating. The decreased fluorescence for the 16%-anthracene Ag@polyACA may have been due to the proximity of the polymer to the nanoparticle surface. The fluorescence

Fig. 6.5 Schematic illustration of LSPR-enhanced TTA-UC from post-modified polymeric acceptors. (**a**) Illustration of the post-polymerization modification of the random copolymer PMMA-*co*-GMA with the anthracene acceptor to produce the polymeric acceptor (polyACA). (**b**) Illustration of silver nanoparticle attached to polyACA to produce polymer-grafted silver nanoparticles (Ag@polyACA). (**c**) Illustration of TTA-UC in solution containing Ag@polyACA as the acceptor and PtOEP (pink squares) as the sensitizer. Reprinted with permission from Fig. 2, ***Dalton Trans.***, 2018, **47**, 8638–8645

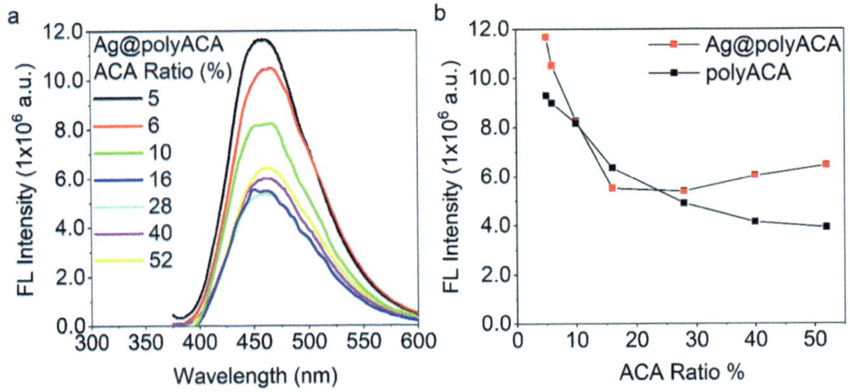

Fig. 6.6 (**a**) Direct fluorescence emission (excited at 380 nm) of the Ag@polyACA series with varying anthracene (ACA) percentage. (**b**) Fluorescence emission intensity as a function of the ACA percentage (equal molar ratios of acceptor to sensitizer for all) in the Ag@polyACA series (red) compared to the polyACA series (black). Reprinted with permission from Fig. 5, ***Dalton Trans.***, 2018, **47**, 8638–8645

enhancement of noble metal nanoparticles is distance dependent and as mentioned previously, fluorescence is quenched when chromophores are less than 5 nm from the surface of the nanoparticle [10, 11]. The 16%-anthracene Ag@polyACA might have been a crossover point at which concentration quenching was still dominating and the anthracene units in polyACA were less than 5 nm from the surface of the nanoparticles, so electronic interaction between anthracene and nanoparticle increased non-radiative decay and decreased fluorescence.

The quenching of fluorescence in the lower anthracene percentage samples due to the proximity to the surface of the nanoparticles was not as pronounced as the reduced concentration quenching, and therefore fluorescence was enhanced. In the higher anthracene percentage samples, even though concentration quenching was expected to dominate, the fluorescence intensity increased with increasing anthracene percentage. This enhancement was attributed to the nanoparticle LSPR. The polymers may extend out past 5 nm from the nanoparticle surface, which allows for plasmon-enhanced fluorescence.

Next, the goal of one polymer chain per nanoparticle was set to help distinguish between intramolecular TTA and intermolecular TTA. Since the Dexter-type energy transfer of TTA requires diffusion of the chromophores, if the chromophores can diffuse more easily to one another, there is an increase in the TTA efficiency and thus the TTA-UC intensity. It was hypothesized that with one polymer chain per nanoparticle, the diffusion of the entire nanocomposite unit would be so slow that if intermolecular TTA was the only pathway, the TTA-UC intensity would be severely reduced due to diminished TTA between the adjacent nanocomposites. However, if the intramolecular TTA is enhanced by the nanoparticle, then the TTA-UC intensity would increase over the polyACA alone.

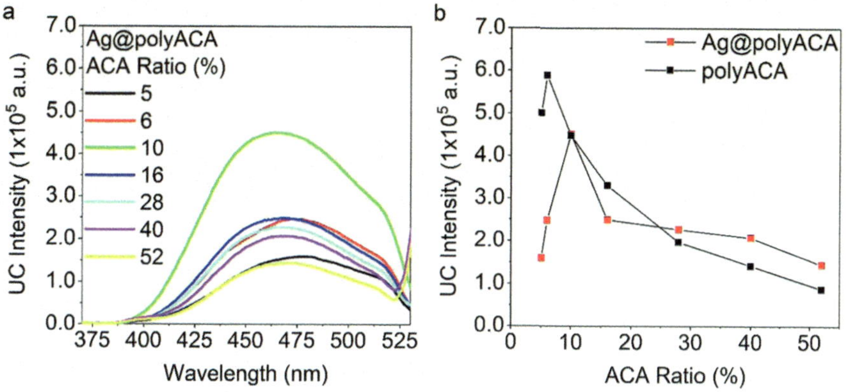

Fig. 6.7 (**a**) TTA-UC of the Ag@polyACA series with specified ACA percentage (excited with a 532 nm laser). (**b**) TTA-UC intensity as a function of ACA percentage, at equal molar ratios of acceptor to sensitizer, in the Ag@polyACA series (red) compared to the polyACA series (black). Reprinted with permission from Fig. 6, ***Dalton Trans.***, 2018, **47**, 8638–8645

It was observed that the upconversion intensities of Ag@polyACA with higher anthracene percentage (28% and higher) were generally enhanced due to the presence of the AgNPs (Fig. 6.7). However, the low anthracene percentage Ag@polyACA had significantly diminished upconversion intensities, as compared to the polyACA alone. These were likely cases in which the intermolecular TTA was reduced due to the slow diffusion of the very bulky nanocomposite. With the lower anthracene percentages, intramolecular TTA was less significant since the distance between acceptor units was much larger than the higher anthracene percentage samples. So, the plasmon enhancement of TTA did not sufficiently compensate for fewer TTA events occurring.

The 10%-anthracene Ag@polyACA and 10%-anthracene polyACA had nearly identical upconversion intensities. Therefore, the intensity threshold (I_{th}) provided a direct comparison of the TTA-UC efficiency with and without AgNPs. The I_{th} is a figure-of-merit parameter for TTA-UC that describes the efficiency of the TTA-UC system, dependent on excitation intensity. The I_{th} for polyACA was 352.5 mW·cm^{-2} and that for Ag@polyACA was 341.3 mW·cm^{-2}. The smaller I_{th} for Ag@polyACA as compared to that for polyACA indicated that the TTA-UC process was more efficient for Ag@polyACA, likely due to the nanoparticle LSPR enhancement.

Interestingly, a log-log plot of the phosphorescence intensity of PtOEP in these two systems versus the excitation intensity showed an increase in the phosphorescence intensity of PtOEP for the Ag@polyACA system, as compared to polyACA (Fig. 6.8). The phosphorescence was likely also enhanced in the presence of AgNPs. This enhancement has been observed previously [16] and in this case was likely due to the spin-orbit coupling between the silver of the nanoparticles and PtOEP, which enhanced the formation of triplet states via ISC. In the absence of enhanced energy

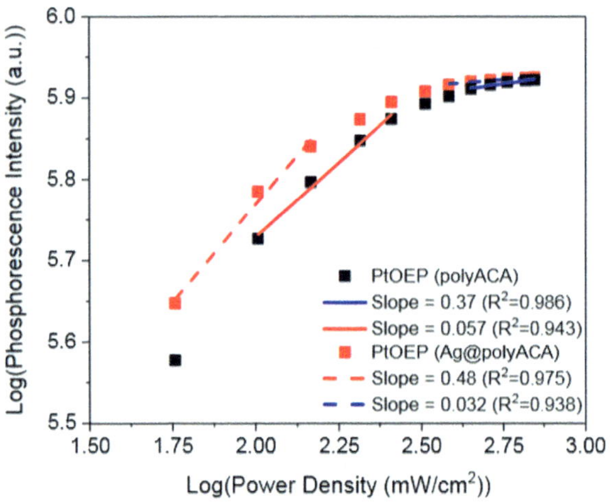

Fig. 6.8 Power dependence log-log plot of the PtOEP phosphorescence intensity versus the excitation power density for the polyACA series (black squares) and the Ag@polyACA series (red squares). Reprinted with permission from Fig. S4, ***Dalton Trans.***, 2018, **47**, 8638–8645

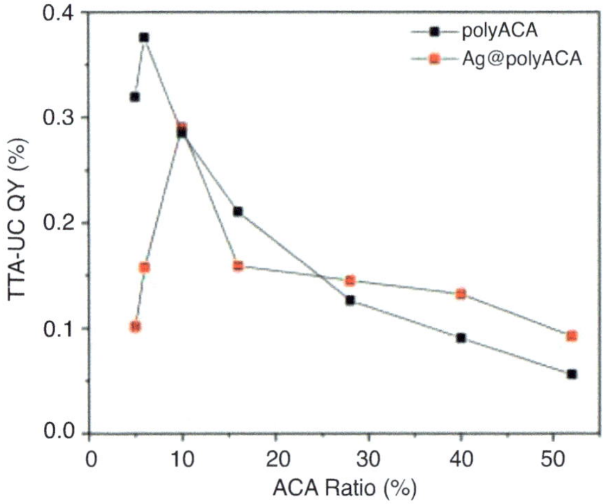

Fig. 6.9 Φ_{UC} of the polyACA series (black squares) and the Ag@polyACA series (red squares), using Rose Bengal as the standard fluorophore reference. Reprinted with permission from Fig. S3, ***Dalton Trans.***, 2018, **47**, 8638–8645

transfer, since the LSPR band did not overlap with the PtOEP absorption, more efficient triplet-state formation produced greater phosphorescence.

Finally, the effect of the LSPR on the Φ_{UC} was investigated by comparing the two systems. With higher anthracene percentage, the presence of AgNPs increased Φ_{UC} (Fig. 6.9). In fact, the 52%-anthracene Ag@polyACA had Φ_{UC} nearly double that of 52%-anthracene polyACA. This produced a 1.6-fold TTA-UC enhancement for 52%-anthracene Ag@polyACA over 52%-anthracene polyACA. This enhancement was attributed to the overlap of the AgNP LSPR band and the fluorescence emission of the polymeric anthracene acceptor, which enabled plasmon-enhanced fluorescence and TTA. The enhancement was generally less pronounced than that observed for plasmon-enhanced excitation rate and energy transfer when the LSPR band overlapped with the sensitizer absorption.

A separate investigation using a solid-state polymer thin film embedded with chromophores and silver nanoclusters of different sizes was reported to elucidate the effect of LSPR on the various steps of TTA-UC [18]. It is well known that the size of a noble metal nanoparticle causes the wavelength shift of the LSPR; as the nanoparticle size increases, the wavelength of the LSPR band increases [19]. The appropriate nanocluster was chosen with LSPR that overlapped the absorption or phosphorescence emission of the sensitizer, palladium(II) tetraphenyl-tetrabenzoporphyrin (Pd-TPTBP), or fluorescence emission of the acceptor, 9,10-bis-(phenylethynyl)anthracene (BPEA). When the LSPR band overlapped with the fluorescence of the acceptor but not the excitation wavelengths of the sensitizer, there was an increase in the fluorescence decay rate that caused TTA-UC enhancement of up to 3.7 times. When the LSPR band overlapped with the excitation of the sensitizer's Q band, the excitation efficiency was enhanced, which led to a TTA-UC enhancement of up to 8.0 times. Additionally, the I_{th} was reduced from 3500 mW cm^{-2} to 240 mW cm^{-2}. The LSPR effectively increased the excitation intensity, leading to the observed decrease in I_{th}. When the LSPR band overlapped with the phosphorescence wavelengths of the sensitizer, there was enhanced energy transfer from the sensitizer to the nanoparticle, strongly quenching the TTA-UC by 85%.

As a summary, in order to enhance homogeneous TTA-UC and heterogeneous TTA using the LSPR of plasmonic nanostructures, the LSPR band should overlap with the absorption of the sensitizer or the fluorescence of the acceptor, and should not overlap with the phosphorescence of the sensitizer.

References

1. T. Ming, H. Chen, R. Jiang, Q. Li, J. Wang, Plasmon-controlled fluorescence: beyond the intensity enhancement. J. Phys. Chem. Lett. **3**(2), 191–202 (2012). https://doi.org/10.1021/jz201392k
2. R.F. Oulton, V.J. Sorger, T. Zentgraf, R.M. Ma, C. Gladden, L. Dai, G. Bartal, X. Zhang, Plasmon lasers at deep subwavelength scale. Nature **461**(7264), 629–632 (2009). https://doi.org/10.1038/nature08364
3. K.A. Willets, R.P. Van Duyne, Localized surface plasmon resonance spectroscopy and sensing. Annu. Rev. Phys. Chem. **58**(1), 267–297 (2007). https://doi.org/10.1146/annurev.physchem.58.032806.104607
4. F. Tam, G.P. Goodrich, B.R. Johnson, N.J. Halas, Plasmonic enhancement of molecular fluorescence. Nano Lett. **7**(2), 496–501 (2007). https://doi.org/10.1021/nl062901x
5. D.V. Guzatov, S.V. Vaschenko, V.V. Stankevich, A.Y. Lunevich, Y.F. Glukhov, S.V. Gaponenko, Plasmonic enhancement of molecular fluorescence near silver nanoparticles: theory, modeling, and experiment. J. Phys. Chem. C **116**(19), 10723–10733 (2012). https://doi.org/10.1021/jp301598w
6. O. Siiman, A. Lepp, Protonation of the methyl orange derivative of aspartate adsorbed on colloidal silver: a surface-enhanced resonance Raman scattering and fluorescence emission study. J. Phys. Chem. **88**(12), 2641–2650 (1984). https://doi.org/10.1021/j150656a043
7. J.C. Rubim, I.G.R. Gutz, O. Sala, Surface-Enhanced Raman Scattering (SERS) and fluorescence spectra from mixed copper(I)/pyridine/iodide complexes on a copper electrode. Chem. Phys. Lett. **111**, 117–122 (1984)

8. J.R. Lakowicz, Radiative decay engineering 5: metal-enhanced fluorescence and plasmon emission. Anal. Biochem. **337**(2), 171–194 (2005). https://doi.org/10.1016/j.ab.2004.11.026
9. R.M. Bakker, H.K. Yuan, Z. Liu, V.P. Drachev, A.V. Kildishev, V.M. Shalaev, R.H. Pedersen, S. Gresillon, A. Boltasseva, Enhanced localized fluorescence in plasmonic nanoantennae. Appl. Phys. Lett. **92**(4), 043101 (2008). https://doi.org/10.1063/1.2836271
10. S. Eustis, M.A. El-Sayed, Why gold nanoparticles are more precious than pretty gold: noble metal surface plasmon resonance and its enhancement of the radiative and nonradiative properties of nanocrystals of different shapes. Chem. Soc. Rev. **35**(3), 209–217 (2006). https://doi.org/10.1039/B514191E
11. K.G. Thomas, P.V. Kamat, Chromophore-functionalized gold nanoparticles. Acc. Chem. Res. **36**(12), 888–898 (2003). https://doi.org/10.1021/ar030030h
12. S. Baluschev, F. Yu, T. Miteva, S. Ahl, A. Yasuda, G. Nelles, W. Knoll, G. Wegner, Metal-enhanced up-conversion fluorescence: effective triplet-triplet annihilation near silver surface. Nano Lett. **5**(12), 2482–2484 (2005). https://doi.org/10.1021/nl0517969
13. K. Poorkazem, A.V. Hesketh, T.L. Kelly, Plasmon-enhanced triplet-triplet annihilation using silver nanoplates. J. Phys. Chem. C **118**(12), 6398–6404 (2014). https://doi.org/10.1021/jp412223m
14. H. Yonemura, Y. Naka, M. Nishino, H. Sakaguchi, S. Yamada, Effect of gold nanoparticle on photon upconversion based on sensitized triplet–triplet annihilation in polymer films. Mol. Cryst. Liq. Cryst. **654**(1), 196–200 (2017). https://doi.org/10.1080/15421406.2017.1358044
15. Ł. Bujak, K. Narushima, D.K. Sharma, S. Hirata, M. Vacha, Plasmon enhancement of triplet exciton diffusion revealed by nanoscale imaging of photochemical fluorescence upconversion. J. Phys. Chem. C **121**(45), 25479–25486 (2017). https://doi.org/10.1021/acs.jpcc.7b08495
16. X. Cao, B. Hu, R. Ding, P. Zhang, Plasmon-enhanced homogeneous and heterogeneous triplet-triplet annihilation by gold nanoparticles. Phys. Chem. Chem. Phys. **17**(22), 14479–14483 (2015). https://doi.org/10.1039/c5cp01876e
17. E.G. Westbrook, P. Zhang, Plasmon-enhanced triplet-triplet annihilation upconversion of post-modified polymeric acceptors. Dalton Trans. **47**(26), 8638–8645 (2018). https://doi.org/10.1039/c8dt00269j
18. S. Jin, K. Sugawa, N. Takeshima, H. Tahara, S. Igari, S. Yoshinari, Y. Kurihara, S. Watanabe, M. Enoki, K. Sato, W. Inoue, K. Tokuda, T. Akiyama, R. Katoh, K. Takase, H. Ozawa, T. Okazaki, T. Watanabe, J. Otsuki, Precise control of localized surface plasmon wavelengths is needed for effective enhancement of triplet-triplet annihilation-based upconversion emission. ACS Photonics. **5**(12), 5025–5037 (2018). https://doi.org/10.1021/acsphotonics.8b01292
19. E. Petryayeva, U.J. Krull, Localized surface plasmon resonance: nanostructures, bioassays and biosensing—a review. Anal. Chim. Acta **706**(1), 8–24 (2011). https://doi.org/10.1016/j.aca.2011.08.020

Part II
Molecular Oxygen and Triplets: Photophysics and Protective Strategies

Chapter 7
Molecular Oxygen in Photoresponsive Organic Materials

Mikkel Bregnhøj and Peter R. Ogilby

Abstract The presence of molecular oxygen in organic materials designed for use in photoresponsive devices (e.g., solar cells, photon up-converters) can adversely influence the performance of the device in several ways. The most important of these is arguably through reactions that oxygenate and/or oxidize the organic components and thereby change properties relevant for a functioning device. The ground electronic state of molecular oxygen is a spin triplet, $O_2(X^3\Sigma_g^-)$. As such, it behaves as a biradical in its chemical reactions, trapping adventitious organic free radicals to yield reactive peroxyl radicals. The lowest excited electronic state of molecular oxygen is a spin singlet, $O_2(a^1\Delta_g)$, and can be formed in appreciable yield by energy transfer from a photoexcited organic molecule to $O_2(X^3\Sigma_g^-)$. Functional groups common to molecules used in photoresponsive materials (e.g., double bonds, sulfides) can react with $O_2(a^1\Delta_g)$ to form peroxides, which likewise are reactive and propagate disruption. The quenching of an excited-state organic component by $O_2(X^3\Sigma_g^-)$ also points to other ways in which oxygen can influence device performance. For example, the oxygen-mediated deactivation of comparatively long-lived triplet states can adversely influence the energy fusion process essential for some up-conversion devices. Likewise, electron transfer from an excited-state organic molecule to $O_2(X^3\Sigma_g^-)$ and/or $O_2(a^1\Delta_g)$ can not only interfere with desired charge movement (e.g., in a photovoltaic device), but it will also produce the superoxide radical ion that, in turn, can contribute to oxygenation reactions. Thus, to exploit fully the functional capabilities of photoresponsive organic materials and to prolong device longevity, and in lieu of completely excluding oxygen, it is necessary to monitor, understand, and ultimately control the behavior of oxygen in such systems.

Keywords Oxidative degradation · Photosensitizer · Singlet oxygen · Superoxide anion · Antioxidants

---

M. Bregnhøj · P. R. Ogilby (✉)
Department of Chemistry, Aarhus University, Aarhus, Denmark
e-mail: mibr@chem.au.dk; progilby@chem.au.dk

J. S. Lissau, M. Madsen (eds.), *Emerging Strategies to Reduce Transmission and Thermalization Losses in Solar Cells*, https://doi.org/10.1007/978-3-030-70358-5_7

7.1 Introduction

Light-mediated interactions between molecular oxygen, O_2, and organic molecules, M, are commonplace on earth. The material published on this subject is extensive, and covers almost all of the scientific disciplines ranging from biology and medicine to polymer physics. However, even with the current emphasis on interdisciplinary interactions, a common and accurate understanding of the photochemistry and photophysics of oxygen-dependent processes is lacking. This is perhaps surprising given that the central character in this story is a simple homonuclear diatomic molecule. Nevertheless, the study of oxygen-dependent processes continues to provide challenges of fundamental relevance as well as for practical applications.

We do not intend to provide an exhaustive review of interactions between oxygen and organic molecules pertinent for photoresponsive materials. Such an endeavor would arguably require a book in itself. Rather, we will attempt to provide a synopsis of selected light-mediated processes that involve molecular oxygen. As such, our goal with this chapter is to provide a commonly accepted foundation for continued work in this field.

The Cast of Characters

This chapter could arguably be rendered irrelevant simply by precluding contact between oxygen and the photoresponsive material. Indeed, given much of what we present below, this approach is often followed to prepare functional devices. In turn, the relevant discussion becomes one of encapsulation and oxygen scavenging [1, 2]. However, even in the absence of oxygen, one can still encounter undesired reactions, many of which derive from the same intermediates integral to the function of the photoresponsive device (e.g., electronic excited states, radical ions). As with any material, trace impurities can also give rise to undesired reactions. A relevant example in this regard is the radical-mediated degradation of solar cells that occurs upon irradiation of residual amounts of 1,8-diiodooctane commonly used to control the morphology of the active layer [3]. In the presence of oxygen, many of these intermediates undergo different reactions that are kinetically competitive. Thus, for a given device, encapsulation may not be a panacea, but it can be a useful approach.

As for any process in which an organic molecule absorbs light, the pertinent nascent intermediates are electronic excited states. Although excitation via a resonant one-photon process is generally the rule, certainly for many systems based on solar excitation, two-photon excitation using pulsed lasers can be a useful mechanistic tool (Fig. 7.1) [4].

For ground states with singlet spin, $^1M_0$, commonly encountered with organic molecules, the excited state of initial concern is likewise a singlet (i.e., the fluorescent state, $^1M_1$). This state often has a lifetime, ~1 ns, that is short enough

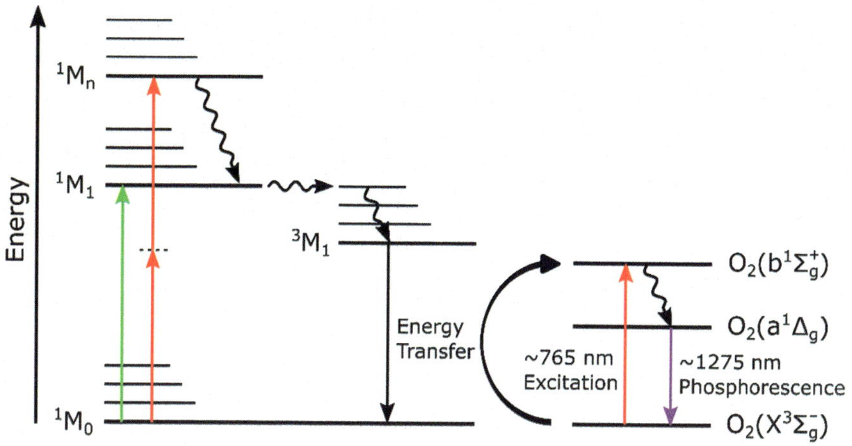

Fig. 7.1 Diagram showing the creation of excited states of a molecule M by one- and two-photon absorption (green and red arrows, respectively). Also shown are the triplet-state photosensitized production of $O_2(b^1\Sigma_g{}^+)$, with the subsequent nonradiative process to yield $O_2(a^1\Delta_g)$; the 765 nm transition to produce $O_2(b^1\Sigma_g{}^+)$ upon direct irradiation of oxygen; and the 1275 nm phosphorescence of $O_2(a^1\Delta_g)$. For molecules M in which the triplet-state energy is less than the $O_2(b^1\Sigma_g{}^+)$ excitation energy, energy transfer to $O_2(X^3\Sigma_g{}^-)$ may still produce $O_2(a^1\Delta_g)$

to preclude diffusion-dependent bimolecular interactions, including those with oxygen. Relevant exceptions are for devices constructed to exploit a diffusion-independent process of static quenching. For example, to facilitate electron transfer in organic photovoltaic solar cells, techniques are employed to optimize interpenetrating nanoscale networks of donor and acceptor moieties [5, 6]. Under these conditions, where the concentration of an active solute can be quite high, static quenching of a short-lived singlet fluorescent state by oxygen can also occur and thus lead to undesired effects.

The triplet state, $^3M_1$, often produced efficiently upon intersystem crossing (Fig. 7.1), can have a lifetime that is long enough that bimolecular interactions become important, even in viscous materials where mass diffusion is inhibited. One relevant example of such a "bimolecular" reaction, triplet-triplet annihilation, is the foundation for devices designed to up-convert incident light to a higher frequency [2, 7, 8]. In this case, depending on the morphology of the active material, the required annihilation reaction may occur as a consequence of exciton diffusion rather than mass diffusion. In either case, annihilation will kinetically compete with other triplet-state reactions that, depending on the functional groups present, can produce a plethora of reactive intermediates (e.g., free radicals) [9]. In systems that contain even a small amount of oxygen, quenching by oxygen will likely dominate the competition for triplet-state removal. Such quenching is at the core of our discussion on the photooxidative stability of a given material.

For systems containing molecular oxygen, exposure to light can likewise result in the production of oxygen excited electronic states. This can occur through energy transfer from the organic molecule excited state in a photosensitized process, or because of M-induced transitions in oxygen itself (Fig. 7.1). These processes can be deleterious in two ways: (a) the quenching of an M excited state by oxygen can adversely affect the M-dependent photoresponse (e.g., triplet-triplet annihilation in an up-conversion device) and (b) the oxygen excited state can oxygenate and/or oxidize M such as to irreversibly degrade the material. Moreover, photoinduced electron transfer from M to O_2 often competes with energy transfer, and the resultant radical ions, $M^{+\cdot}$ and $O_2^{-\cdot}$, can likewise have adverse effects on the device of interest.

In the context of oxygen-mediated adverse effects, the moniker "reactive oxygen species," ROS, is often used. In addition to an excited state of oxygen itself, other ROS of importance are the products of oxygenation reactions. These include a variety of peroxides and peroxyl radicals.

We elaborate on these reactions and reactive intermediates below.

7.2 Relevant Electronic States of Oxygen

Molecular oxygen has three low-energy electronic states, which are commonly denoted by their term symbols (Fig. 7.1) [10]. The ground state, $O_2(X^3\Sigma_g^-)$, is a spin triplet, whereas the two low-lying excited states are spin singlets, $O_2(a^1\Delta_g)$ and $O_2(b^1\Sigma_g^+)$. The moniker "singlet oxygen" is often used to denote only the lowest energy singlet state, $O_2(a^1\Delta_g)$. However, recent experiments have established the need to distinguish between the $O_2(a^1\Delta_g)$ and $O_2(b^1\Sigma_g^+)$ singlet states in many condensed-phase systems. As such, we rely on the term symbols.

The Triplet Ground State of Oxygen

The fact that oxygen's ground electronic state, $O_2(X^3\Sigma_g^-)$, is a spin triplet is a classic example of Hund's rule of maximum spin multiplicity; degenerate orbitals are singly occupied. As such, it is effectively a biradical and will readily react with other free radicals (i.e., it is a good "radical trap"). Thus, if a carbon-based free radical is produced upon homolytic bond cleavage in an excited state of M, for example, $O_2(X^3\Sigma_g^-)$ can trap this radical to form an alkyl peroxide. This can then start well-established propagating reactions that ultimately result in the oxidative degradation of M [11].

As described further in Sect. 7.3 below, $O_2(X^3\Sigma_g^-)$ can also play a role as a light absorber under selected conditions. The relevant prerequisite in this case is the formation of a very weakly bound complex between $O_2(X^3\Sigma_g^-)$ and $^1M_0$, arguably best characterized as a collision complex. The issue here is not that light no longer becomes available for the operation of the photoresponsive device, but rather that adverse ROS can be formed as a consequence.

The First Excited Singlet State of Oxygen

The energy difference between the $O_2(X^3\Sigma_g^-)$ and $O_2(a^1\Delta_g)$ states, ~94 kJ/mol, depends only slightly on the surrounding environment, with the spectroscopic transition occurring in the range 1270–1280 nm [12–14]. With very few exceptions, organic molecules have triplet-state energies that exceed 94 kJ/mol [15], and thus are energetically capable of producing $O_2(a^1\Delta_g)$ in a photosensitized process. However, despite these energetics, other processes often kinetically compete and reduce the yields of $O_2(a^1\Delta_g)$ upon collision of $^3M_1$ and $O_2(X^3\Sigma_g^-)$ (e.g., electron transfer and/or charge-transfer oxygen-mediated $^3M_1$ deactivation) [13].

The most relevant feature of $O_2(a^1\Delta_g)$ is that it reacts with common organic functional groups (e.g., double bonds, polycyclic aromatic hydrocarbons, sulfides) [16] to yield peroxides that, in turn, likewise contribute to the degradation of M [11]. Many of these functional groups are often integral to photoresponsive materials. Indeed, it is well established that $O_2(a^1\Delta_g)$ adversely affects many photoresponsive materials [2, 17–19]. Considering its importance in this regard, much of our discussion in this chapter focuses on the behavior of $O_2(a^1\Delta_g)$.

The Second Excited Singlet State of Oxygen

Unlike $O_2(a^1\Delta_g)$, there is no evidence that $O_2(b^1\Sigma_g^+)$ undergoes chemical reactions that result in the oxygenation of M [20]. Rather, once formed in condensed-phase systems, $O_2(b^1\Sigma_g^+)$ rapidly decays to produce $O_2(a^1\Delta_g)$ with almost unit efficiency (Fig. 7.1) [13]. In this way, $O_2(b^1\Sigma_g^+)$ is important solely as a precursor for $O_2(a^1\Delta_g)$, either upon direct irradiation of $O_2(X^3\Sigma_g^-)$ at 765 nm or in an M-sensitized process when the energy of the sensitizer triplet state exceeds the $O_2(b^1\Sigma_g^+)$ excitation energy of ~157 kJ/mol (Fig. 7.1).

7.3 Singlet Oxygen Production

Photosensitized Production

In photoresponsive materials, sensitization is arguably the most important process by which $O_2(a^1\Delta_g)$ is produced. Specifically, an excited electronic state of the organic component, M, is quenched by $O_2(X^3\Sigma_g^-)$, and energy transfer occurs to produce $O_2(a^1\Delta_g)$. Because the transition in oxygen is spin forbidden, this process must occur through a collision between M and O_2. In short, orbital overlap is required and we then treat the problem as a transition within the M-O_2 collision complex [10], not as a through-space Förster process.

The collision-dependent quenching of the $^1M_1$ state by $O_2(X^3\Sigma_g^-)$ occurs at the diffusion-controlled limit in liquids with a rate constant of ~3 × 10^{10} s^{-1} M^{-1} [21]. Nevertheless, for systems based solely on the physical displacement of M and O_2, the concentration of oxygen in a given sample is generally not large enough to render such quenching competitive with the unimolecular deactivation of $^1M_1$ through fluorescence, internal conversion, and intersystem crossing. Even so, $^1M_1$ states can still be quenched, certainly if one considers (a) static quenching by $O_2(X^3\Sigma_g^-)$ fortuitously placed near $^1M_1$, and/or (b) "movement" of $^1M_1$ via exciton diffusion [22]. In this case, $O_2(X^3\Sigma_g^-)$-induced $^1M_1 \rightarrow {}^1M_0$ internal conversion is a rare event [23], and the process is dominated by $O_2(X^3\Sigma_g^-)$-induced $^1M_1 \rightarrow {}^3M_1$ intersystem crossing [24, 25]. If the $^1M_1$-$^3M_1$ energy gap is larger than the excitation energy of $O_2(a^1\Delta_g)$, then energy transfer can occur and $O_2(a^1\Delta_g)$ can be produced [13].

The collision-dependent quenching of the $^3M_1$ state by $O_2(X^3\Sigma_g^-)$ in liquids occurs with rate constants of ~1–5 × 10^9 s^{-1} M^{-1}, depending on the molecule and solvent [15]. In amorphous polymers, where the diffusion coefficients of both M and O_2 are reduced, this rate constant for bimolecular interaction decreases by approximately two orders of magnitude [26]. Of course, $O_2(X^3\Sigma_g^-)$-mediated $^3M_1$ deactivation becomes more probable when the site of excitation can move via exciton diffusion. The key factor in all cases is that unimolecular channels for $^3M_1$ deactivation are sufficiently improbable that they generally do not compete with quenching by $O_2(X^3\Sigma_g^-)$, even with comparatively low concentrations of oxygen. As such, triplet-state photosensitized $O_2(b^1\Sigma_g^+)$ and $O_2(a^1\Delta_g)$ production is a common phenomenon and, depending on M, can occur with quantum efficiencies of unity. The magnitude of the latter depends on a variety of parameters including the extent of charge transfer in the M-O_2 complex and whether $^3M_1$ is energetically resonant with either the $O_2(b^1\Sigma_g^+)$ or the $O_2(a^1\Delta_g)$ state [13].

Although the active components in a given photoresponsive material readily form $^1M_1$ and $^3M_1$ states as part of the desired function (e.g., up-conversion via triplet-triplet annihilation), one cannot neglect corresponding excited states formed from adventitious impurities in the material. Indeed, the photophysics in a given material may be such that impurities could be a significant source of photosensitized $O_2(a^1\Delta_g)$ production.

To recapitulate, in solid materials/films where the active photoresponsive compound may be present in high concentration, the notion of an isolated and localized $^1M_1$ or $^3M_1$ state may not be accurate. Rather, we should consider that the excitation energy may readily "move" (e.g., exciton transport). Such movement may facilitate the energy transfer process from M to $O_2(X^3\Sigma_g^-)$, overcoming limitations for M and O_2 orbital overlap that arise from the inhibited translation/diffusion of the molecules themselves.

Optical Transitions in Oxygen

Although absorption cross sections for organic chromophores can be quite large, certainly for photoresponsive materials, one should not neglect light absorption by oxygen itself. As an isolated molecule, the optical transitions between the $O_2(X^3\Sigma_g^-)$, $O_2(a^1\Delta_g)$, and $O_2(b^1\Sigma_g^+)$ states are not very probable [27]. However, the perturbations associated with a collision between M and O_2 can be sufficient to make these processes important [10, 14].

As shown in Fig. 7.1, the $O_2(X^3\Sigma_g^-) \rightarrow O_2(b^1\Sigma_g^+)$ transition occurs at 765 nm, a wavelength that is present in many light-driven systems. We have shown that upon irradiation of an oxygen-containing liquid at 765 nm, appreciable amounts of $O_2(a^1\Delta_g)$ can indeed be rapidly produced upon the decay of the $O_2(b^1\Sigma_g^+)$ state [28, 29]. However, because this transition also results in a pronounced atmospheric absorption band in the solar spectrum [9], it may not play an important role for solar cells, for example, at the earth's surface. On the other hand, it is a very useful mechanistic tool for the study of M-O_2 photophysics and $O_2(a^1\Delta_g)$-mediated chemical reactions [30, 31], partly because many organic chromophores do not absorb at 765 nm and partly because the $O_2(a^1\Delta_g) \rightarrow O_2(X^3\Sigma_g^-)$ phosphorescence at ~1275 nm is sufficiently red-shifted to preclude interference from the 765 nm pump wavelength.

Although $O_2(a^1\Delta_g)$ can likewise be produced upon irradiation of $O_2(X^3\Sigma_g^-)$ at ~1275 nm [32–34], this is not as convenient a mechanistic tool, principally because one cannot easily monitor the $O_2(a^1\Delta_g) \rightarrow O_2(X^3\Sigma_g^-)$ phosphorescence at the same wavelength (i.e., the Stokes shift is very small).

M-O_2 Charge-Transfer Absorption

Complementing M-perturbed transitions in oxygen, we must also consider a cooperative transition that is unique to the M-O_2 complex. In this case, the ground-state M-O_2 collision complex absorbs light at wavelengths longer than (a) inherent transitions in M alone (e.g., $\pi \rightarrow \pi^*$ processes), and (b) fully allowed transitions in oxygen alone (i.e., the Herzberg and Schumann-Runge UV absorption bands) [35]. With the high concentrations of M found in both liquid solvents and solid polymers, such light absorption can be important. This is illustrated in Fig. 7.2 for cyclohexane, an arguably benign molecule.

This transition is due to the creation of an M-O_2 charge-transfer, CT, state often represented as a radical ion pair ($M^{+\cdot}O_2^{-\cdot}$) [35]. For many organic molecules, the one-photon process generally occurs in the UV region of the spectrum, as illustrated in Fig. 7.2 for cyclohexane. However, the same transition can be affected via a two-photon process with visible light [37]. The latter may thus be important under certain circumstances in which high-intensity lasers are used (e.g., two-photon up-conversion devices).

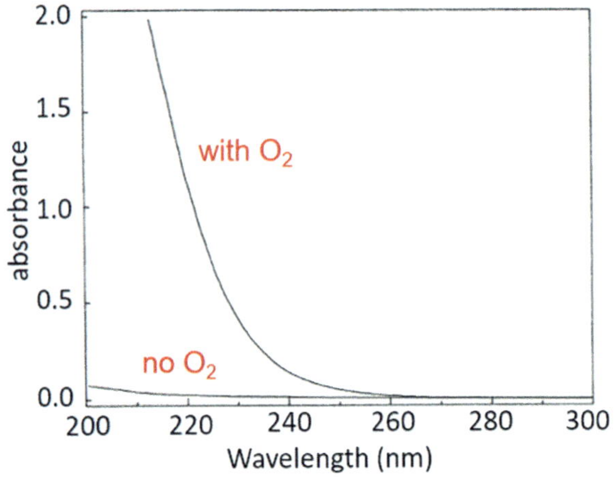

Fig. 7.2 Absorption spectra of N_2- and O_2-saturated liquid cyclohexane recorded over a 1 cm path length. Analogous spectra have also been recorded from a range of amorphous polymers [36]

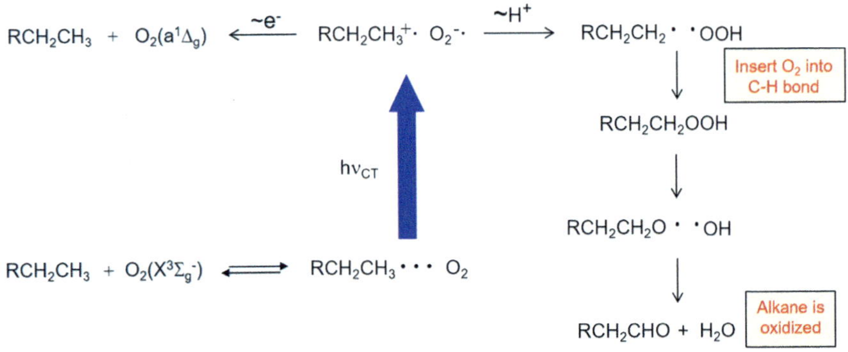

Fig. 7.3 Processes that can occur upon irradiation into the M-O_2 CT band, as illustrated for an alkane. Proton transfer within the CT state to produce a caged radical pair kinetically competes with the back electron transfer that produces $O_2(a^1\Delta_g)$. Both cases represent entry points for propagating degradation reactions

Populating the M-O_2 CT state can have adverse effects. First, rapid back electron transfer within the CT state can produce $O_2(a^1\Delta_g)$ with quantum yields as large as ~0.2, even for a "benign" molecule such as cyclohexane [38]. As such, for a suitable $O_2(a^1\Delta_g)$ reactant, either as a neat material or dissolved in a solvent, this is an entry point to the world of $O_2(a^1\Delta_g)$-mediated oxygenation reactions (vide infra) [36, 39]. Second, proton transfer within the CT state kinetically competes with the electron transfer reaction to make $O_2(a^1\Delta_g)$. As illustrated in Fig. 7.3, this ultimately results in the insertion of oxygen into a C-H bond and the start of common peroxide degradation reactions [40]. In this way, a molecule that does not chemically react with $O_2(a^1\Delta_g)$ (i.e., an alkane) can still degrade to produce products that both sensitize $O_2(a^1\Delta_g)$ formation and react with $O_2(a^1\Delta_g)$.

Chemical Generation

For an article on photoresponsive materials, it may appear nonsensical to discuss the production of $O_2(a^1\Delta_g)$ through chemical reactions that occur in the absence of light. However, peroxyl radicals formed upon the reaction of $O_2(X^3\Sigma_g^-)$ and an organic free radical, for example, can combine to yield a ground-state tetroxide that can fragment to yield $O_2(a^1\Delta_g)$ in non-negligible yields (the so-called Russell mechanism, Fig. 7.4) [41, 42].

Carrying this point further, the ground-state products of some light-initiated reactions involving $O_2(a^1\Delta_g)$ may, themselves, act as thermal/chemical precursors to $O_2(a^1\Delta_g)$ in subsequent processes that occur in the dark. Examples include the decomposition of aromatic endoperoxides [43, 44] and dioxetanes [45, 46] (Fig. 7.5). The latter may occur via an intermediary triplet-state carbonyl that, in turn, could sensitize the production of $O_2(a^1\Delta_g)$.

Thus, $O_2(a^1\Delta_g)$ may be a relevant reactive intermediate in (a) common radical-mediated auto-oxidation processes, and (b) processes that occur long after light exposure has ceased.

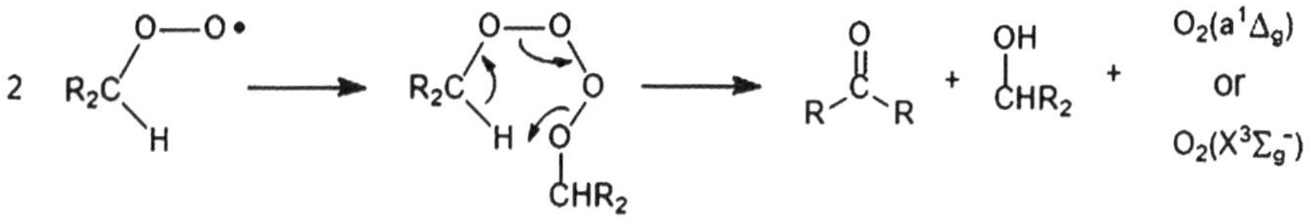

Fig. 7.4 Illustration of the Russell mechanism in which the combination of two peroxyl radicals can result in $O_2(a^1\Delta_g)$ production

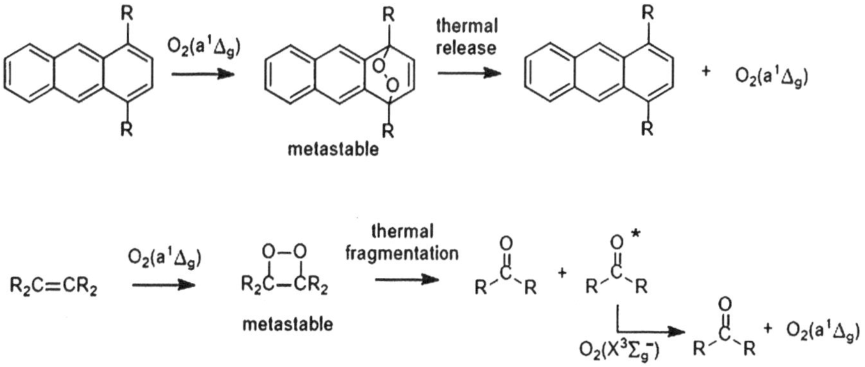

Fig. 7.5 Examples of $O_2(a^1\Delta_g)$ reactions that produce metastable species that can subsequently yield $O_2(a^1\Delta_g)$ via thermal fragmentation reactions. (Top) The retro 2 + 4 cycloaddition of an aromatic endoperoxide. (Bottom) The fragmentation of a ground-state dioxetane to yield an electronically excited-state carbonyl that, in turn, can sensitize the production of $O_2(a^1\Delta_g)$ from $O_2(X^3\Sigma_g^-)$

7.4 Singlet Oxygen Reactions

Thus far, we have indicated that $O_2(a^1\Delta_g)$ will react with organic functional groups commonly found in many photoresponsive materials. It is sufficient for our present discussion to elaborate on only some aspects of this point, leaving those interested in the details of these reactions to explore the extensive literature on this topic [16, 47].

First, $O_2(a^1\Delta_g)$ is an electrophile. Rate constants for the chemical reactions of $O_2(a^1\Delta_g)$ and, independently, for the nonreactive quenching of $O_2(a^1\Delta_g)$ (vide infra) are greatest for electron-rich species [48]. This information alone facilitates rational design in the construction of functional devices that are more stable to oxidative degradation.

Second, and arguably most importantly, although the acronym ROS is commonly applied to $O_2(a^1\Delta_g)$, it has led to a misinterpretation of $O_2(a^1\Delta_g)$'s behavior. Specifically, many investigators refer to $O_2(a^1\Delta_g)$ as a "highly reactive species." This is not correct. In the strict kinetic sense, $O_2(a^1\Delta_g)$ is selective, not reactive. It is this feature of $O_2(a^1\Delta_g)$'s behavior that makes it important as a synthetic reagent, for example [16, 49]. Bimolecular rate constants for the reactions of $O_2(a^1\Delta_g)$ with common organic functional groups rarely exceed ~5 × 10^7 s^{-1} M^{-1} in liquid solvents [48], a number that is well below that of ~3 × 10^{10} s^{-1} M^{-1} for a diffusion-controlled process that characterizes a true "reactive" intermediate. Admittedly, the rate constants for such reactions far removed from the diffusion-controlled limit may increase when carried out in a solid polymer matrix, as opposed to a liquid solvent, due to matrix caging effects [50]. However, these increases are generally quite small [50].

The general "lack of singlet oxygen's reactivity," as reflected in rate constants that are not large, is an important feature that allows for exploitation; $O_2(a^1\Delta_g)$ quenchers added as stabilizers can kinetically compete with adverse degradation reactions. We discuss this in a separate section below.

7.5 Nonreactive Deactivation of Singlet Oxygen

The chemical reactions of $O_2(a^1\Delta_g)$ kinetically compete with processes that deactivate $O_2(a^1\Delta_g)$ back to $O_2(X^3\Sigma_g{}^-)$. This is expressed in Eq. 7.1, where k_Δ is the first-order rate constant for total $O_2(a^1\Delta_g)$ removal, and is the reciprocal of the $O_2(a^1\Delta_g)$ lifetime, τ_Δ:

$$k_\Delta = \tau_\Delta^{-1} = k_{\mathrm{rxn}}\,[\mathrm{R}] + k_{\mathrm{deact}}\,[\mathrm{R}] + k_{\mathrm{q}}\,[\mathrm{Q}] + k_{\mathrm{solv}}\,[\mathrm{S}] + k_{\mathrm{O}} \tag{7.1}$$

Removal of $O_2(a^1\Delta_g)$ by chemical reaction with some reactant R is quantified by the product k_{rxn}[R]. However, R may also simply promote $O_2(a^1\Delta_g)$ deactivation to $O_2(X^3\Sigma_g{}^-)$, and we denote this with the bimolecular rate constant k_{deact}. In the literature, these processes are often denoted as chemical and physical quenching, respectively.

A key concept for multicomponent systems, or for systems into which an antioxidant has been added, is that a second quencher molecule, denoted by Q, will compete with R for $O_2(a^1\Delta_g)$ removal. In Eq. 7.1, we use the rate constant k_q to denote the sum of Q-mediated reactive and nonreactive $O_2(a^1\Delta_g)$ removal. If R and Q are solutes in a solvent or matrix, the latter can have a noticeable effect on $O_2(a^1\Delta_g)$ removal, particularly because the concentration of this solvent, [S], is likely much greater than that of R or Q. Although k_{solv} is not large (~10^3 s^{-1} M^{-1}), it depends strongly on the solvent in question and, as such, the product k_{solv} [S] is likewise highly solvent dependent [14]. Finally, the unimolecular decay of $O_2(a^1\Delta_g)$ to $O_2(X^3\Sigma_g{}^-)$, quantified by k_0, is negligible in most condensed-phase systems.

In discussing nonreactive removal of $O_2(a^1\Delta_g)$ mediated by some molecule M, be it a solute or the solvent, one must also distinguish radiative from nonradiative deactivation.

In the radiative process, the excitation energy of $O_2(a^1\Delta_g)$ is released as light (i.e., phosphorescence, because a spin change is involved). This transition, which occurs at ~1275 nm, is a very useful and characteristic probe for $O_2(a^1\Delta_g)$, and is discussed in Sect. 7.7 below. In condensed organic systems, the bimolecular rate constant for M-mediated radiative $O_2(a^1\Delta_g)$ decay, k_{rad}, is much smaller than the rate constant for nonradiative decay, k_{nr} [13, 14]. Even though both rate constants depend strongly on the collision partner M [14], the inequality $k_{rad} < k_{nr}$ holds and the phosphorescence quantum efficiency of $O_2(a^1\Delta_g)$ is very small (~10^{-6}–10^{-9}).

In contrast, rate constants for the collision-induced nonradiative $O_2(a^1\Delta_g) \rightarrow O_2(X^3\Sigma_g{}^-)$ transition can be quite large, and in many cases are greater than the rate constants for reactive removal of $O_2(a^1\Delta_g)$ [48]. For most collision partners, the excitation energy of $O_2(a^1\Delta_g)$ is lost as heat (e.g., energy transferred into the vibrational modes of the collision partner). However, for selected molecules whose triplet-state energy is less than the excitation energy of $O_2(a^1\Delta_g)$, of which carotenoids provide the best and most prevalent example, the mechanism of $O_2(a^1\Delta_g)$ deactivation can involve energy transfer to populate this excited state of the quenching molecule [13, 51]. The key point here is that selected molecules can be added to a system to kinetically compete with, and thus mitigate, the reactions of $O_2(a^1\Delta_g)$ (e.g., Q in Eq. 7.1).

7.6 Singlet Oxygen as a Diffusible Reagent

In this section, we illustrate the notion of $O_2(a^1\Delta_g)$ as a diffusible reagent with numerical exercises that show the importance of the $O_2(a^1\Delta_g)$ lifetime and oxygen's diffusion coefficient in a given medium. It is useful to start this discussion focusing on liquids where accurate data are more readily available. In this regard, water is an interesting medium simply because the lifetime of $O_2(a^1\Delta_g)$ in H_2O is comparatively short [14]. Although, like oxygen, water is often considered to be an undesired molecule in photoresponsive devices [52], aqueous heterogeneous systems have been examined to exploit the benefits of molecular confinement [2] and are relevant for the application of selected photoresponsive devices in life science applications [53].

Given the kinetic definition of a lifetime [9], a population of $O_2(a^1\Delta_g)$ molecules will be reduced by a factor of ~150 over a period of $t = 5\tau_\Delta$. In neat liquid H_2O at room temperature, τ_Δ is 3.5 μs [14]. Thus, a population of $O_2(a^1\Delta_g)$ produced in H_2O could initiate reactions with some molecule R for a period of ~18 μs. With a diffusion coefficient, D, of 2×10^{-5} cm^2 s^{-1} in water [54], the radial diffusion distance d of $O_2(a^1\Delta_g)$ over this period t is ~460 nm ($d = (6tD)^{1/2}$). Under conditions where the $O_2(a^1\Delta_g)$ lifetime is longer, a common occurrence in many organic systems, the radial diffusion distance of $O_2(a^1\Delta_g)$ can be appreciably larger. Even in glassy polymers, the decrease in D is often not enough to preclude reactions of $O_2(a^1\Delta_g)$ at distances remote from its point of production (illustrated below in a numerical exercise). In short, $O_2(a^1\Delta_g)$ is a diffusable "reagent."

In liquid solutions, values of k_q for the carotenoids are ~3×10^{10} s^{-1} M^{-1} (i.e., quenching occurs at the diffusion-controlled limit) [48, 51, 55, 56]. Molecules that have easily donated electrons (e.g., amines with a low oxidation potential) are likewise efficient quenchers of $O_2(a^1\Delta_g)$, and values of k_q can be as large as ~1×10^{10} s^{-1} M^{-1} in liquid solutions [48]. For these latter molecules, the mechanism of quenching involves CT-mediated coupling between the M-$O_2(a^1\Delta_g)$ and M-$O_2(X^3\Sigma_g^-)$ states of the M-O_2 complex; for a molecule M with a low oxidation potential, the $M^{+\cdot}O_2^{-\cdot}$ CT state is stabilized and comes closer in energy to the M-$O_2(a^1\Delta_g)$ state [10].

For many photoresponsive systems, however, one must consider rate constants and diffusion coefficients characteristic of a solid material (e.g., an amorphous polymer), and concentrations of reactive components that can be much higher than in liquid solvents. In this case, $O_2(a^1\Delta_g)$ may be created in the immediate vicinity of a reactive component, and one no longer has to take diffusion-based encounter into account (i.e., we enter the domain of static quenching). Alternatively, and as discussed above, exciton diffusion can also be important in certain solid materials. However, when mass diffusion is involved, the change from a liquid solvent to a solid polymer generally results in a leveling effect on the magnitude of bimolecular rate constants: large rate constants characteristic of the diffusion-controlled limit become smaller, and small rate constants characteristic of the so-called pre-equilibrium limit become larger [50, 57]. The latter phenomenon likely reflects the caging properties of a solid matrix.

With these points in mind, it is useful to present a numerical exercise that illustrates features of $O_2(a^1\Delta_g)$ as a diffusable reagent and the effect that an added antioxidant/quencher can have in a solid polymer matrix. In amorphous glassy polystyrene, the polymer-mediated rate constant for $O_2(a^1\Delta_g)$ removal is ~5×10^4 s^{-1} (i.e., τ_Δ ~ 20 μs) [58], and oxygen's diffusion coefficient, D, is ~2×10^{-7} cm^2 s^{-1} [26, 59]. Thus, over a period of $5\tau_\Delta$, the radial diffusion distance of $O_2(a^1\Delta_g)$ is ~110 nm. For an efficient $O_2(a^1\Delta_g)$ quencher, such as a carotenoid or a readily oxidized amine that removes $O_2(a^1\Delta_g)$ with a rate constant of ~1×10^{10} s^{-1} M^{-1} in liquid solutions, the removal rate constant, k_q, in glassy polystyrene is reduced to ~1×10^8 s^{-1} M^{-1} (i.e., the rate constant for a diffusion-controlled reaction gets smaller) [50]. If this quencher Q is homogeneously distributed in the polystyrene matrix at a concentration of 1×10^{-3} M, the $k_q[Q]$ term reduces the $O_2(a^1\Delta_g)$ lifetime to 6.7 μs. In turn, over a period of $5\tau_\Delta$, the radial diffusion distance of $O_2(a^1\Delta_g)$ is reduced to 63 nm. Thus, even in the presence of a comparatively high antioxidant concentration, $O_2(a^1\Delta_g)$ can still diffuse over an appreciable distance and initiate oxidation reactions.

Diffusion coefficients can be reduced by lowering the temperature. However, in many cases the lifetime of $O_2(a^1\Delta_g)$ correspondingly increases because the nonradiative $O_2(a^1\Delta_g) \rightarrow O_2(X^3\Sigma_g^-)$ transition is an activated process [30, 60]. Thus, depending on the system and the reactants involved, cooling alone may not solve the problem of $O_2(a^1\Delta_g)$-mediated sample degradation.

The comparatively long lifetime and large diffusion distance of $O_2(a^1\Delta_g)$ can be beneficial for systems in which $O_2(a^1\Delta_g)$ is used to mediate the up-conversion of near-IR light [61]. In this case, $O_2(a^1\Delta_g)$ transports energy from a photoexcited sensitizer and deposits this energy in a second molecule that can subsequently undergo triplet-triplet annihilation.

7.7 Detecting and Monitoring the Behavior of Singlet Oxygen

Unlike many ROS that can influence the behavior and properties of photoresponsive systems, $O_2(a^1\Delta_g)$ has the unique feature that it can be directly detected in time-resolved optical experiments. Moreover, these $O_2(a^1\Delta_g)$ experiments are performed in spectral and temporal domains that generally facilitate discrimination from optical processes inherent to the function of the photoresponsive system.

Singlet Oxygen Phosphorescence

The $O_2(a^1\Delta_g) \rightarrow O_2(X^3\Sigma_g^-)$ phosphorescent transition has a characteristic spectral signature: a narrow band centered at ~1275 nm whose position is only weakly influenced by the surrounding medium [14]. Although the $O_2(a^1\Delta_g)$ phosphorescence quantum efficiency is small, this spectrum can readily be detected in steady-state experiments using high-throughput monochromators and standard near-IR detectors. Spectral contamination usually appears as the tail of the fluorescence and/or phosphorescence spectrum from other components in the system whose intensity decreases over the range from 1250 to 1300 nm in which the distinct $O_2(a^1\Delta_g)$ band appears.

Although spectral discrimination is important in isolating the $O_2(a^1\Delta_g)$ phosphorescence signal, the combined use of spectral and temporal discrimination can be quite useful. Specifically, because the $O_2(a^1\Delta_g)$ lifetime is generally much longer than both the fluorescence and phosphorescence lifetimes of typical organic molecules, certainly in the presence of oxygen that will quench these excited states, one can "gate out" these faster decaying luminescence signals. Such temporal discrimination usually also applies for any scattered light transmitted through the collection optics and/or that causes near-IR luminescence from the collection optics [62]. Nevertheless, detector saturation at time = 0 and the concomitant "afterpulsing" are common problems [37, 63], and compromises must be made between the intensity of the exciting laser and the time spent signal averaging.

Time-resolved detection of the 1275 nm $O_2(a^1\Delta_g)$ phosphorescence facilitates appreciable mechanistic insight, as can be inferred from the discussion in preceding sections. For some systems, where $O_2(a^1\Delta_g)$ decay follows single exponential kinetics, $O_2(a^1\Delta_g)$ lifetimes and quenching rate constants can be obtained directly from the time-resolved data. In other cases, where the $O_2(a^1\Delta_g)$ decay follows multi-exponential kinetics, one can begin to infer about the extent of sample inhomogeneity in the spatial domain through which $O_2(a^1\Delta_g)$ moves in a given time period. Time-resolved $O_2(a^1\Delta_g)$ signals have been detected from a wide range of thin films, including films directly relevant for photoresponsive systems (Fig. 7.6) [64].

Depending on the system and experimental conditions, the time-resolved $O_2(a^1\Delta_g)$ phosphorescence signal can also provide information about the kinetics of the $O_2(a^1\Delta_g)$ precursor, not just the kinetics of $O_2(a^1\Delta_g)$ itself. Translating

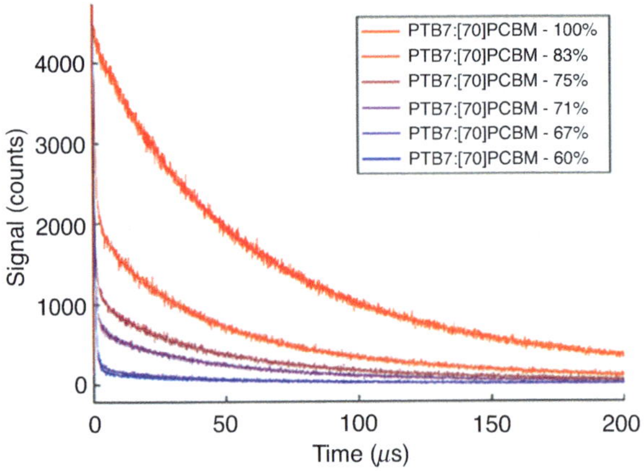

Fig. 7.6 Time-resolved 1275 nm $O_2(a^1\Delta_g)$ phosphorescence traces recorded from ~200 nm thick films prepared from blends of a functionalized derivative of C_{70} and PTB7, a thiophene derivative. Such blends are used as the active material in organic photovoltaic devices. The weight percentage of the fullerene in the film is given in the legend. Reprinted with permission from Bregnhøj et al. [64]

the illustration of $^3M_1$-photosensitized $O_2(a^1\Delta_g)$ production shown in Fig. 7.1 into kinetic expressions provides an informative example. In Eq. 7.2, we write the appropriate differential equation, where k_{form} is the bimolecular rate constant for diffusion-dependent energy transfer from $^3M_1$ to $O_2(X^3\Sigma_g{}^-)$:

$$\frac{d\left[O_2\left(a^1\Delta_g\right)\right]}{dt} = k_{form}\left[O_2\left(X^3\Sigma_g^-\right)\right]\left[^3M_1\right] - k_\Delta\left[O_2\left(a^1\Delta_g\right)\right] \quad (7.2)$$

Accounting for the fact that $^3M_1$ is itself a transient, integrating this differential equation yields an expression that includes a difference of two exponential functions (Eq. 7.3) [4]:

$$\left[O_2\left(a^1\Delta_g\right)\right]_t = \frac{k_{form}\left[O_2\left(X^3\Sigma_g^-\right)\right]\left[^3M_1\right]_0}{k_\Delta - k_T}\left\{\exp\left(-k_T t\right) - \exp\left(-k_\Delta t\right)\right\} \quad (7.3)$$

A key parameter in Eq. 7.3 is k_T, the first-order rate constant that accounts for all processes of $^3M_1$ removal, only one of which is $O_2(a^1\Delta_g)$ formation as quantified in the product $k_{form}[O_2(X^3\Sigma_g{}^-)]$. The difference of two exponential functions shown in Eq. 7.3 allows one to model the appearance and then disappearance of $O_2(a^1\Delta_g)$. When $k_T \sim k_\Delta$ this is manifested as a distinct rising component followed by a distinct falling component in the time-resolved $O_2(a^1\Delta_g)$ phosphorescence kinetic trace [65]. Under such conditions, it is often necessary to independently quantify k_T (e.g., through a time-resolved triplet-triplet absorption experiment) to obtain accurate information about k_Δ [65].

Time-resolved $O_2(a^1\Delta_g)$ phosphorescence signals are often recorded under conditions where $O_2(a^1\Delta_g)$ formation is fast relative to $O_2(a^1\Delta_g)$ removal (i.e., $k_T \gg k_\Delta$). This is exemplified by the data shown in Fig. 7.6, and could represent conditions of high $O_2(X^3\Sigma_g{}^-)$ concentration, facile $O_2(X^3\Sigma_g{}^-)$ and $^3M_1$ diffusion, and/or the facile diffusion of a triplet exciton to meet any number of quenchers including $O_2(X^3\Sigma_g{}^-)$. Carrying this point forward, one can apply the general format of Eq. 7.3 to the formation of $O_2(a^1\Delta_g)$ upon irradiation into the M-O_2 CT band (Fig. 7.3). The key point here is that, under these conditions, $O_2(a^1\Delta_g)$ formation is likewise fast relative to $O_2(a^1\Delta_g)$ removal. This is illustrated in the time-resolved $O_2(a^1\Delta_g)$ trace shown as a solid line in Fig. 7.7a obtained upon the irradiation of the CT band of oxygenated polystyrene glass.

We can now illustrate how the time-resolved 1275 nm $O_2(a^1\Delta_g)$ phosphorescence signal can be used as a sensitive kinetic probe of events that result in the photooxidative degradation of a sample. Consider, for example, the reactions shown in Figs. 7.3, 7.4, and 7.5 where oxidative degradation results in the formation of a carbonyl moiety that, in turn, could act as an efficient $O_2(a^1\Delta_g)$ photosensitizer. The latter could add an independent slower component to $O_2(a^1\Delta_g)$ formation that

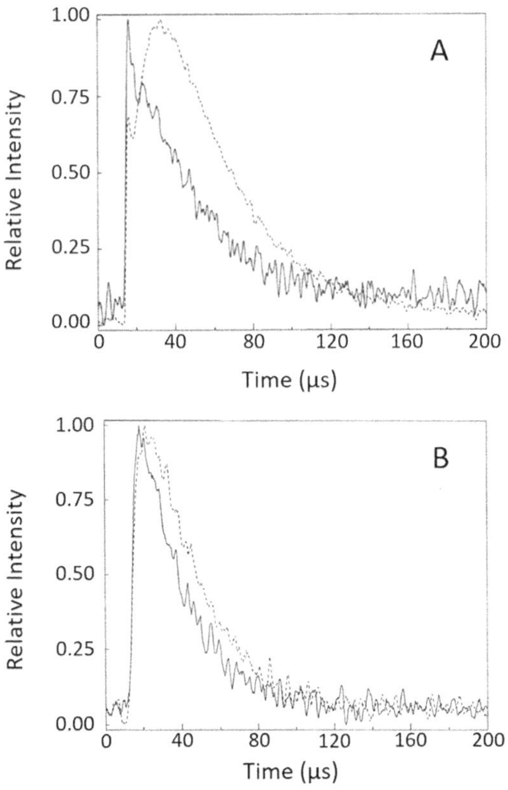

Fig. 7.7 Examples of how the time-resolved 1275 nm $O_2(a^1\Delta_g)$ phosphorescence can be used as a sensitive kinetic probe for the presence of impurities that sensitize the production of $O_2(a^1\Delta_g)$. The traces shown were recorded upon pulsed laser excitation of polystyrene glasses. (**a**) Data were recorded upon excitation at 309 nm. The trace shown with a solid line was recorded from a sample prepared via low-temperature free radical polymerization. It shows the rapid formation of $O_2(a^1\Delta_g)$ expected upon excitation of the M-O_2 CT state in a "clean" sample containing no moieties that can sensitize the production of $O_2(a^1\Delta_g)$. The trace shown with a dashed line was recorded from a sample prepared by molding the polystyrene at a high temperature. The kinetics reflect the formation of functional groups from thermal degradation that, in turn, sensitize $O_2(a^1\Delta_g)$ production (i.e., $k_T \sim k_\Delta$ in Eq. 7.3). (**b**) Data were recorded upon excitation at 341 nm. The trace shown with a solid line was recorded from a freshly prepared sample, and shows the fast rise expected upon excitation of the M-O_2 CT state in a "clean" sample. The trace shown with a dashed line was recorded from the same sample after it had been subjected to prolonged photolysis at 341 nm. As in panel (**a**), the kinetics indicate the presence of moieties that sensitize the production of $O_2(a^1\Delta_g)$. These figures are reprinted from Ogilby et al. [36]

depends appreciably on the diffusion-dependent encounter of $O_2(X^3\Sigma_g^-)$ with the newly created and localized $^3M_1$, and would be clearly manifested in the time-resolved $O_2(a^1\Delta_g)$ trace. This is illustrated using the traces shown with dashed lines in Fig. 7.7 in which both the thermal and photochemical production of $O_2(a^1\Delta_g)$ sensitizers are documented. In this way, a time-resolved $O_2(a^1\Delta_g)$ phosphorescence trace could complement experiments in which the products of sample oxidation are directly characterized by FTIR, for example (vide infra).

Compared to steady-state $O_2(a^1\Delta_g)$ spectroscopic experiments, and experiments based on other indirect approaches (see below), the time-resolved kinetic measurements are generally more demanding. In the least, they require sophisticated equipment (e.g., pulsed lasers for excitation, and fast and sensitive near-IR detectors).

Fluorescent Probes for Singlet Oxygen

Under certain conditions, it may be possible to add a molecule to the system which, upon reaction with $O_2(a^1\Delta_g)$, gives a characteristic fluorescent signal in the visible region of the spectrum. The advantage of this approach is that the fluorescence quantum efficiency of the probe will likely be much greater than the phosphorescence quantum efficiency of $O_2(a^1\Delta_g)$, and the light emitted will be in a spectral region where optical detectors are more responsive. The disadvantage is that it is an indirect measurement for the presence of $O_2(a^1\Delta_g)$. Specifically, other ROS may trigger the fluorescent response that would otherwise be assigned to $O_2(a^1\Delta_g)$. Moreover, the $O_2(a^1\Delta_g)$ formation and decay kinetics that would be accessible from a direct time-resolved $O_2(a^1\Delta_g)$ measurement would not be readily available from the indirect fluorescent signal of the probe. Given the importance of $O_2(a^1\Delta_g)$ to a broad range of disciplines, these issues are relevant to a large cross section of the scientific community. To this end, a large number of such probes have been developed and examined [66–70], some of which have been used in studies of photoresponsive materials [18, 71]. Although some probes are more useful than others, depending on the system being studied, the ideal general probe for $O_2(a^1\Delta_g)$ has yet to be developed.

Although the selectivity and accuracy with which a given probe responds to $O_2(a^1\Delta_g)$ are clearly important, one must also consider whether the probe compromises the system being studied. Thus, for a photoresponsive system, the presence of the probe and its oxidized products may interfere with the desired photoinduced electron transfer, triplet-triplet annihilation, or light absorption. The clear inference is that luminescent probes for $O_2(a^1\Delta_g)$ must be used with caution and in conjunction with relevant control experiments.

Characterizing Reaction Products

A corollary to the use of a probe that gives a fluorescent signal upon reaction with $O_2(a^1\Delta_g)$ is the direct characterization of oxidation products and ascertaining whether such products can be assigned to a reaction with $O_2(a^1\Delta_g)$. This can be a challenging process, particularly when sample amounts are small. Moreover,

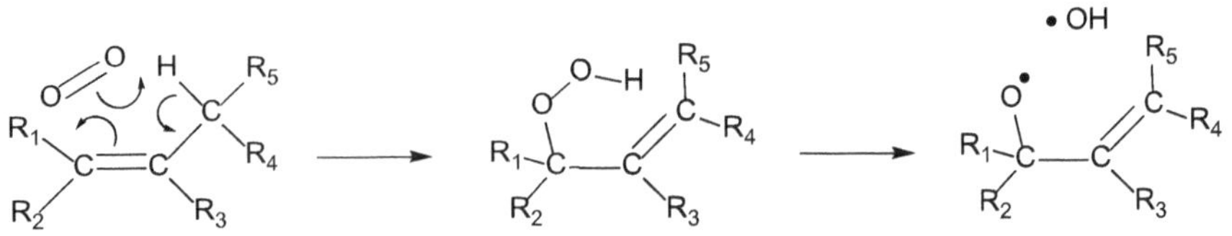

Fig. 7.8 Example of one reaction of $O_2(a^1\Delta_g)$ that leads to other ROS. In this case, the ene reaction results in the formation of an allylic hydroperoxide that, in turn, can cleave to yield the hydroxyl radical and an alkoxyl radical. These radicals can propagate to result in the oxidative degradation of the material

although many of the nascent oxidation products may indeed be characteristic of $O_2(a^1\Delta_g)$ (e.g., an allylic hydroperoxide formed via the ene reaction, Fig. 7.8, or a dioxetane formed via a 2 + 2 cycloaddition reaction, Fig. 7.5) [16], such products often evolve to yield other molecules and thereby blur the direct connection to $O_2(a^1\Delta_g)$. To help resolve this issue, one could examine the relative amounts of products formed as a function of experimental conditions that uniquely influence the rates of $O_2(a^1\Delta_g)$ reactions. This includes examining solvent/matrix isotope effects that influence k_{solv} (Eq. 7.1) or adding $O_2(a^1\Delta_g)$ quenchers, as discussed in the next section.

Although the isolation and characterization of products using chromatography and NMR, for example, may be difficult, in situ characterization using other spectroscopic techniques may be useful. For example, FTIR has been used to monitor oxidation products in conducting polymers and light-emitting diodes [72, 73] as well as in materials used for solar cells [19]. Of course, caution must again be exercised when assigning the observed products to a reaction with $O_2(a^1\Delta_g)$, particularly since carbonyls are often monitored and this functional group is not a nascent product of known $O_2(a^1\Delta_g)$ reactions.

Judicious Use of Additives and Isotope Effects

As indicated in Eq. 7.1, the rate of $O_2(a^1\Delta_g)$ removal is sensitive to the presence of certain quenchers and to the nature of the solvent/matrix. As such, through the judicious use of an additive or a change in the "matrix environment" to which $O_2(a^1\Delta_g)$ is exposed, this kinetic competition will be manifested in (a) the intensity and kinetics of the $O_2(a^1\Delta_g)$ phosphorescence signal, (b) the fluorescent response of a $O_2(a^1\Delta_g)$ probe, and (c) the nature and amounts of oxidation products.

Perhaps the most definitive test for the presence/participation of $O_2(a^1\Delta_g)$ is to exploit the effect of the solvent H/D isotope effect on the magnitude of the $k_{solv}[S]$ term in Eq. 7.1. Replacing C-H and O-H bonds in the solvent with C-D and O-D bonds results in an appreciable decrease in the magnitude of k_{solv} (i.e., the solvent isotope effects on τ_Δ in the neat solvent fall in the range of 10–30) [48]. Although an accurate description of the origins of this phenomenon can be convoluted [30], it is generally described as a change in the solvent's ability to accept the electronic excitation energy of $O_2(a^1\Delta_g)$ through the process of electronic-to-vibrational energy transfer. The key point is that, with few exceptions [74], the replacement of H by D in the *solvent* has not been found to have comparable effects on other photophysical and photochemical processes of *solutes* in oxygenated systems. Thus, H/D isotope effects observed in the amount of oxygenated products formed, for example, are a good indication that $O_2(a^1\Delta_g)$ was involved in the process. Of course, many photoresponsive materials/devices do not have a "solvent" in the traditional sense. Thus, to exploit this isotope effect as a mechanistic tool in a photovoltaic solar cell, for example, one would need to work with a perdeuterated electron donor and/or acceptor. An added complication in this case would be that the H/D effect could also be manifested in the photophysics that results in the yield of sensitized $O_2(a^1\Delta_g)$ production (e.g., the lifetime of the pertinent triplet state quenched by $O_2(X^3\Sigma_g^-)$).

The addition of certain quenchers, Q, is often thought to be a sufficiently accurate test for the presence of $O_2(a^1\Delta_g)$. For example, NaN_3 is acknowledged to be an efficient quencher of $O_2(a^1\Delta_g)$ [48]. Thus, upon the addition of NaN_3 to a given system, a decrease in (a) the amount of oxygenated products formed, or (b) the intensity of the $O_2(a^1\Delta_g)$ phosphorescence signal is often attributed to the fact that NaN_3 has directly quenched $O_2(a^1\Delta_g)$ in the system. However, one must be careful in interpreting such data because NaN_3 will also quench the triplet state of many organic molecules [75]. If that triplet state also happened to be the precursor of $O_2(a^1\Delta_g)$, then the observed effect of NaN_3 may not reflect the quenching of $O_2(a^1\Delta_g)$ itself. The latter could only be discerned by the effect of NaN_3 on the kinetics of the time-resolved $O_2(a^1\Delta_g)$ phosphorescence signal. This problem also extends to molecules added as a $O_2(a^1\Delta_g)$ probe; they too may quench the $O_2(a^1\Delta_g)$ precursor and thereby influence the amount of $O_2(a^1\Delta_g)$ produced [76].

7.8 Mitigating Degradation Mediated by Singlet Oxygen in Photoresponsive Materials

When it is ascertained that $O_2(a^1\Delta_g)$ is indeed a component in the degradation of photoresponsive materials, it should now be evident that steps can be taken to mitigate this process.

Our use of the word "mitigate" is arguably more accurate than the word "eliminate." Even if one appreciably reduces (a) the amount of $O_2(a^1\Delta_g)$ produced, and/or (b) the rate of $O_2(a^1\Delta_g)$-mediated oxygenation reactions, a low nascent concentration of oxygenated product can still propagate to a point of being disruptive. The latter includes the fact that products such as carbonyl-containing compounds (e.g., Fig. 7.5) can be good photosensitizers that only escalate the yield of $O_2(a^1\Delta_g)$ production. Thus, a realistic goal is "mitigation sufficient to yield a useful device."

Molecular Modification of the Photoresponsive Material

An inert/benign photoresponsive material would be the ideal solution to the problem. To this end, we consider two questions. Can one create a functional photoresponsive material that (1) is a sufficiently poor $O_2(a^1\Delta_g)$ sensitizer and, independently, (2) lacks functional groups that readily react with $O_2(a^1\Delta_g)$? Many of the fundamental mechanistic issues needed to answer these questions have already been investigated and resolved.

To make an unreactive material, one can capitalize on the fact that $O_2(a^1\Delta_g)$ is an electrophile. Thus, by using compounds that do not readily donate electrons, materials that do not rapidly react with $O_2(a^1\Delta_g)$ can be prepared. In some cases, the solution is as simple as adding an electron-withdrawing group (e.g., a cyano moiety) to a double bond that would otherwise react with $O_2(a^1\Delta_g)$, as has been demonstrated with conducting polymers used in light-emitting diodes [77].

Features that characterize a good $O_2(a^1\Delta_g)$ sensitizer are also known, and include a triplet-state energy that is close to the excitation energy of $O_2(a^1\Delta_g)$ or $O_2(b^1\Sigma_g^+)$, a comparatively long triplet-state lifetime to facilitate the required collision with $O_2(X^3\Sigma_g^-)$, and a minimal amount of CT character in the sensitizer-oxygen complex [13]. With respect to the latter, however, one should be careful about designing a system based solely on the inclusion of more M-O_2 CT character. First, by choosing molecules that would lower the energy of the $M^{+\cdot}O_2^{-\cdot}$ state, one runs the risk of enabling electron transfer chemistry at the expense of the energy transfer reaction that produces $O_2(a^1\Delta_g)$ [78]. Second, a readily oxidized molecule may not efficiently produce $O_2(a^1\Delta_g)$, but it would more easily react with any $O_2(a^1\Delta_g)$ that was produced (vide supra). Of course, for up-conversion devices based on triplet-triplet annihilation, the combination of (1) the desired long triplet-state lifetimes and (2) triplet-state energies that invariably exceed the $O_2(a^1\Delta_g)$ excitation energy can make it difficult to preclude $O_2(a^1\Delta_g)$ production, and other strategies should be employed (vide infra) [2, 53].

Exploiting the Benefits of an Added Quencher or Antioxidant

The addition of a molecule that could quench both $O_2(a^1\Delta_g)$ and/or the $O_2(a^1\Delta_g)$ precursor could sufficiently mitigate $O_2(a^1\Delta_g)$-mediated degradation to yield a useful device. We have already discussed fundamental features of this approach, including selected molecules that could be used.

For additives that remove $O_2(a^1\Delta_g)$, a nonreactive molecule that simply promotes the $O_2(a^1\Delta_g) \rightarrow O_2(X^3\Sigma_g^-)$ transition is preferred (e.g., NaN_3 or selected amines). In the ideal limit, such an additive remains active indefinitely and irreversible chemical changes do not occur. In contrast, a $O_2(a^1\Delta_g)$ scavenger (i.e., a molecule that reacts with $O_2(a^1\Delta_g)$) is less desired. In this case, not only does the concentration of the sacrificial scavenger decline over time, but also the oxidation products may promote degradation in themselves (e.g., start a free radical chain reaction) or may evolve to become good $O_2(a^1\Delta_g)$ sensitizers (e.g., a carbonyl as a reaction product).

When using carotenoids as added antioxidants, one must consider three separate modes of action. First, because the energy of the carotenoid triplet state is invariably lower than the $O_2(a^1\Delta_g)$ excitation energy of 94 kJ/mol, energy transfer from $O_2(a^1\Delta_g)$ to the carotenoid provides an efficient channel for $O_2(a^1\Delta_g)$ removal [55, 79–81]. This process requires a collision between $O_2(a^1\Delta_g)$ and the carotenoid, and the process can be optimized upon the judicious placement of the carotenoid such as to preclude/minimize a dependence on diffusion (e.g., carotenoid placement in the photosynthetic unit of a plant such as to enable static quenching). Second, the same energetic inequalities that make a carotenoid a good $O_2(a^1\Delta_g)$ quencher also make it a good quencher of the $O_2(a^1\Delta_g)$ precursor, the triplet state of an organic molecule [15]. In this way, the carotenoid could reduce the yield of photosensitized $O_2(a^1\Delta_g)$ production, again through a collision-dependent process of energy transfer. One could distinguish between these two processes by examining the effect of the carotenoid on k_Δ and k_T, respectively. As discussed in Sect. 7.7, k_Δ can be obtained through time-resolved $O_2(a^1\Delta_g)$ phosphorescence experiments and k_T through time-resolved triplet-triplet absorption and/or $^3M_1 \rightarrow {}^1M_0$ phosphorescence experiments. A corollary is that the carotenoid could also quench an excited state critical to the operation of the photoresponsive material (e.g., a triplet state in an up-conversion device). One solution to this problem could be the use of molecular compartmentalization (vide infra) to separate the photoresponsive material from the carotenoid. For oxidative stabilization to work with this approach, one would rely on a slow reaction between $O_2(a^1\Delta_g)$ and the photoresponsive material combined with the facile diffusion of $O_2(a^1\Delta_g)$ into the domain housing the carotenoid. Third, the carotenoid can act as a sacrificial reactant to inhibit radical reactions initiated by the $O_2(a^1\Delta_g)$-mediated oxygenation of an organic molecule [56]. Although this can be beneficial in the short run, it eventually results in carotenoid depletion and, as such, is an undesired process. One solution for this problem would be to incorporate other additives that specifically work as radical traps and that would kinetically compete with the carotenoid to inhibit radical propagation, allowing the carotenoid to work as a physical quencher of $O_2(a^1\Delta_g)$ for a longer period of time. It has been shown that carotenoids can indeed quench $O_2(a^1\Delta_g)$ in photovoltaic materials and devices and, as such, prolong the functional lifetime of the device [64, 82].

Like carotenoids, nickel chelates are multifaceted antioxidants. They are efficient photophysical quenchers of $O_2(a^1\Delta_g)$ [50, 83] and, independently, organic triplet states that can photosensitize the production of $O_2(a^1\Delta_g)$ [84]. They are also sacrificial quenchers of free radicals [85]. Such chelates have likewise been shown to mitigate the $O_2(a^1\Delta_g)$-mediated degradation of photovoltaic materials and prolong device lifetimes [86].

Exploiting the Benefits of Phase Separation and Molecular Confinement

As stated at the outset, we will not address the large-scale limit of confinement (i.e., encapsulation of the entire photoresponsive device). Rather, we focus on molecular scale features that can influence the local concentration of $O_2(X^3\Sigma_g^-)$, the local lifetime of $O_2(a^1\Delta_g)$, and the magnitude of rate constants for the reactions of $O_2(a^1\Delta_g)$ within a local domain.

Although reactive scavengers could be added to the system to remove $O_2(X^3\Sigma_g^-)$ (e.g., the oxidation of sodium sulfite) [2, 87], they are limited by the fact that, over time, the scavenger is consumed. This approach was originally presented as a tool to deoxygenate the hydrophobic domain of a micellar system that housed the photoresponsive molecules [87]. Specifically, the reaction of $O_2(X^3\Sigma_g^-)$ and sodium sulfite in the surrounding water would create an oxygen concentration gradient across the oil-water interface and force more $O_2(X^3\Sigma_g^-)$ out of the hydrophobic domain. As such, to solve the problem of scavenger depletion here, one could arguably design a system that replenished the surrounding water with more sodium sulfite.

Another approach to control the local concentration of $O_2(X^3\Sigma_g^-)$ is to put the photoresponsive system itself in water and exploit the fact that the solubility of $O_2(X^3\Sigma_g^-)$ in water is approximately ten times less than in a hydrocarbon medium [88]. This would also require a judicious choice of photoactive molecules whose excited-state lifetimes are short enough to preclude quenching by $O_2(X^3\Sigma_g^-)$ at these concentrations. A further advantage of this approach is that H_2O itself is a good physical deactivator of $O_2(a^1\Delta_g)$ (i.e., the lifetime of $O_2(a^1\Delta_g)$ in neat H_2O is ~3.5 μs) [14]. As such, for any $O_2(a^1\Delta_g)$ that is produced in this system, one might be able to sufficiently mitigate the degradation of photoresponsive molecules by exploiting the kinetic competition between $O_2(a^1\Delta_g)$ reaction and $O_2(a^1\Delta_g)$ deactivation simply by decreasing the concentration of the material that would react. Although this approach to handle the presence of $O_2(X^3\Sigma_g^-)$ is not ideal for an up-conversion device based on triplet-triplet annihilation, for example, it could nevertheless be useful when combined with other methods of mediating the effects of oxygen.

With increasingly sophisticated structures that both confine molecular ensembles and provide a barrier to oxygen [2], one can better control adverse effects by controlling local $O_2(X^3\Sigma_g^-)$ concentrations, diffusional encounter of an organic molecule triplet state and $O_2(X^3\Sigma_g^-)$, and encounter between $O_2(a^1\Delta_g)$ and a reactive substrate. Despite the elegance of many such approaches, many polymeric barriers will only mitigate but not preclude oxygen transport; diffusion coefficients are not zero. Thus, one cannot neglect the importance of ensuring that the kinetics of the competing processes disfavor the undesired effects of oxygen. This general point has been clearly made for the process of up-conversion via triplet-triplet annihilation [7]. Nevertheless, as has long been known for cyclodextrin enclosures, for example, the inhibitory effects on oxygen transport can be appreciable for a photophysical process involving the triplet states of organic molecules [89]. Of course, one can mitigate the effects of $O_2(a^1\Delta_g)$ diffusion through a barrier by incorporating functional groups into the barrier that will promote the $O_2(a^1\Delta_g) \rightarrow O_2(X^3\Sigma_g^-)$ transition. One should also consider that processes of electron and/or energy transfer can happen through barriers. One relevant example is the H_2/D_2 isotope effect observed for the quenching of $O_2(a^1\Delta_g)$ through the walls of a fullerene [90]. Another example involves photoinduced electron transfer between a protein-encapsulated chromophore and a molecule in the surrounding solvent [91, 92].

7.9 Other ROS and the Superoxide Radical Anion

The production of $O_2(a^1\Delta_g)$ can result in the formation of other ROS that can lead to the degradation and failure of a photoresponsive device. Perhaps the most prevalent in this regard are peroxides and oxygen-based radicals that are formed as products of a reaction between $O_2(a^1\Delta_g)$ and an organic functional group (Fig. 7.8). In this way, the photoinitiated production of $O_2(a^1\Delta_g)$ leads to a plethora of radical propagation steps that occur in the absence of light. These reactions have long been the focus of the organic polymer community [93], and an extensive industry of producing appropriate antioxidants and stabilizers exists.

In many photo-initiated processes that involve oxygen, electron transfer from an organic molecule excited state to $O_2(X^3\Sigma_g^-)$ will kinetically compete with the energy transfer process that produces $O_2(a^1\Delta_g)$ [94]. In this way, one obtains the radical cation of the organic component and the superoxide radical anion, $O_2^{-\cdot}$. The moniker "superoxide" can be misleading in that $O_2^{-\cdot}$ is not reactive under many conditions [95, 96]. However, when protonated, the resultant hydroperoxyl radical, $HOO^\cdot$, is quite reactive, providing yet another entry point to adverse radical chain reactions [11]. Given the pK_b of the superoxide radical anion, such protonation occurs at pH values less than about 6 [97]. Thus, for aqueous systems to which we have inferred in our discussion above, even though the equilibrium between $HOO^\cdot$ and $O_2^{-\cdot} + H^+$ favors the unreactive superoxide at neutral pH, $HOO^\cdot$ will be constantly removed via reaction. Through this equilibrium, the superoxide ion is best viewed as a benign precursor to more reactive species.

With its excitation energy of 94 kJ/mol (i.e., ~1 eV), $O_2(a^1\Delta_g)$ will be more easily reduced than $O_2(X^3\Sigma_g^-)$. Thus, if there is a suitable reductant in a given system, the superoxide ion can also be formed even when the nascent photoproduct is $O_2(a^1\Delta_g)$.

7.10 Summary and Conclusions

Because we live in a world filled with light, organic molecules, and oxygen, the scientific investigation of the interaction between these three entities is almost as old as science itself. The photo-initiated production and subsequent removal of $O_2(a^1\Delta_g)$ is a cornerstone of this system, and has likewise been the subject of intense investigation over the years. As in many fields, the advent of pulsed lasers and associated microscopes has particularly led to a better understanding of space- and time-resolved events that involve $O_2(a^1\Delta_g)$. In turn, these studies facilitate a better understanding and ultimately the control of oxygen-dependent processes that influence many photoresponsive materials.

Many photoresponsive devices are currently deemed "functional" only under the condition that the active material is encapsulated and oxygen is excluded. This stipulation can put severe restrictions on the implementation of many devices. However, given the current knowledge base of relevant molecular events, it is reasonable to expect that many photoresponsive materials and devices could soon be rendered functional even in the absence of encapsulation.

References

1. J. Ahmad, K. Bazaka, L.J. Anderson, R.D. White, M.V. Jacob, Materials and methods for encapsulation of OPV: a review. Renew. Sust. Energ. Rev. **27**, 104–117 (2013)
2. M.A. Filatov, S. Baluschev, K. Landfester, Protection of densely populated excited triplet state ensembles against deactivation by molecular oxygen. Chem. Soc. Rev. **45**, 4668–4689 (2016)
3. I.E. Jacobs, F. Wang, Z.I.B. Valdez, A.N.A. Oviedo, D.J. Bilsky, A.J. Moule, Photoinduced degradation from trace 1,8-diiodooctane in organic photovoltaics. J. Mater. Chem. C **6**, 219–225 (2018)
4. P.R. Ogilby, Singlet oxygen: there is indeed something new under the sun. Chem. Soc. Rev. **39**, 3181–3209 (2010)
5. M.T. Fontana, H. Kang, P.Y. Yee, Z. Fan, S.A. Hawks, L.T. Schelhas, S. Subramaniyan, Y.-J. Hwang, S.A. Jenekhe, S.H. Tolbert, B.J. Schwartz, Low-vapor-pressure solvent additives function as polymer swelling agents in bulk heterojunction organic photovoltaics. J. Phys. Chem. C **122**, 16574–16588 (2018)
6. C. Yan, S. Barlow, Z. Wang, H. Yan, A.K.-Y. Jen, S.R. Marder, X. Zhan, Non-fullerene acceptors for organic solar cells. Nat. Rev. Mater. **3**, 18003 (2018)
7. T.W. Schmidt, F.N. Castellano, Photochemical upconversion: the primacy of kinetics. J. Phys. Chem. Lett. **5**, 4062–4072 (2014)
8. T.N. Singh-Rachford, F.N. Castellano, Photon upconversion based on sensitized triplet-triplet annihilation. Coord. Chem. Rev. **254**, 2560–2573 (2010)
9. P. Klán, J. Wirz, *Photochemistry of Organic Compounds* (Wiley, Chichester, 2009)
10. M.J. Paterson, O. Christiansen, F. Jensen, P.R. Ogilby, Overview of theoretical and computational methods applied to the oxygen-organic molecule photosystem. Photochem. Photobiol. **82**, 1136–1160 (2006)
11. C. Walling, in *Active Oxygen in Chemistry*, ed. by C. S. Foote, J. S. Valentine, A. Greenberg, J. F. Liebman, (Blackie Academic and Professional, Glasgow, 1995), pp. 24–65
12. P.R. Ogilby, Solvent effects on the radiative transitions of singlet oxygen. Acc. Chem. Res. **32**, 512–519 (1999)
13. C. Schweitzer, R. Schmidt, Physical mechanisms of generation and deactivation of singlet oxygen. Chem. Rev. **103**, 1685–1757 (2003)
14. M. Bregnhøj, M. Westberg, B.F. Minaev, P.R. Ogilby, Singlet oxygen photophysics in liquid solvents: converging on a unified picture. Acc. Chem. Res. **50**, 1920–1927 (2017)

15. M. Montalti, A. Credi, L. Prodi, M.T. Gandolfi, *Handbook of Photochemistry* (CRC Press, Boca Raton, 2006)
16. E.L. Clennan, A. Pace, Advances in singlet oxygen chemistry. Tetrahedron **61**, 6665–6691 (2005)
17. R.D. Scurlock, B. Wang, P.R. Ogilby, J.R. Sheats, R.L. Clough, Singlet oxygen as a reactive intermediate in the photodegradation of an electroluminescent polymer. J. Am. Chem. Soc. **117**, 10194–10202 (1995)
18. Y.W. Soon, H. Cho, J. Low, H. Bronstein, I. McCulloch, J.R. Durrant, Correlating triplet yield, singlet oxygen generation and photochemical stability in polymer/fullerene blend films. Chem. Commun. **49**, 1291–1293 (2013)
19. A. Perthue, I.F. Dominguez, P. Verstappen, W. Maes, O.J. Duatel, G. Wantz, A. Rivaton, An efficient and simple tool for assessing singlet oxygen involvement in the photo-oxidation of conjugated materials. Sol. Energy Mater. Sol. Cells **176**, 336–339 (2018)
20. R.D. Scurlock, B. Wang, P.R. Ogilby, Chemical reactivity of singlet sigma oxygen ($b^1\Sigma g^+$) in solution. J. Am. Chem. Soc. **118**, 388–392 (1996)
21. W.R. Ware, Oxygen quenching of fluorescence in solution: an experimental study of the diffusion process. J. Phys. Chem. **66**, 455–458 (1962)
22. L. Lüer, H.-J. Egelhaaf, D. Oelkrug, G. Cerullo, G. Lanzani, B.-H. Huisman, D. de Leeuw, Oxygen-induced quenching of photoexcited states in polythiophene films. Org. Electron. **5**, 83–89 (2004)
23. F.M. Cabrerizo, J. Arnbjerg, M.P. Denofrio, R. Erra-Balsells, P.R. Ogilby, Fluorescence quenching by oxygen: "debunking" a classic rule. ChemPhysChem **11**, 796–798 (2010)
24. C.S. Parmenter, J.D. Rau, Fluorescence quenching in aromatic hydrocarbons by oxygen. J. Chem. Phys. **51**, 2242–2246 (1969)
25. R. Potashnik, C.R. Goldschmidt, M. Ottolenghi, Triplet state formation in the quenching of fluorescence by molecular oxygen. Chem. Phys. Lett. **9**, 424–425 (1971)
26. L. Poulsen, P.R. Ogilby, Oxygen diffusion in glassy polymer films: effects of other gases and changes in pressure. J. Phys. Chem. A **104**, 2573–2580 (2000)
27. G. Herzberg, *Molecular Spectra and Molecular Structure. I. Spectra of Diatomic Molecules* (Van Nostrand Reinhold, New York, 1950)
28. M. Bregnhøj, A. Blázquez-Castro, M. Westberg, T. Breitenbach, P.R. Ogilby, Direct 765 nm optical excitation of molecular oxygen in solution and in single mammalian cells. J. Phys. Chem. B **119**, 5422–5429 (2015). (See correction: J. Phys. Chem. B **123**, 10569 (2019))
29. M. Bregnhøj, M.V. Krægpøth, R.J. Sørensen, M. Westberg, P.R. Ogilby, Solvent and heavy-atom effects on the $O_2(X^3\Sigma_g^-)$ - $O_2(b^1\Sigma_g^+)$ absorption transition. J. Phys. Chem. A **120**, 8285–8296 (2016)
30. M. Bregnhøj, M. Westberg, F. Jensen, P.R. Ogilby, Solvent-dependent singlet oxygen lifetimes: temperature effects implicate tunneling and charge-transfer interactions. Phys. Chem. Chem. Phys. **18**, 22946–22961 (2016)
31. M. Westberg, M. Bregnhøj, C. Banerjee, A. Blázquez-Castro, T. Breitenbach, P.R. Ogilby, Exerting better control and specificity with singlet oxygen experiments in live mammalian cells. Methods **109**, 81–91 (2016)
32. F. Anquez, I.E. Yazidi-Belkoura, S. Randoux, P. Suret, E. Courtade, Cancerous cell death from sensitizer free photoactivation of singlet oxygen. Photochem. Photobiol. **88**, 167–174 (2012)
33. A.A. Krasnovsky, Y.V. Roumbal, A.A. Strizhakov, Rates of $^1O_2$ ($^1\Delta_g$) production upon direct excitation of molecular oxygen by 1270 nm laser radiation in air-saturated alcohols and micellar aqueous dispersions. Chem. Phys. Lett. **458**, 195–199 (2008)
34. A. Blázquez-Castro, Direct $^1O_2$ optical excitation: a tool for redox biology. Redox Biol. **13**, 39–59 (2017)
35. H. Tsubomura, R.S. Mulliken, Molecular complexes and their spectra. XII. Ultraviolet absorption spectra caused by the interaction of oxygen with organic molecules. J. Am. Chem. Soc. **82**, 5966–5974 (1960)
36. P.R. Ogilby, M. Kristiansen, R.L. Clough, Singlet oxygen formation in a solid organic polymer upon irradiation of the oxygen-polymer charge-transfer band. Macromolecules **23**, 2698–2704 (1990)

37. M. Bregnhøj, P.R. Ogilby, Two-photon excitation of neat aerated solvents with visible light produces singlet oxygen. J. Phys. Chem. A **123**, 7567–7575 (2019)
38. R.D. Scurlock, P.R. Ogilby, Singlet molecular oxygen ($a^1\Delta_g$) formation upon irradiation of an oxygen ($X^3\Sigma_g^-$)-organic molecule charge-transfer absorption band. J. Phys. Chem. **93**, 5493–5500 (1989)
39. P.-G. Jensen, J. Arnbjerg, L.P. Tolbod, R. Toftegaard, P.R. Ogilby, Influence of an intermolecular charge-transfer state on excited-state relaxation dynamics: solvent effect on the methylnaphthalene-oxygen system and its significance for singlet oxygen production. J. Phys. Chem. A **113**, 9965–9973 (2009)
40. K. Onodera, G. Furusawa, M. Kojima, M. Tsuchiya, S. Aihara, R. Akaba, H. Sakuragi, K. Tokumaru, Mechanistic considerations on photoreaction of organic compounds via excitation of contact charge transfer complexes with oxygen. Tetrahedron **41**, 2215–2220 (1985)
41. J.A. Howard, K.U. Ingold, The self-reaction of *sec*-butylperoxy radicals. Confirmation of the Russell mechanism. J. Am. Chem. Soc. **90**, 1056–1058 (1968)
42. Q. Niu, G.D. Mendenhall, Structural effects on the yields of singlet molecular oxygen ($^1\Delta_gO_2$) from alkylperoxyl radical recombination. J. Am. Chem. Soc. **112**, 1656–1657 (1990)
43. J.M. Aubry, C. Pierlot, J. Rigaudy, R. Schmidt, Reversible binding of oxygen to aromatic compounds. Acc. Chem. Res. **36**, 668–675 (2003)
44. N.J. Turro, M.-F. Chow, J. Rigaudy, Mechanism of thermolysis of endoperoxides of aromatic compounds. Activation parameters, magnetic field, and magnetic isotope effects. J. Am. Chem. Soc. **103**, 7218–7224 (1981)
45. S. Wilsey, F. Bernardi, M. Olivucci, M.A. Robb, S. Murphy, W. Adam, The thermal decomposition of 1,2-dioxetane revisited. J. Phys. Chem. A **103**, 1669–1677 (1999)
46. C.S. Foote, T.R. Darling, Decomposition of dioxetanes: a unique probe into mechanism and energy transfer processes. Pure Appl. Chem. **41**, 495–506 (1975)
47. W. Fudickar, T. Linker, in *Singlet Oxygen: Applications in Biosciences and Nanosciences*, ed. by S. Nonell, C. Flors, (Royal Society of Chemistry, Cambridge, 2016), pp. 431–446
48. F. Wilkinson, W.P. Helman, A.B. Ross, Rate constants for the decay and reactions of the lowest electronically excited singlet state of molecular oxygen in solution. An expanded and revised compilation. J. Phys. Chem. Ref. Data **24**, 663–1021 (1995)
49. A.A. Ghogare, A. Greer, Using singlet oxygen to synthesize natural products and drugs. Chem. Rev. **116**, 9994–10034 (2016)
50. R.D. Scurlock, M. Kristiansen, P.R. Ogilby, V.L. Taylor, R.L. Clough, Singlet oxygen reactions in a glassy polystyrene matrix. Polym. Degrad. Stab. **60**, 145–159 (1998)
51. P.F. Conn, W. Schalch, T.G. Truscott, The singlet oxygen and carotenoid interaction. J. Photochem. Photobiol. B Biol. **11**, 41–47 (1991)
52. D. Yu, Y.-Q. Yang, Z. Chen, Y. Tao, Y.-F. Liu, Recent progress on thin-film encapsulation technologies for organic electronic devices. Opt. Commun. **362**, 43–49 (2016)
53. S.H.C. Askes, S. Bonnet, Solving the oxygen sensitivity of sensitized photon upconversion in life science applications. Nat. Rev. **2**, 437–452 (2018)
54. M. Tsushima, K. Tokuda, T. Ohsaka, Use of hydrodynamic chronocoulometry for simultaneous determination of diffusion coefficients and concentrations of dioxygen in various media. Anal. Chem. **66**, 4551–4556 (1994)
55. R. Edge, D.J. McGarvey, T.G. Truscott, The carotenoids as anti-oxidants: a review. J. Photochem. Photobiol. B Biol. **41**, 189–200 (1997)
56. R. Edge, T.G. Truscott, Singlet oxygen and free radical reactions of retinoids and carotenoids—a review. Antioxidants **7**, 5 (2018)
57. P.R. Ogilby, M.P. Dillon, M. Kristiansen, R.L. Clough, Quenching of singlet oxygen in solid organic polymers. Macromolecules **25**, 3399–3405 (1992)
58. R.D. Scurlock, D.O. Mártire, P.R. Ogilby, V.L. Taylor, R.L. Clough, Quantum yield of photosensitized singlet oxygen ($a^1\Delta_g$) production in solid polystyrene. Macromolecules **27**, 4787–4794 (1994)
59. Y. Gao, A.M. Baca, B. Wang, P.R. Ogilby, Activation barriers for oxygen diffusion in polystyrene and polycarbonate glasses: effects of low molecular weight additives. Macromolecules **27**, 7041–7048 (1994)

60. R.L. Jensen, J. Arnbjerg, P.R. Ogilby, Temperature effects on the solvent-dependent deactivation of singlet oxygen. J. Am. Chem. Soc. **132**, 8098–8105 (2010)
61. B. Fückel, D.A. Roberts, Y.Y. Cheng, R.G.C.R. Clady, R.B. Piper, N.J. Ekins-Daukes, M.J. Crossley, T.W. Schmidt, Singlet oxygen mediated photochemical upconversion of NIR light. J. Phys. Chem. Lett. **2**, 966–971 (2011)
62. R.D. Scurlock, K.-K. Iu, P.R. Ogilby, Luminescence from optical elements commonly used in near infrared spectroscopic studies: the photosensitized formation of singlet molecular oxygen ($^1\Delta_gO_2$) in solution. J. Photochem. **37**, 247–255 (1987)
63. M. Bregnhøj, S. Rodal-Cedeira, I. Pastoriza-Santos, P.R. Ogilby, Light scattering versus plasmon effects: optical transitions in molecular oxygen near a metal nanoparticle. J. Phys. Chem. C **122**, 15625–15634 (2018)
64. M. Bregnhøj, M. Prete, V. Turkovic, A.U. Petersen, M.B. Nielsen, M. Madsen, P.R. Ogilby, Oxygen-dependent photophysics and photochemistry of prototypical compounds for organic photovoltaics: inhibiting degradation initiated by singlet oxygen at a molecular level. Methods Appl. Fluoresc. **8**, 014001 (2020)
65. J.W. Snyder, E. Skovsen, J.D.C. Lambert, L. Poulsen, P.R. Ogilby, Optical detection of singlet oxygen from single cells. Phys. Chem. Chem. Phys. **8**, 4280–4293 (2006)
66. K. Tanaka, T. Miura, N. Umezawa, Y. Urano, K. Kikuchi, T. Higuchi, T. Nagano, Rational design of fluorescein-based fluorescence probes. Mechanism-based design of a maximum fluorescence probe for singlet oxygen. J. Am. Chem. Soc. **123**, 2530–2536 (2001)
67. B. Song, G. Wang, M. Tan, J. Yuan, A Europium(III) complex as an efficient singlet oxygen luminescence probe. J. Am. Chem. Soc. **128**, 13442–13450 (2006)
68. S. Kim, M. Fujitsuka, T. Majima, Photochemistry of singlet oxygen sensor green. J. Phys. Chem. B **117**, 13985–13992 (2013)
69. S.K. Pedersen, J. Holmehave, F.H. Blaikie, A. Gollmer, T. Breitenbach, H.H. Jensen, P.R. Ogilby, Aarhus sensor green: a fluorescent probe for singlet oxygen. J. Org. Chem. **79**, 3079–3087 (2014)
70. R. Ruiz-Gonzalez, A.L. Zanocco, in *Singlet Oxygen: Applications in Biosciences and Nanosciences*, ed. by S. Nonell, C. Flors, (Royal Society of Chemistry, Cambridge, 2016), pp. 103–120
71. M. Manceau, A. Rivaton, J.-L. Gardette, Involvement of singlet oxygen in the solid-state photochemistry of P3HT. Macromol. Rapid Commun. **29**, 1823–1827 (2008)
72. M. Yan, L.J. Rothberg, F. Papadimitrakopoulos, M.E. Galvin, T.M. Miller, Defect quenching of conjugated polymer luminescence. Phys. Rev. Lett. **73**, 744–747 (1994)
73. B.H. Cumpston, I.D. Parker, K.F. Jensen, In situ characterization of the oxidative degradation of a polymeric light emitting device. J. Appl. Phys. **81**, 3716–3720 (1997)
74. C.A. Davis, K. McNeill, E.M.-L. Janssen, Non-singlet oxygen kinetic solvent isotope effects in aquatic photochemistry. Environ. Sci. Technol. **52**, 9908–9916 (2018)
75. I. Kraljic, L. Lindqvist, Laser photolysis study of triplet eosin and thionine reactions in photosensitized oxidations. Photochem. Photobiol. **20**, 351–355 (1974)
76. M. Bregnhøj, L. Dichmann, C.K. McLoughlin, M. Westberg, P.R. Ogilby, Uric acid: a less-than-perfect probe for singlet oxygen. Photochem. Photobiol. **95**, 202–210 (2019)
77. N. Dam, R.D. Scurlock, B. Wang, L. Ma, M. Sundahl, P.R. Ogilby, Singlet oxygen as a reactive intermediate in the photodegradation of phenylenevinylene oligomers. Chem. Mater. **11**, 1302–1305 (1999)
78. M. Westberg, M. Etzerodt, P.R. Ogilby, Rational design of genetically encoded singlet oxygen photosensitizing proteins. Curr. Opin. Struct. Biol. **57**, 56–62 (2019)
79. C.S. Foote, R.W. Denny, Chemistry of singlet oxygen. VII. Quenching by β-carotene. J. Am. Chem. Soc. **90**, 6233–6235 (1968)
80. A. Farmilo, F. Wilkinson, On the mechanism of quenching of singlet oxygen in solution. Photochem. Photobiol. **18**, 447–450 (1973)
81. R. Schmidt, Deactivation of $O_2(^1\Delta_g)$ singlet oxygen by carotenoids: internal conversion of excited encounter complexes. J. Phys. Chem. A **108**, 5509–5513 (2004)

82. V. Turkovic, M. Prete, M. Bregnhøj, L. Inasaridze, D. Volyniuk, F.A. Obrezkov, J.V. Grazulevicius, S. Engmann, H.-G. Rubahn, P.A. Troshin, P.R. Ogilby, M. Madsen, Biomimetic approach to inhibition of photooxidation in organic solar cells using beta-carotene as an additive. ACS Appl. Mater. Interfaces **11**, 41570–41579 (2019)
83. D.J. Carlsson, G.D. Mendenhall, T. Suprunchuk, D.M. Wiles, Singlet oxygen ($^1\Delta_g$) quenching in the liquid phase by metal(II) chelates. J. Am. Chem. Soc. **94**, 8960–8962 (1972)
84. A. Adamczyk, F. Wilkinson, Quenching of triplet states by schiff base nickel(II) complexes. J. Chem. Soc. Faraday Trans. 2. **68**, 2031–2041 (1972)
85. N. Grassie, G. Scott, *Polymer Degradation and Stabilisation* (Cambridge University Press, Cambridge, 1985)
86. M. Salvador, N. Gasparini, J.D. Perea, S.H. Paleti, A. Distler, L.N. Inasaridze, P.A. Troshin, L. Luer, H.-J. Egelhaaf, C. Brabec, Suppressing photooxidation of conjugated polymers and their blends with fullerenes through nickel chelates. Energy Environ. Sci. **10**, 2005–2016 (2017)
87. M.E. Diaz Garcia, A. Sanz-Medel, Fast chemical deoxygenation of micellar solutions for room temperature phosphorescence. Anal. Chem. **58**, 1436–1440 (1986)
88. R. Battino, T.R. Rettich, T. Tominaga, The solubility of oxygen and ozone in liquids. J. Phys. Chem. Ref. Data **12**, 163–178 (1983)
89. N.J. Turro, G.S. Cox, X. Li, Remarkable inhibition of oxygen quenching of phosphorescence by complexation with cyclodextrins. Photochem. Photobiol. **37**, 149–153 (1983)
90. J. Lopez-Gejo, A.A. Marti, M. Ruzzi, S. Jockusch, K. Komatsu, F. Tanabe, Y. Murata, N.J. Turro, Can H_2 inside C_{60} communicate with the outside world? J. Am. Chem. Soc. **129**, 14554–14555 (2007)
91. C.C. Page, C.C. Moser, X. Chen, P.L. Dutton, Natural engineering principles of electron tunneling in biological oxidation-reduction. Nature **402**, 47–52 (1999)
92. C. Shih, A.K. Museth, M. Abrahamsson, A.M. Blaco-Rodriguez, A.J. Di Bilio, J. Sudhamsu, B.R. Crane, K.L. Ronayne, M. Towrie, A. Vlcek, J.H. Richards, J.R. Winkler, H.B. Gray, Tryptophan-accelerated electron flow through proteins. Science **320**, 1760–1762 (2008)
93. R. L. Clough, N. C. Billingham, K. T. Gillen (eds.), *Polymer Durability: Degradation, Stabilization, and Lifetime Protection* (ACS, Washington, 1996)
94. C.S. Foote, Definition of type I and type II photosensitized oxidation. Photochem. Photobiol. **54**, 659 (1991)
95. D.T. Sawyer, J.S. Valentine, How super is superoxide? Acc. Chem. Res. **14**, 393–400 (1981)
96. D. Sawyer, T. Oxygen, *Chemistry* (Oxford University Press, New York, 1991)
97. B. Bielski, D. Cabelli, R. Arudi, A. Ross, Reactivity of HO_2/O_2 radicals in aqueous solutions. J. Phys. Chem. Ref. Data **14**, 1041–1101 (1985)

Chapter 8
Protective Strategies Toward Long-Term Operation of Annihilation Photon Energy Upconversion

Stanislav Baluschev

Abstract The process of triplet-triplet annihilation photon energy upconversion (TTA-UC) performed in soft-matter environment relays on optically created densely populated organic triplet ensembles. In soft-matter matrix, it is a diffusion-controlled process, simultaneously demonstrating essential dependences on the environmental parameters such as matrix temperature, matrix viscosity, and presence of molecular oxygen, dissolved into the solvent or adsorbed on the polymer film. It is important to notice that all these environmental parameters are strongly interrelated and their impact on the temporal evolution of the densely populated triplet ensembles is not a linear combination of its partial impacts. If the TTA-UC process is applied toward solar energy storage and/or conversion technologies, the influence of the singlet oxygen generation generally leads to lower quantum yield (QY) and simultaneously to accelerated aging of the upconversion device. The generation of singlet oxygen is much more harmful for the studied object, if the process of TTA-UC is used as a sensing mechanism for probing of the local temperature/local oxygen content in cell cultures. In this chapter, the development of an effective protection strategy against quenching by molecular oxygen and protection against the subsequent photooxidation caused by singlet oxygen is discussed.

Keywords Annihilation upconversion · Sunlight excitation · Sacrificial singlet oxygen scavenger

The TTA-UC possesses a chain of crucial technological advantages: First of all, the TTA-UC is the only upconversion method that has been experimentally demonstrated to operate with noncoherent low-intensity illumination such as moderate

S. Baluschev (✉)
Optics and Spectroscopy Department, Faculty of Physics, Sofia University "St. Kliment Ochridski", Sofia, Bulgaria

Max-Planck-Institute for Polymer Research, Mainz, Germany
e-mail: balouche@phys.uni-sofia.bg; balouche@mpip-mainz.mpg.de

J. S. Lissau, M. Madsen (eds.), *Emerging Strategies to Reduce Transmission and Thermalization Losses in Solar Cells*, https://doi.org/10.1007/978-3-030-70358-5_8

concentrated sunlight [1, 2]. Secondly, the spectral power density of the excitation source prerequisite for effective TTA-UC is comparable with that of the terrestrial sunlight. Sustainable and efficient TTA-UC was demonstrated under extremely low excitation intensity of 1 Sun [3] and corresponding with solar spectral power density less than 120 $\mu W \times cm \times nm^{-1}$.

The process of triplet-triplet annihilation photon energy upconversion (TTA-UC) performed in fluidic organic systems (volatile or nonvolatile organic solvents) [4–10] or in hydrophobic soft-matter environment relies on optically created densely populated organic triplet ensembles. The driving force for TTA-UC is the *inter*-molecular triplet energy transfer, including the subsequent processes of triplet-triplet transfer (TTT) and triplet-triplet annihilation (TTA). In soft-matter matrix both processes are diffusion controlled, simultaneously demonstrating essential dependences on the material and environmental parameters such as degree of overlapping of the interacting energy states, matrix temperature, matrix viscosity, and presence of molecular oxygen, dissolved into the solvent or adsorbed on the polymer film. It is important to notice that all these materials and environmental parameters are strongly interrelated and their impact on the densely populated triplet ensembles is not a linear combination of its partial impacts.

The TTA-UC takes place in multi-chromophore systems consisting of energetically optimized pairs of emitter molecules (typically aromatic hydrocarbons) and sensitizer molecules (usually, metalled macrocycles, such as porphyrins and phthalocyanines), as shown in Fig. 8.1. Shortly, the photon energy absorbed by a sensitizer is stored in its metastable triplet state, formed in the process of intersystem crossing (ISC). Further, this energy is transferred to an emitter triplet state via the process of triplet-triplet transfer (TTT). Next, the excited triplet states of two emitter molecules undergo triplet-triplet annihilation (TTA), in which one emitter molecule returns back to its singlet ground state and the other molecule gains the energy of both triplet states and is excited to the higher singlet state. As the singlet-state emitter decays radiatively back to the ground state, a delayed fluorescence photon (the orange arrow, Fig. 8.1, marked further as **dF**) bearing higher energy than that of the excitation photons is emitted. Simultaneously, if the energy overlap between the triplet manifolds of the emitter and sensitizer molecules is not optimal, the sensitizer triplet state will not be completely depopulated and therefore residual sensitizer phosphorescence (the dark violet arrow, Fig. 8.1, marked further as **rPh**) will be observed [11].

In the presence of even vanishing concentrations of molecular oxygen, the energetic scheme of the process of TTA-UC becomes more complicated (please refer to Fig. 8.1, including the energy levels of the molecular oxygen). In oxygen-contaminated environment the energy stored in the excited states of the triplet ensembles is being actively dissipated: thus, well-observable competition occurs between one site, the population of the oxygen singlet state, and other site, the emissive (phosphorescence) and the non-emissive (triplet to triplet) energy transfer processes. Sometimes, if the oxygen concentration is high enough, all other relaxation channels, except generation of singlet oxygen, could be completely

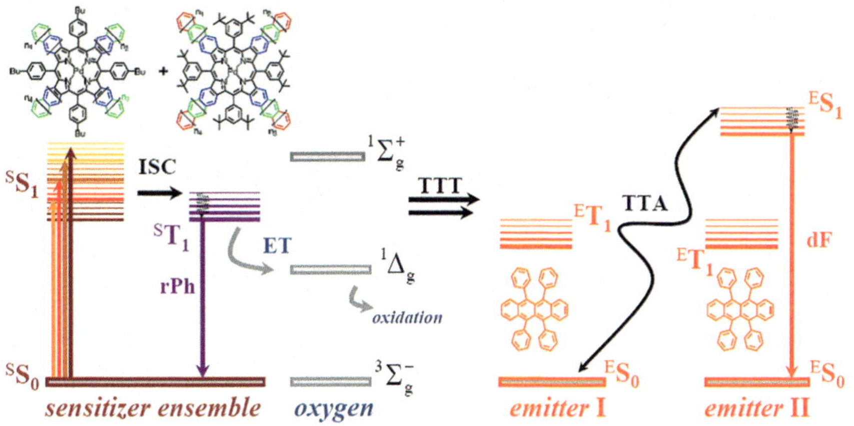

Fig. 8.1 Simplified energetic scheme of the TTA-UC process in multicomponent organic system in molecular oxygen-rich environment. **Insets**: Upper—general structures of the sensitizer ensemble (mixed benzo-naphthoporphyrin family and mixed naphtho-anthraporphyrin family [from [3]]), right—structure of the emitter (rubrene)

quenched (Fig. 8.1). Correspondingly, neither residual sensitizer phosphorescence nor delayed emitter fluorescence could be observed. Immediately, it must be pointed out explicitly that the quenching of the TTT process or rPh process is not a final stage, but even more a transitional phase of the TTA-UC system, where the duration and the strength of the quenching depend on a variety of material parameters.

The physical process behind these observable quenching processes is the process of TTA now, between the excited triplet state of the sensitizer molecule and the ground triplet state of the oxygen molecule: as a result, the sensitizer molecule returns back to its singlet ground state and the oxygen molecule gains the energy of both triplet states and is excited to the metastable singlet state ($^1\Delta_g$, further called singlet oxygen or $^{O2}S_1$):

$$^{S}T_1 + {}^{O_2}T_0 \overset{TTA}{\rightarrow} {}^{S}S_0 + {}^{O_2}S_1$$

Furthermore, quenching of the excited triplet state of the emitter molecule by the oxygen is also possible, but with significant lower probability:

$$^{E}T_1 + {}^{O_2}T_0 \overset{TTA}{\rightarrow} {}^{E}S_0 + {}^{O_2}S_1$$

The possibility to generate other reactive oxygen species [12–14] during the optical excitation along the studied TTA-UC objects is neglected on purpose: firstly, in order to simplify this representation and especially to reveal the complex dynamical properties of singlet oxygen generation and the optical response of the TTA-UC organic system.

Singlet oxygen is a highly reactive species [12–14], leading to real-time oxidation of the photoactive molecules (sensitizer and/or emitter) even at room temperatures. Further, the singlet oxygen formed during the optical excitation takes part in a Diels-Alder-type process on a parent sensitizer molecule, which leads to a loss of conjugation in the π-system and to a loss of the Q-band absorption. Similar process is also involved in photo-bleaching of the emitter chromophores, e.g., anthracenes, perylenes, terrylenes, and terrylenediimides [15–19]. Thus, the TTA-UC process suffers doubly by the presence of molecular oxygen—first, because corresponding to the oxygen concentration and/or the rate of molecular oxygen diffusion significant amounts of optically excited sensitizer triplet states are lost, and second (which is an even more destructive consequence) a considerable amount of the photoactive molecules are being damaged and they fail out from the UC process.

If the TTA-UC process is applied toward solar energy storage and/or conversion technologies, the influence of the singlet oxygen generation leads only to lower quantum yield (QY) and simultaneously to accelerated aging of the upconversion device. Generation of singlet oxygen is much more harmful for the studied object, if the process of TTA-UC is used as a sensing mechanism for probing of the local temperature/local oxygen content in cell cultures [20].

Therefore, in order to demonstrate the advantages of the TTA-UC technology for different photonic applications, the development of an effective protection strategy against quenching by molecular oxygen and protection against the subsequent photodamages caused by singlet oxygen is of crucial importance. It must be pointed out that the adopted oxygen protection techniques must be carefully tailored for the specific photonic application since the key optical parameters for the various TTA-UC applications strongly differ.

There is a vast number of reported and experimentally demonstrated [11–14] material compositions leading to the effective protection of the optically excited triplet ensembles from photoinduced oxidative damage. Mainly, two types of protection strategies can be recognized: (1) **passive protection**, based on the application of barrier materials for packaging, sealing, or encapsulation of the active substances, which prevent oxygen molecules from penetration into the UC media or significantly slow down the oxygen diffusion, thus preventing the physical contact of molecular oxygen with the excited triplet states and (2) **active protection**, based on the application of singlet oxygen scavenger (SOS) moieties, which react chemically with the singlet oxygen, and thus minimize the amount of oxygen available for oxidative processes, causing degradation of the photoactive molecules. Each protection strategy has its advantages but also its principal weaknesses:

For instance, passive protection techniques, i.e., decreasing oxygen permeability by means of encapsulation of the UC-active substances in polymer films [1] or nano- [21, 22] and micro-carriers [23, 24]—although effective for some minute or hours—are insufficient for technological application due to their short lifetime. The use of glass-packaged samples, prepared and sealed in a glove box, restricts the application potential of the UC devices considerably, decreasing significantly the ability to integrate the UC device into the desired application layout. Additionally, desired properties of the TTA-UC devices such as mechanical flexibility and/or miniaturization are not compatible with glass-type sealing. The straightforward active protection strategy—namely, addition of sacrificial oxygen-scavenging species to the UC matrix—results in molar concentrations of the oxygen-scavenging additives comparable with the molar concentrations of the protected species. As a consequence relatively short device lifetime in order of hours or single days was achieved.

Therefore, it is essential to identify the optimal combination of passive and active oxygen protection for the specific TTA-UC application. Of a significant importance is to find materials, working effectively as TTA-UC media, and simultaneously, these materials have to ensure the required sacrificial oxygen-scavenging properties: for instance, hyperbranched unsaturated poly(phosphoester)s [25] or linear polyunsaturated triterpene [26]. The oxidation of squalene (structure (**4**) from Table 8.2) by singlet oxygen and free radicals was revealed in detail in [26]. In both examples, the protective solvents introduce a huge excess of "sacrificial" double bonds, combined with low viscosity, essential for high molecular mobility of the UC molecules and hence relatively high QY of the TTA-UC process. In order to choose the proper protective media, it is essential to follow the empirical rules, stated below:

1. To choose the proper optical selection criterion for comparison and characterization of the oxygen protection efficiency—it can be the QY of the TTA-UC process, or the dynamical characteristics of the TTA-UC process (including the aging of the sample).
2. The identified oxygen protection media must ensure efficient scavenging of oxygen in excited singlet state, but do not react with the molecular oxygen in the triplet ground state. This behavior is crucially important if sensing applications of the TTA-UC are targeted. This feature ensures that the TTA-UC sensing tool can be used as "on-shelf" manner, and the possibility for that long-term storage.
3. Binding of singlet oxygen by the sacrificial singlet oxygen scavengers (SSOS) is in strong competition with the photooxidation of the active TTA-UC moieties. Therefore, the presence of sacrificial scavenger in a concentration lower than some threshold value does not lead to exhaustive protection. Consequently, the total concentration of the SSOS materials must be proportionally higher than the amount of present molecular oxygen (at the starting moment of the optical excitation) and the amount of oxygen, penetrating toward the access (reaction) optical spot for the desired duration (lifetime) of the TTA-UC process.

The setup for registration of the dynamical parameters of the TTA-UC in oxygen-rich environment is shown in Fig. 8.2. As excitation sources' temperature and current are stabilized, single-mode continuous-wave (*cw*) diode lasers, mounted on laser heads (TCLDM9, Thorlabs Inc.) and operating at different wavelengths, are used (Fig. 8.2). All lasers are collimated to a mutually uniform spot with an aspect ratio of ~1.1 by applying cylindrical telescopes. The temperature and current control of all diode lasers is performed by PRO8000/8-Slot Modular Rack Chassis (Thorlabs Inc.), supporting laser current control units LDC8002/LDC8010 and laser diode temperature controller modules TED8040 (Thorlabs Inc.).

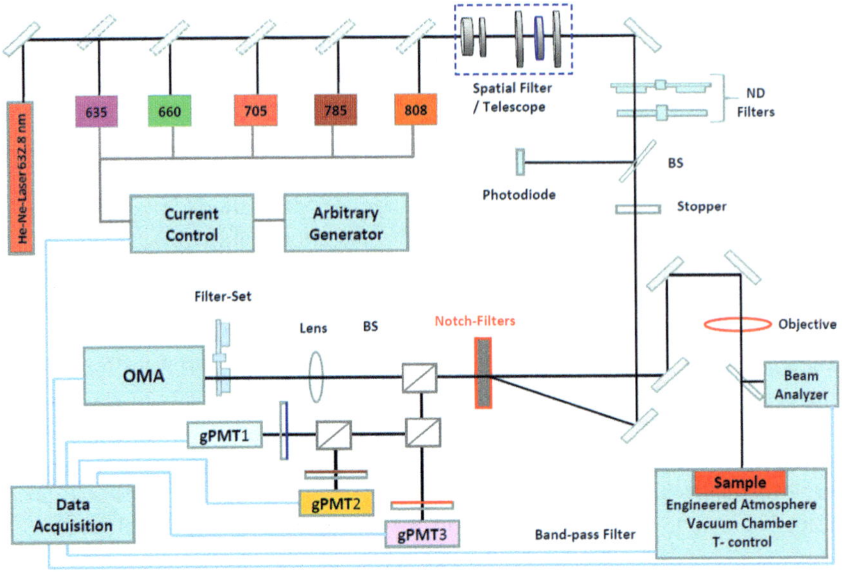

Fig. 8.2 Setup for registration of the dynamical parameters of the TTA-UC in oxygen-rich environment. Depending on the excitation laser wavelength the central wavelength of the notch filter is changed accordingly. There is possibility to measure (1) sealed liquid samples, at an atmospheric pressure of 1 bar, and (2) not sealed samples—films, in vacuum chamber with 1×10^{-5} torr dynamical pressure or engineered atmosphere. The samples are measured in environments with active temperature control and stabilization. In order to minimize the effect of solvent convection, the liquid samples are measured in a horizontal position

Each laser beam passes through the spatial filter in order to reach nearly TEM_{00} transversal intensity distribution. A series of reflective neutral density (ND) filters (Thorlabs Inc.) placed on revolving optical holder (in order to keep the optical axes of the experiment unchanged) were used to attenuate the beam power smoothly. Afterward, the beam passed through the system of ultra-broadband mirrors (MaxMirror, Semrock Inc.) and was finally focused by apochromatic lens (NA = 0.24) onto the sample. The excitation spot diameter is permanently controlled by beam profiler (BP104-VIS, Thorlabs Inc.).

The optical density of all the samples at the excitation wavelength is nearly 0.1; therefore the attenuation of the pump light can be neglected for the given thickness of the optical samples (1000 μm). The luminescence emission generated by the sample was collected with the same apochromatic lens; thus the excitation and observation spots are completely spatially overlapped. The emission of the excitation lasers was rejected by the appropriate notch filters: for instance by application of a notch filter designed for $\lambda = 633$ nm (FWHM ~29 nm, NF 03-633E-25, Semrock Inc.) rejection better than 10^6 times can be achieved. Other notch filters are $\lambda = 658$ nm (NF03-658E-25), $\lambda = 785$ nm (NF03-785E-25), and $\lambda = 808$ nm (NF03-808E-25).

The *cw*-emission spectra were registered by optical multichannel analyzer (Hamamatsu Inc.) or fiber spectrometer (C10083CA, Hamamatsu Inc.) with absolute wavelength calibration and corrected spectral response.

For the time-resolved measurements of the sensitizer residual phosphorescence and delayed emitter UC fluorescence additional removable beam splitters were installed in order to reflect the sample emission toward the time-resolved registration unit. The used laser heads (TCLDM9, Thorlabs Inc.) allow current modulation via modulation input, depending on the maximal current supplied, and the modulation frequency varies from 50 kHz up to 200 kHz. Therefore, the laser intensity was modulated by arbitrary pulse generator (TGA1244, *TTi Inc.*) with caring frequency of 40 MHz. The obtained laser diode pulses have almost rectangular time shape, with rise and decay slopes less than 2.5×10^{-6} s.

The time-response function of the used setup is better than 50×10^{-9} s which is much lower than the characteristic timescale of the processes of TTA and TTT (usually, in order of $10 \div 100 \times 10^{-6}$ s). The time interval between the excitation pulses can be produced to be enough large, and no technical limits apply (arbitrary pulse generator is used). Therefore, the cross talk between the optically excited molecular ensembles is completely avoided. One edge-beam splitter (FF01-694/SP-25, Semrock Inc.) was used to separate spatially the sensitizer residual phosphorescence and delayed emitter UC fluorescence, registered by gPMT1 and gPMT2, accordingly. Additional edge filters, namely a short-pass (SP01-633RS-25, Semrock Inc.) filter with cutting wavelength $\lambda = 650$ nm and long-pass (HQ700LP, Chroma GmbH) filter with cutting wavelength $\lambda = 700$ nm, were installed in front of the registration optoelectronic devices at the delayed-fluorescence branch and phosphorescence branch, respectively.

The optical signals were detected by gated photomultiplier tubes H1156-01-NF (Hamamatsu Photonics Inc., gPMT1 and gPMT2, Fig. 8.5) and gated photomultiplier tube H1156-20-NF with extended red multialkali photocathode with enhanced sensitivity (Hamamatsu Photonics Inc., gPMT3, Fig. 8.5). The gPMTs are working in analogous mode with a rise and decay times less than 30×10^{-9} s. The gated photomultiplier tubes are set in "Normally OFF" mode.

The gating pulses should be with an amplitude of 3.5 V $\div$ 5 V applied on 10 kΩ load, with maximal repetition rate of 10 kHz up to ∞. The gPMT operation delay is less than 0.2×10^{-6} s.

Each gPMT is equipped with a wide-bandwidth amplifier unit C9663 (DC up to 150 MHz), where the current input signal from the photomultiplier tube is directly converted into a voltage output signal with 50 Ω load at a conversion factor of 4 mV/μA. The electrical signals were registered by multichannel digital storage oscilloscope (54622A, Agilent Inc.) with a bandwidth of 100 MHz and 12-bit resolution.

The sample chamber is equipped with input and output optical windows; thus possible parasitic reflections for both excitation and information optical signals caused by metal walls of the chamber are completely avoided. The optical windows possess broadband antireflection coating, keeping the unwanted reflections on levels less than 0.6% for all used wavelengths up to 1.1 μm. The sample chamber can be evacuated up to a dynamical pressure of 1×10^{-5} mbar by application of a turbo vacuum pump hiCUBE, Pfeiffer Vacuum GmbH. Alternatively after evacuation engineered oxygen-containing atmosphere can be used. In the case of usage of not sealed samples (liquid NC suspensions) the sample chamber possesses a septum input, allowing introducing of a water-based sample into the engineered atmosphere, preventing the noncontrolled evaporation of the biological sample.

The sample is the subject of active temperature control performed by Peltier element equipped with a hole with a diameter of 4×10^{-3} m. Thus, possible parasitic emissions/reflections caused by the ceramic surface of the Peltier element are completely avoided. The Peltier element current is actively controlled and stabilized by applying a negative-feedback electronic loop; thus any temperature at the interval of $\Delta T \sim 4\ °C \div 64\ °C$ with an accuracy of 0.1 °C can be realized and kept unchanged for long enough period. Stable operation of the Peltier element temperature determination is achieved by appropriate remit of the parasitic heat created by means of recirculation chiller Ministat CC3, Huber GmbH.

8.1 Selection Criteria for Singlet Oxygen Protection Efficiency

The Quantum Yield of the TTA-UC

The quantum yield ($\eta_{\text{TTA-UC}}$) of the TTA-UC process is certainly of decisive importance. The IUPAC definition of quantum yield (QY) for arbitrary emissive process is given by $\eta_{\text{TTA-UC}} = N^{\text{photons}}_{\text{emitted}} / N^{\text{photons}}_{\text{absorbed}}$. It has to be pointed out that such a classical term (i.e., QY) is attributed to a complex process like TTA-UC, where $N^{\text{photons}}_{\text{absorbed}}$ represents the number of photons absorbed by the UC-sensitizer molecules and $N^{\text{photons}}_{\text{emitted}}$ represents the number of photons emitted by the UC-emitter molecules. The main outcome of such a classical definition is that clear and non-contradictive knowledge about the ratio of absorbed excitation photons to emitted UC photons could be derived. Any other definition (such as internal efficiency [27] or amount of participating emitter triplet states [28, 29]) can only lead the reader to inconsistency of the real UC-photon flux expected.

It was mentioned explicitly that TTA-UC process is a diffusion-controlled process. Therefore it is important to consider also the diffusion of excited triplet states throughout the virtual borders of the excitation spot, especially if the QY of the TTA-UC system is estimated at steady -state excitation conditions. The diffusion of excited triplet states leads to the well-measurable dependence of the QY on the diameter of the excitation spot, shown in Fig. 8.3. Specifically, if the excitation sources for TTA-UC are lasers the precise definition and control of the excitation spot parameters throughout the complete experiment are from decisive importance.

As shown in Fig. 8.3, the QY of the TTA-UC fluorescence at constant molar concentration of the optically excited triplet states depends substantially on the

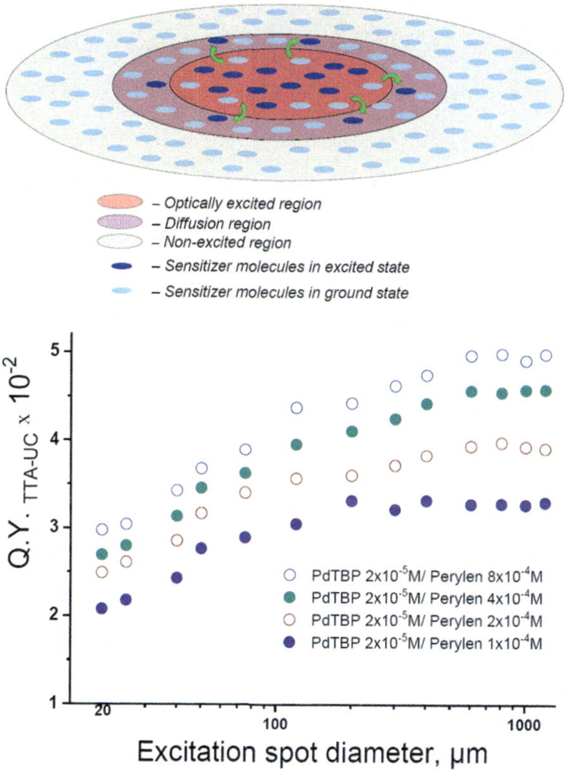

Fig. 8.3 (**a**) Dependence of the QY of the TTA-UC fluorescence on the excitation spot diameter. The excitation intensity across each spot is constant, 32 mWcm^{-2}. *Experimental conditions*: UC couple, PdTBP (2×10^{-5} M)/perylene (shown on the graph); room temperature; sample thickness 1000 μm; excitation laser HeNe, $\lambda_{exc} = 633$ nm. Vitrotube® glass sample, sealed in nitrogen-filled glove box, residual oxygen <2 ppm; solvent 98 vol% toluene/2 vol% squalene; (**b**) cartoon elucidating the diffusion of the excited triplets outside the borders of the excitation spot

diameter of the excitation spot. For low-viscosity volatile solvent such as toluene the QY can be changed more than 80%, depending on the excitation beam diameter. As shown in Fig. 8.3 the molar concentration ratio C_S/C_E plays a significant role in the QY excitation beam dependence: at low molar ratio $C_S/C_E = 1/5$ (Fig. 8.3, the violet-filled circles) the QY dependence reaches a saturation at $r_{EXC} = 200$ μm; oppositely, at high molar ratio $C_S/C_E = 1/40$ (Fig. 8.3, the blue open circles) the QY dependence even at $r_{EXC} > 1100$ μm does not reach saturation. Special attention was paid to keep all other experimental conditions constant: (1) there was no molecular oxygen present (samples were prepared, and sealed in glove box, less than 2 ppm residual oxygen) and (2) samples were kept at a constant temperature $T = 22$ °C (active temperature stabilization).

In Fig. 8.4 the dependence of the QY of the phosphorescence on the excitation beam diameter is shown. Oppositely to the process of TTA-UC, higher concentration of excited sensitizer triplet states leads to lower QY of the phosphorescence; therefore its efficiency decreases substantially with the increase of the excitation beam diameter.

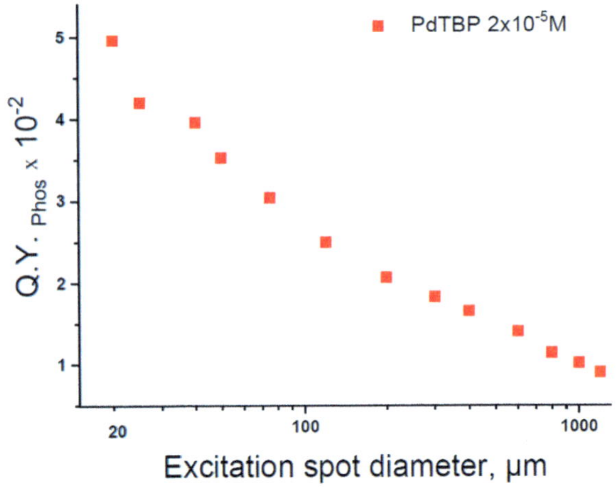

Fig. 8.4 Dependence of the QY of the phosphorescence of PdTBP (2×10^{-5} M) on the excitation spot diameter. The excitation intensity across each spot is constant, 32 mWcm^{-2}. The excitation intensity across each spot is constant, 32 mWcm^{-2}. *Experimental conditions*: room temperature; sample thickness 1000 μm; excitation laser HeNe, $\lambda_{exc} = 633$ nm. Vitrotube® glass sample, sealed in nitrogen-filled glove box, residual oxygen <2 ppm; solvent 98 vol% toluene/2 vol% squalene

Rise Time of the TTA-UC

As evident from Fig. 8.5, in oxygen-free environment the rise-up time of the TTA-UC fluorescence does not depend on the excitation intensity. In fact, for time less than 100 ms the UC fluorescence reaches a steady-state condition.

It is also well seen that the delayed emitter fluorescence practically is constant during the complete excitation period; small signal changes can be attributed to small (less than 0.2%) instabilities of the intensity of the excitation laser.

The sub-quadratic dependence of the UC fluorescence on the excitation intensity, demonstrated for many UC systems [1, 21, 30, 31] dissolved in organic solvents in oxygen-free conditions, is reproduced here for the UC systems dissolved in organophosphates (Fig. 8.6). The experimental data are well approximated with a power law $I_{UC} = a \times I_{exc}^{b}$, where the parameter $\boldsymbol{b} = 1.18$.

Alternatively, in Fig. 8.7 the rise-up dynamic of the TTA-UC fluorescence in oxygen-rich environment for different excitation intensities is reported. The TTA-UC molecules (the same UC system, as reported in Fig. 8.5) are dissolved in a SSOS solvent (phenyl-di-(3-buten-1-yl)-phosphate structure (**4**) from Table 8.2), bearing a huge excess of terminal double bonds, and working effectively as a sacrificial singlet oxygen scavenger. Additionally, low viscosity of the SSOS solvent ensures high mobility of the UC molecules, and thus high efficiency of the TTA-UC process.

As evident from Fig. 8.7b, in oxygen-rich environment the necessary time to reach steady-state UC fluorescence (τ_{UC}^{SS}) depends substantially on the excitation intensity. For instance, at high excitation intensity (~1.5 W $\times$ cm^{-2}) this time is

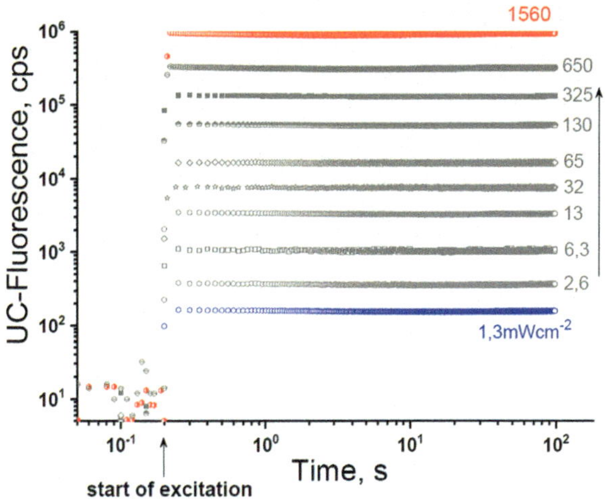

Fig. 8.5 Rise-up dynamic of the TTA-UC fluorescence at different excitation intensities (shown on the graph). *Experimental conditions*: UC couple, PdTBP (2×10^{-5} M)/perylene (4×10^{-4} M); ratio sensitizer/emitter $C_S/C_E = 1/20$; room temperature; sample thickness 1000 μm; single-mode laser diode, $\lambda_{exc} = 635$ nm; optical registration via gated PMT; integration time 50 ms; excitation beam diameter $r_{EXC} = 1200$ μm; Vitrotube® glass sample, sealed in nitrogen-filled glove box, residual oxygen <2 ppm; solvent 98 vol% phenyl-di-(3-buten-1-yl)-phosphate/2 vol% toluene

Fig. 8.6 Dependence of the UC intensity on the excitation intensity. The data are obtained at moment $t = 1800$ ms (2000 ms after the start of the excitation). The same experimental conditions as in Fig. 8.5

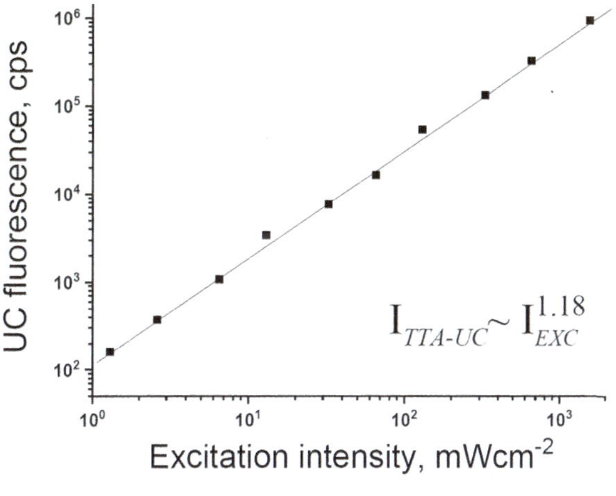

$\tau_{UC}^{SS} \approx 100 \times 10^{-3}$s, and at low excitation intensities (in order of a few mW $\times$ cm^{-2}) this time increases up to $\tau_{UC}^{SS} = 20$s. In contrast, in oxygen-free conditions the rise time for the same UC couple is nearly $\tau_{UC}^{SS} < 100 \times 10^{-3}$s even at 1.3 mW $\times$ cm^{-2} excitation intensity.

These experimental data support the hypothesis: in oxygen-saturated solvents embedding metallated macrocycles, generation of singlet oxygen is the dominant process during the optical excitation. In this case the sensitizer triplet state is

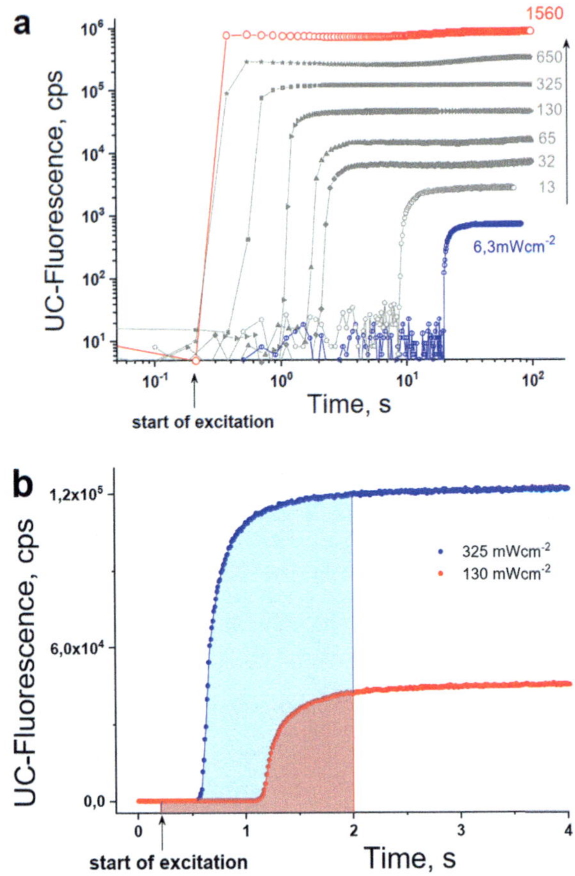

Fig. 8.7 (**a**) Rise-up dynamic of the TTA-UC fluorescence at different excitation intensities (shown on the graph). *Experimental conditions*: UC couple, PdTBP (2×10^{-5} M)/perylene (4×10^{-4} M); ratio sensitizer/emitter $C_S/C_E = 1/20$; room temperature; sample thickness 1000 μm; single-mode laser diode, $\lambda_{exc} = 635$ nm; optical registration via gated PMT; excitation beam diameter $r_{EXC} = 1200$ μm; integration time 50 ms; solvent 98 vol% phenyl-di-(3-buten-1-yl)-phosphate/2 vol% toluene; Vitrotube® glass sample, steered in ambient environment and sealed; molar concentration of the dissolved oxygen estimated to be 2×10^{-4} M. (**b**) Rise-up dynamic for two specific excitation intensities

effectively quenched, and the process of TTA-UC is not observed, unless the concentration of the molecular oxygen reaches some lower lying threshold amount. This explains the experimental data that at oxygen-saturated starting conditions, initially no TTA-UC is observed (Fig. 8.7a, see for instance the blue curve, excitation intensity ~6.3 mW × cm^{-2}). However, as the molecular oxygen concentration is reduced within the time (Fig. 8.5) the UC fluorescence builds up and steady-state UC emission is reached. Therefore, only after consumption (oxidizing the terminal double bonds of the organophosphate solvent) the process of TTA-UC becomes the prevalent relaxation path for the excited sensitizer triplet states. Consistently with this hypothesis, the higher the excitation intensity, the shorter the time needed for the complete scavenging of the molecular oxygen which was initially present in the optically excited volume (Fig. 8.7a) and therefore the faster the rise time of the upconverted emission.

In Fig. 8.7b the rise-up dynamics for two excitation intensities, namely $I_{exc} = 325$ mW × cm^{-2} (Fig. 8.7b, the blue-filled circles) and $I_{exc} = 130$ mW × cm^{-2} (Fig. 8.7b, the red-filled circles), are shown. The rise-up times in oxygen-saturated environments are, respectively, $\tau_{UC}^{SS} \approx 450 \times 10^{-3}$s and $\tau_{UC}^{SS} \approx 900 \times 10^{-3}$s. If the registration of the intensity of the delayed emitter fluorescence is registered "after reaching the steady-state condition," the measured UC intensities will be much higher than if an integrating light detector (with integration time of, for instance, 2 s) were used: the light blue surface area in Fig. 8.7b corresponds to the registered UC signal at $I_{exc} = 325$ mW $\times$ cm^{-2} and the light red surface area to $I_{exc} = 130$ mW $\times$ cm^{-2}. It must be mentioned explicitly that straightforward definition of the time necessary to reach steady-state UC emission is not possible: for instance, for low excitation intensities ($I_{exc} < 6$ mW $\times$ cm^{-2}, for the demonstrated TTA-UC system) this time is definitely higher than 20s (Fig. 8.7a). Nevertheless, this experimental fact does not mean that no TTA-UC exists even at this excitation intensity.

More importantly, at low excitation intensities this can lead to the often observed "nonlinearity" of the UC signals on the excitation intensity: actually, this is due to the fact that the UC signals are compared in environments differently contaminated with oxygen. At low total oxygen contamination the sensitizer and the emitter molecules itself can serve as chemical traps; indeed, some amounts (not controlled) of the optically active molecules will be irrevocably lost for the TTA-UC process, but the rest of the UC ensemble will demonstrate upconversion with lower than possible efficiency.

8.2 Sacrificial Singlet Oxygen Scavengers (SSOS)

SSOS Selection Criteria

The desired singlet oxygen scavenger molecules, designed to support the process of TTA-UC in bulk or in confined environment must fulfill a chain of complex requirements:

1. The SSOS molecules must bind chemically **preferably** the singlet oxygen present at the optically excited region. The formed oxidation products must be chemically stable for the temperature region of interest (T < 50 °C). This requirement is of critical importance—if the sacrificial oxygen scavenging material works as a common antioxidant, i.e., chemically binding even the molecular oxygen in ground state (at the given temperature), some of the sensing features of TTA-UC process will be lost.
2. The SSOS molecules must show pronounced hydrophobicity and be well miscible with organic solvents/matrices. This requirement is basic, if the SSOS is used in a macroscopic (bulk) matrix: the TTA-UC process is demonstrated only in strongly hydrophobic environment (the UC moieties are hydrophobic). Oppositely, in nanoconfined environment, i.e., TTA-UC couples embedded in colloidal micelles or nano-containers [32], it is possible to add hydrophilic SSOS in the continuous phase of the dispersion.

3. If the SSOS materials are directly applied into the UC-embedding matrix viscosity and melting temperature of the resulting material are changed. In order to fit these material parameters to specific requirements, it is possible to prepare blends of SSOS materials with low melting temperature and those with high melting temperature; thus the optimal matrix viscosity and melting temperate can be found: for instance, UC nanoparticles based on natural wax/natural oils were demonstrated to serve as TTA-UC temperature and oxygen sensors [20].
4. Last but not least, the SSOS molecules must be biocompatible and nontoxic. This selection criterion is unavoidable if applications targeting temperature/oxygen content sensing in cell cultures or medical treatment are targeted.

There is a large bunch of literature [33] dealing with the oxidative damage (please see Table 8.2) and application of essential oils (please see Table 8.1) as oxygen scavengers [34]. The ability of different antioxidants to chemically bind oxygen radicals is given by the ORAC value, where oxygen radical absorption capacity (ORAC) assay measures the degree of inhibition of peroxy-radical-induced oxidation by the compounds of interest in a chemical milieu. It measures the value as Trolox equivalents and includes both inhibition time and extent of inhibition of oxidation [35] (Tables 8.2 and 8.3).

5. **In nanoconfined environment**, the process of TTA-UC encounters additional complications: namely the process of oxygen diffusion throughout the 3D interface of the NC combined with the relatively low amount of SSOS materials embedded in the NC core. The oxygen diffusion is modulated by the presence of nonionic surfactants and a mixture of hydrophobic NC-core materials with partial amphiphilicity (natural oils and waxes). It is assumed that the equilibrium concentration between the oxygen dissolved in the continuous phase (water environment) and oxygen dissolved in the dispersed phase (nanoparticle core) is reached only for long interaction times. As a consequence, one can expect that at certain experimental conditions it will be possible to observe mutually steady-state UC emission: in this case, the rate of oxygen diffusion through the NC shell will be lesser than the rate of chemical binding of the oxygen inside the NC core. In Fig. 8.8, wax/oil TTA-UC nanoparticle loaded with the standard sensitizer/emitter couple is shown. The contamination with oxygen is assumed to be in equilibrium.

The NC core is built from a mixture of carnauba wax and rice bran oil (Fig. 8.8, please see the NC-core composition). This matrix composition allows for significant increase of the molecular rotational diffusion of the optically active molecules for a temperature interval centered at 36 °C. Both products are natural and are defined

Table 8.1 ORAC index for some hydrophobic SSOS

	Clove oil	Citronella oil	Chamomile oil	Soybean oil
ORAC indexμmolTE/100 g	107,870.0	31,200	21,860.0	5409.2

Table 8.2 Structure of some hydrophilic antioxidants, used at the continuous phase of nanoconfined TTA-UC

Short name	Structure
α-Lipoic acid (**1**)	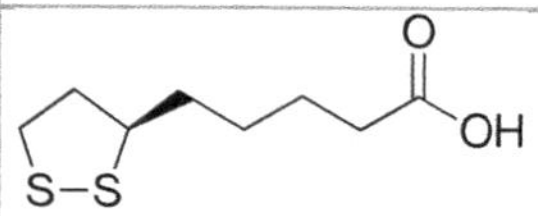
Ascorbic acid (**2**)	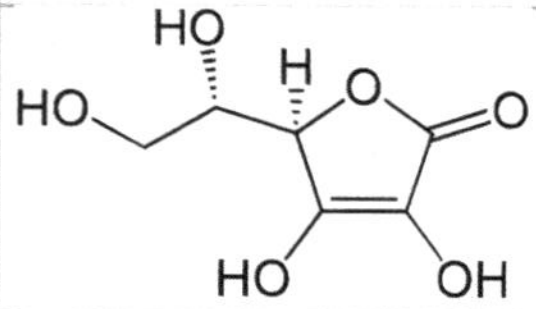
Carotenoid (**3**)	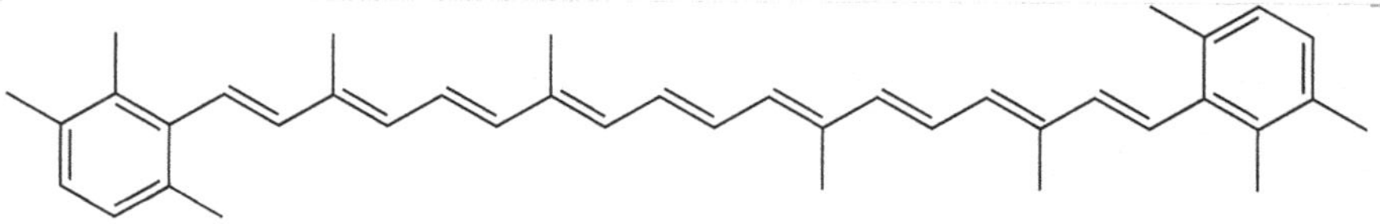
Phenyl-di-(3-buten-1-yl)-phosphate (**4**)	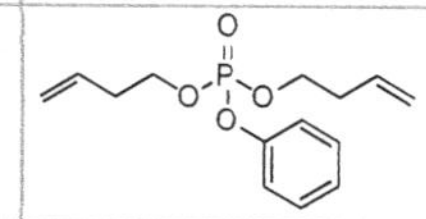
Squalene (**5**)	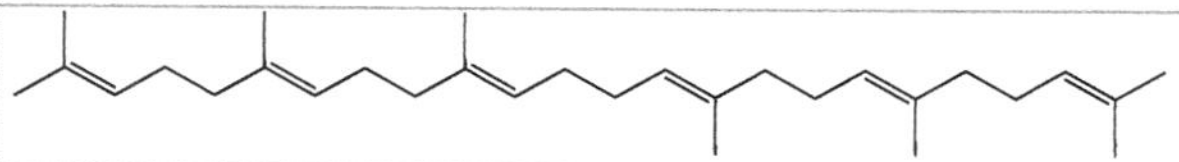

Table 8.3 Structure of amphiphilic antioxidants, slightly soluble in water

Short name	Structure
Polyoxyethanyl-alpha-tocopheryl sebacate	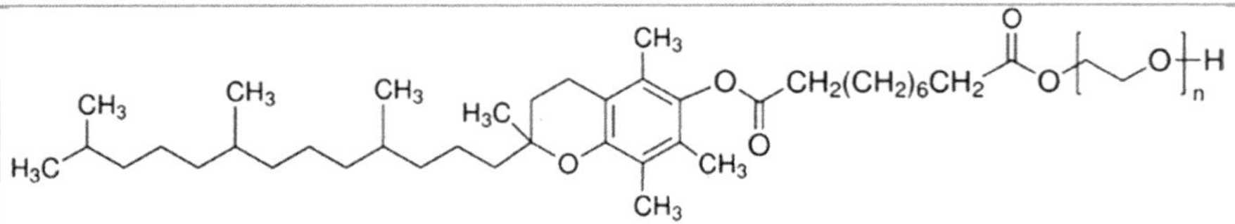
4-(2-Hydroxyethyl)phenol (Tyrosol)	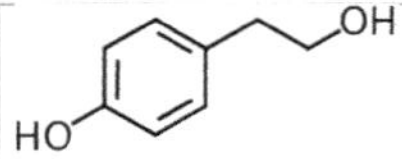

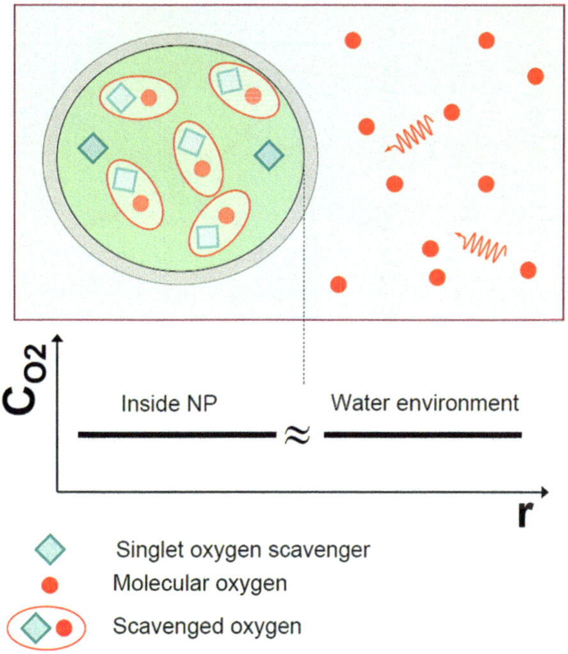

Fig. 8.8 Cartoon elucidating the oxygen diffusion in 3D nanosized object for the case of "NC from oxygen-saturated environment/oxygen-saturated water." After long enough (more than 6 h) equilibration time, it can be assumed that the oxygen concentration inside the NC and the water is equal. *Core material composition*: PdTBP (2×10^{-5} M)/perylene/40% wt. carnauba wax/60% wt. rice bran oil; *surfactant*: Tween 81; ratio sensitizer/emitter $C_S/C_E = 1/20$

as food additives. Additionally, the rice bran oil is known to be a highly potential SSOS material.

In Fig. 8.9, the dynamical property of the normalized UC fluorescence as a function of the excitation intensity is shown. It is evident that for high intensity (Fig. 8.9, red open circles) the oxygen scavenging capacity of the embedded SSOS materials is consumed very fast: steady-state UC emission cannot be observed. On the contrary, if the excitation intensity is too low (Fig. 8.9, blue open circles) the rate of oxygen diffusion throughout the NC shell is mutually equal to the rate of generation of singlet oxygen and its consequent scavenging. Also in this case it is not possible to observe steady-state UC emission even for very long excitation period (as long as 300 s). Nevertheless, for intermediate excitation intensity (Fig. 8.9, dark green open circles) it is possible to observe a nearly steady-state emission for the signals of upconversion for a period of nearly 50 s. Such an experimental result fits well with the requirements stated for minimally invasive sensing in cell culture samples since the TTA-UC process is performed in very mild excitation conditions (~4 mW × cm^{-2}) by application of UC materials embedded in inherently biocompatible matrix.

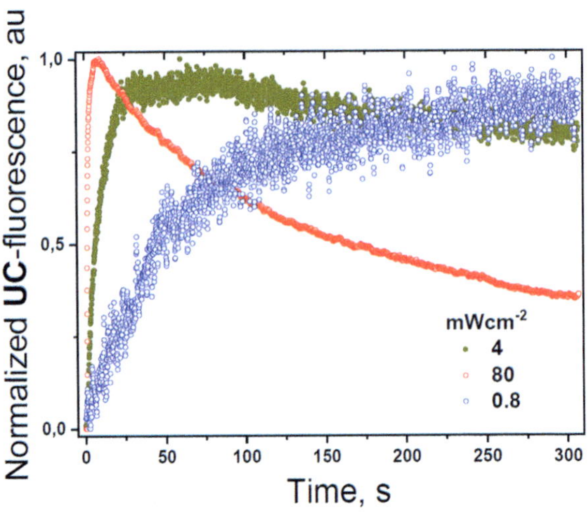

Fig. 8.9 Dynamical properties of the signal of normalized UC fluorescence as a function of the excitation intensity for UC nanoparticles: *Experimental conditions*: water suspension (10% wt. disperse phase); single-mode laser diode, $\lambda_{exc} = 635$ nm; optical registration via gated PMT; acquisition time $t_{AQ} = 100$ ms; air-saturated environment, sample isolated; RT; sample thickness 1000 μm; excitation beam diameter $r_{EXC} = 1200$ μm; each experimental plot is obtained at neat excitation spot, after complete intermixing of the sample. Optical data are obtained only for the duration of the excitation pulse of 300 s; no additional illumination of the sample is performed

References

1. R. Islangulov, J. Lott, C. Weder, F.N. Castellano, Noncoherent low-power upconversion in solid polymer films. J. Am. Chem. Soc. **129**(42), 12652–12653 (2007)
2. S. Baluschev, T. Miteva, V. Yakutkin, G. Nelles, A. Yasuda, G. Wegner, Phys. Rev. Lett. **97**, 143903 (2006)
3. E. Heinrich, Y. Avlasevich, K. Landfester, S. Baluschev, Annihilation upconversion: harvesting the entire deep-red spectral range of the sun irradiation. J. Photonics Energy **8**(2), 022002 (2018). https://doi.org/10.1117/1.JPE.8.022002
4. H. Goudarzi, P.E. Keivanidis, ACS Appl. Mater. Interfaces **9**(1), 845–857 (2017)
5. V. Gray, K. Moth-Poulsen, B. Albinsson, M. Abrahamsson, Coord. Chem. Rev. **362**, 54–71 (2018)
6. J. Pedrini, A. Monguzzi, J. Photonics Energy **8**, 2 (2017)
7. K.A. El Roz, F.N. Castellano, Chem. Commun. **53**, 85 (2017)
8. L.G. von Reventlow, M. Bremer, B. Ebenhoch, M. Gerken, T.W. Schmidt, A. Colsmann, J. Mater. Chem. C **6**, 15 (2018)
9. T. Ogawa, M. Hosoyamada, B. Yurash, T.Q. Nguyen, N. Yanai, N. Kimizuka, J. Am. Chem. Soc. **140**, 28 (2018)
10. Y. Murakami, S.K. Das, Y. Himuro, S. Maeda, Phys. Chem. Chem. Phys. **19**, 45 (2017)
11. M.A. Filatov, S. Baluschev, K. Landfester, Protection of densely populated excited triplet state ensembles against deactivation by molecular oxygen. Chem. Soc. Rev. **45**, 4668–4689 (2016). https://doi.org/10.1039/c6cs00092d
12. P.R. Ogilby, Chem. Soc. Rev. **39**, 3181 (2010)
13. D.B. Papkovsky, R.I. Dmitriev, Chem. Soc. Rev. **42**, 8700 (2013)

14. J. Zhou, Q. Liu, W. Feng, Y. Sun, F. Li, Chem. Rev. **115**, 395 (2015)
15. B.D. Rihterty, M.E. Kenney, W.E. Ford, M.A.J. Rodgers, J. Am. Chem. Soc. **115**, 8146 (1993)
16. R. Schmidt, W. Drews, H.D. Brauer, J. Phys. Chem. **86**, 4909 (1982)
17. W. Fudickar, T. Linker, J. Am. Chem. Soc. **134**, 15071 (2012)
18. T. Christ, F. Kulzer, P. Bordat, T. Basche, Angew. Chem. Int. Ed. **40**, 4192 (2001)
19. K. Naito, T. Tachikawa, S.-C. Cui, A. Sugimoto, M. Fujitsuka, T. Majima, J. Am. Chem. Soc. **128**, 16430 (2006)
20. K. Landfester, Y. Avlasevich, D. Busko, F. Wurm, S. Balouchev, Single all-optical nanosensor device probing simultaneously the local temperature and local oxygen concentration in soft-matter in non-invasive manner, WO/2016/150677/A1/EP3072942-A1 /WO2016150677-A1 EP3274420-A1/US2018106785-A1 (2016)
21. Y. Murakami, H. Kikuchi, A. Kawai, J. Phys, Chem. B **117**, 2487–2494 (2013)
22. J.-H. Kim, F. Deng, F.N. Castellano, J.-H. Kim, Chem. Mater. **24**, 2250–2252 (2012)
23. Q. Liu, T. Yang, W. Feng, F.J. Li, Am. Chem. Soc. **11**, 5390–5397 (2012)
24. K. Tanaka, H. Okada, W. Ohashi, J.-H. Jeon, K. Inafuku, Y. Chujo, Bioorg. Med. Chem. **21**, 2678–2681 (2013)
25. F. Marsico, A. Turshatov, R. Peköz, Y. Avlasevich, M. Wagner, K. Weber, D. Donadio, K. Landfester, S. Baluschev, F.R. Wurm, Hyperbranched unsaturated polyphosphates as protective matrix for long-term photon upconversion in air. J. Am. Chem. Soc. **136**, 11057–11064 (2014). https://doi.org/10.1021/ja5049412
26. N. Shimizu, J. Ito, S. Kato, et al., Oxidation of squalene by singlet oxygen and free radicals results in different compositions of squalene monohydroperoxide isomers. Sci. Rep. **8**, 9116 (2018). https://doi.org/10.1038/s41598-018-27455-5
27. P. Du, R. Eisenberg, Energy upconversion sensitized by a platinum(II) terpyridyl acetylide complex. Chem. Sci. **1**, 502–506 (2010)
28. D.L. Andrews, C. Curutchet, G.D. Scholes, Laser Photonics Rev. **5**(1), 114–123 (2011)
29. Y.Y. Cheng, T. Khoury, R.G.C.R. Clady, M.J.Y. Tayebjee, N.J. Ekins-Daukes, M.J. Crossley, T.W. Schmidt, Phys. Chem. Chem. Phys. **12**, 66–71 (2010)
30. S. Baluschev, T. Miteva, V. Yakutkin, Y.S. Avlasevich, K. Müllen, G. Nelles, S. Chernov, S. Aleshchenkov, A. Cheprakov, A. Yasuda, G. Wegner, Angew. Chem. Int. Ed. **46**(40), 7693–7696 (2007)
31. T.F. Schulze, J. Czolk, Y.-Y. Cheng, B. Fückel, R.W. MacQueen, T. Khoury, M.J. Crossley, B. Stannowski, K. Lips, U. Lemmer, A. Colsmann, T.W. Schmidt, J. Phys. Chem. C **116**, 22794–22801 (2012)
32. S. Baluschev, K. Katta, Y. Avlasevich, K. Landfester, Annihilation upconversion in nanoconfinement: solving the oxygen quenching problem. Mater. Horiz. **3**, 478–486 (2016). https://doi.org/10.1039/c6mh00289g
33. D.K. Gupta, J.M. Palma, F.J. Corpas (ed.), Reactive Oxygen Species and Oxidative Damage in Plants Under Stress (Springer International Publishing, 2015). https://doi.org/10.1007/978-3-319-20421-5
34. D. P. de Sousa (ed.), *Bioactive Essential Oils and Cancer* (Springer International Publishing, 2015). https://doi.org/10.1007/978-3-319-19144-7
35. http://www.ars.usda.gov/nutrientdata

Chapter 9
Additive-Assisted Stabilization Against Photooxidation of Organic and Hybrid Solar Cells

Michela Prete, Um Kanta Aryal, Jonas Sandby Lissau, Horst-Günter Rubahn, Morten Madsen, and Vida Turkovic

Abstract Thin-film organic and perovskite solar cells have attracted the interest of research and industry. The main reason behind this is the low price, high tunability, and great potential of these technologies. Lightweight, color tunability, translucence, flexibility, and ease and low cost of processing are a few of the excellent characteristics of these technologies. However, they have a high sensitivity to light, oxygen, temperature, and humidity, which are common environmental conditions. Upon exposure to these environmental stresses, the chemical structure and morphological arrangement undergo complex changes that result in dramatic reduction of device functionalities over time. This chapter collects the approaches currently reported in the field of additive-assisted stabilization, a method in which solar cells' active layers are blended with additives that promote their stability (e.g., hydrogen donors, radical scavengers, hydroperoxide decomposers, quenchers, UV absorbers).

Keywords Antioxidant · Radical scavenger · Singlet oxygen quencher · Organic solar cells · Perovskite solar cells · Additive-assisted stabilization · Photooxidation

Printable electronics presents a new era of optoelectronic technologies that in the past years had several breakthroughs in the technological advancement of large-area and low-cost processing methods, which have accelerated the field for various applications—from energy harvesting (solar cells) [1], sensors (photodetectors, electrochromic devices) [2, 3], and displays (light-emitting diodes) [4] to electronic circuitry (transistors) [5]. Two of the most prominent types of materials compatible with printing techniques are organic and perovskite semiconductors. These materials found their use thanks to their strong absorption in the visible region [6] (superior

---

M. Prete · U. K. Aryal · J. S. Lissau · H.-G. Rubahn · M. Madsen · V. Turkovic (✉)
Mads Clausen Institute, University of Southern Denmark, Sønderborg, Denmark
e-mail: prete@mci.sdu.dk; engmann@mci.sdu.dk

J. S. Lissau, M. Madsen (eds.), *Emerging Strategies to Reduce Transmission and Thermalization Losses in Solar Cells*, https://doi.org/10.1007/978-3-030-70358-5_9

in organics) and efficient charge generation and transport properties (superior in perovskites) [7, 8]. However, both organic and perovskite semiconductors are susceptible to degradation in contact with oxygen and light [9, 10], resulting in a progressive loss of absorption and transport properties over their lifetime, and eventually the complete loss of their optoelectronic functionality. Different approaches to an effective stabilization have been utilized in the past years, such as (a) multilayer encapsulation of devices consisting of complex barrier materials, oxygen getters, and UV filters; (b) synthesis of intrinsically stable materials; as well as (c) implementation of stabilizing additives in the active layers. The latter method, in the focus of this chapter, imported from the field of insulating polymers [11], has been pioneered by our group in the field of organic solar cells [12], and is now widely applied to limit the photooxidation of organic and perovskite optoelectronic devices [13]. This chapter highlights the latest studies in this direction and describes different mechanisms involved in additive-assisted stabilization of different thin-film solar cell systems. We note that while the focus of this chapter is on prolonging the lifetime of organic and hybrid solar cell devices by stabilizing the active layers themselves using various different additives, such strategies are also very relevant to consider in devices that rely on upconversion and singlet-fission processes to reduce transmission and thermalization losses, as it is the focus in this book. This is for example in cases where the upconverter or singlet fission materials are placed in direct contact with the active-layer materials inside the device stack, where they thus may perturb the involved degradation and stabilization mechanism in question. Furthermore, as triplet states play a central role in upconversion and singlet fission processes, and as these readily interact with molecular oxygen, potential strategies that can mitigate negative effects from this interaction may be very relevant for the stabilization of such systems.

9.1 Organic Solar Cells

Organic solar cells are based on organic thin films that consist of a blend of electron-donating (D) and electron-accepting (A) organic molecules, usually polymers and small organic molecules. This layer, usually referred to as the active layer, is the "heart" of the device, in charge of light absorption, upon which excitons are generated, and split at the D/A interface to form free charges. The active layer is "sandwiched" from both sides by electron-transporting and hole-transporting layers. Once the exciton has split into an electron and a hole, these transport layers will extract these free charges from the active layer to the cathode and anode on top of them. A typical organic solar cell in commonly used inverted architecture is depicted in Fig. 9.1.

Organic solar cells are nowadays reaching record efficiencies of 18%, as reported for single-junction devices [14], but still suffering from relatively poor stability. The organic nature of the technology is a double-sided sword with good characteristics on one side, such as low weight, high flexibility, and low production costs, but on the

Fig. 9.1 OPV cell

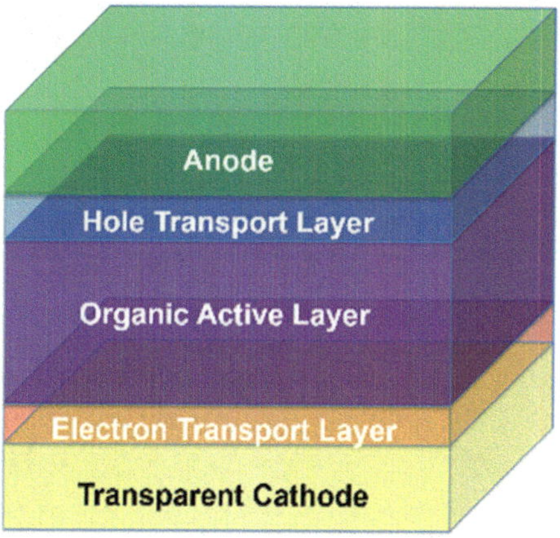

other side, high sensitivity to UV-vis irradiation and oxygen, which causes drastic changes in their optical, electrical, and mechanical properties, and ultimately leads to the loss of device functionality.

A lot of efforts have already been invested into elucidating the degradation mechanisms in organic solar cells [15–17]. The mechanisms are often classified either as extrinsic—caused by oxygen and moisture [18]—or as intrinsic—caused by morphological rearrangements of the molecules within the active layer blend (thermodynamically driven [19], including crystallization by thermal stress [20, 21] or even dimerization by UV-vis irradiation [22]). And although extrinsic degradation could ideally be suppressed by encapsulation, in reality it is impossible to avoid minute amounts of oxygen (present from, e.g., synthesis, or even fabrication in glove box, as well as encapsulation) coming in contact with the active layers, which are sufficient to start the oxidative cycle of the polymer.

The generalized scheme of polymer degradation [23] and possible mechanisms of degradation suppression are depicted in Fig. 9.2. In contact with oxygen and light, the degradation begins by scissioning of the polymer bonds, leading to the formation of a makroalkyl radical R•, which quickly reacts with surrounding oxygen molecules to form peroxy ROO• radical. The peroxy radical readily abstracts hydrogen from the polymer and forms a hydroperoxide ROOH. The labile hydroperoxide is further decomposed into two highly reactive radicals, alkoxy RO• and hydroxyl OH•, which further react with oxygen and polymer and initiate new oxidation chains, until all the radicals have recombined. These processes also take place under elevated temperatures in the absence of light; however instead of light, the energy for the formation of radicals comes from heat.

By addition of specific types of compounds, shown in Fig. 9.2, into the active layer of the devices, one can in different ways prevent or slow down the photooxidation of the organic molecules and thus the decay of the device performance over time. UV absorbers selectively absorb the UV portion of sunlight, thereby shielding the polymer from absorbing it and going through a photooxidative reaction while themselves undergoing rapid intramolecular process upon excitation which returns

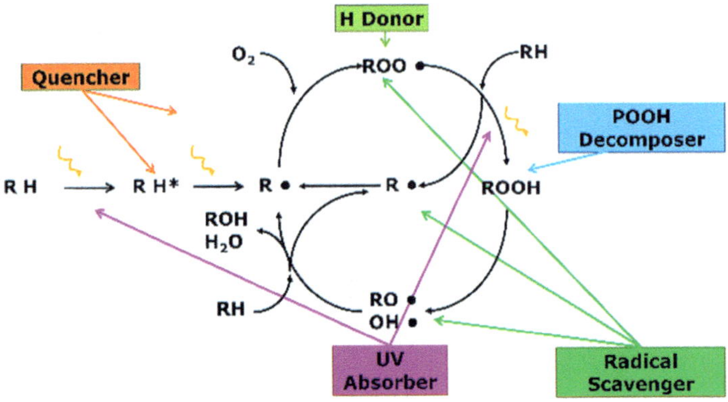

Fig. 9.2 Photooxidation cycle of organic materials

them to original state. Quenchers deactivate excited chromophores in polymers, which otherwise act as (1) sensitizers for the photolysis of hydroperoxides or (2) sensitizers of singlet oxygen, as well as (3) deactivate excited state of oxygen to its less reactive ground state. Hydrogen donors and radical scavengers either (1) transfer a hydrogen from the phenolic or amine moiety to the peroxy radical, thus preventing the abstraction of hydrogen from the polymer (e.g., phenolic antioxidants, aromatic amines), or (2) form radicals (e.g., nitroxyl) which have a very high reaction rate with radicals (e.g., hindered amine stabilizers, HALS) upon which they undergo reactions in which they reform themselves. Hydroperoxide (POOH) decomposers (e.g., organophosphorus and organosulfur compounds) interfere with the homolytic split of the hydroperoxide, a polymer oxidation intermediate, that would otherwise homolytically split into extremely reactive alkoxy and hydroxy radicals, but instead gets split into inert products (e.g., water, alcohols).

As reported in our previous overview chapter [13], there are multiple studies on the stabilization of fullerene-based organic solar cells. These stability studies highlight the ambivalent role of the acceptor molecule in the degradation of OPV. Fullerene can act as (a) a radical scavenger, by addition of radicals to their extended pi system [24–29]; (b) a quencher of the polymer exciton, by electron transfer to fullerene, which reduces the lifetime of the excited singlet state and thus decreases the population of all reactive states populated from it (e.g., long-lived triplet state that can sensitize the highly reactive singlet oxygen) [30–32]; and (c) a UV irradiation screen, which prevents the most destructive part of incident light from reaching the polymers [30]. However, fullerenes can also exert negative influence on the degradation of OPVs: (a) by electron transfer from fullerene anions to molecular oxygen, through which a superoxide radical anion can be created [33, 34] which can initialize the oxidation cycle, or (b) by energy transfer from fullerene triplet, directly or over a polymer triplet state or a polymer:fullerene charge transfer state, to molecular oxygen, in which singlet oxygen is formed, forming hydroperoxides that initialize the oxidation cycle [30, 35–41].

One common polymer:fullerene system where singlet oxygen plays a dominant role in degradation is PTB7:[70]PCBM [32]. It was demonstrated that upon addition of 3% of naturally occurring beta-carotene [42], the presence of singlet oxygen as well as the triplet precursor can be drastically reduced (see Fig. 9.3). These trends were also confirmed for these systems in solution, in case of stabilization using beta-carotene, as well as using a related carotenoid molecule, astaxanthin, as shown in Fig. 9.4 [43]. As a consequence, an improvement in the lifetime of devices, as reflected in a dramatic increase in power-accumulated generation by a factor of 6 measured under ISOS-L-2 protocol [44] on unencapsulated devices, was observed (see Fig. 9.3). Interestingly, the degradation was strongly inhibited in the burn-in period (the initial fast decaying period characteristic of organic solar cells, whose origin is still a topic of controversy), and a direct correlation between singlet oxygen concentration and burn-in in this system was revealed.

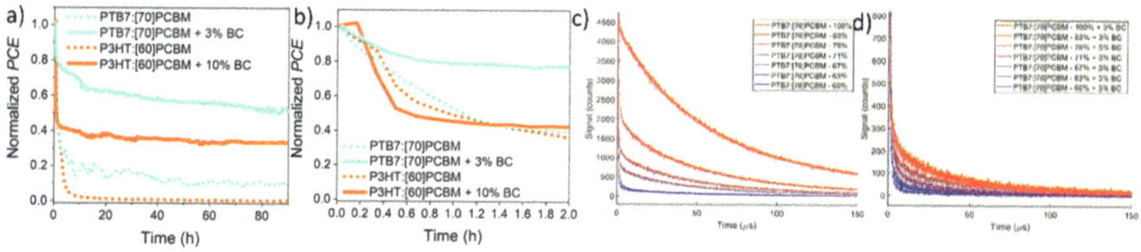

Fig. 9.3 (**a**) Stability of unencapsulated PTB7:[70]PCBM and P3HT:[60]PCBM solar cells without additives and containing 3 and 10% beta-carotene (BC), under 1 Sun irradiation in air according to ISOS-L-2. (**b**) Zoomed-in stability graph in (**a**) on the burn-in region. The values are normalized with respect to initial efficiency (PCE). (**c**) Time-resolved singlet oxygen phosphorescence signals of PTB7:[70]PCBM films with different polymer:fullerene ratios without additives and (**d**) containing 3% beta-carotene. The percentage in the legend refers to the amount of [70]PCBM in each film. Note the different scales in *y*-axes of the respective figures. Republished from Ref. [42]. Copyright 2019, with permission from The American Chemical Society ACS

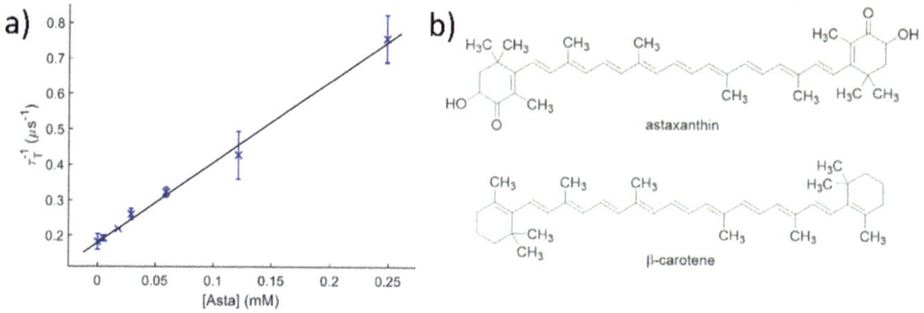

Fig. 9.4 (**a**) The inverse triplet-state lifetime, recorded in time-resolved absorption spectroscopy, as a function of the concentration of added astaxanthin. The solid black line is a linear fit to the data, the slope of which yields the rate constant for triplet PTB7 quenching by astaxanthin. (**b**) Structures of tested additives. Republished from Ref. [43]. Copyright 2020, with permission from The Institute of Physics IOP

Another way in which fullerenes can negatively influence the stability of the devices is by their dimerization in the presence of light [22, 45]. A successful stabilization of this process was reported in P3HT:[60]PCBM system, utilizing an amine class of stabilizers, piperazine [46–49] (see Fig. 9.5). In these studies, conducted in nitrogen glove box, the ability of the piperazine to quench fullerene triplet states that suppress dimerization of fullerene was primarily utilized, which strongly improved the stability of devices under nitrogen atmosphere, although its antioxidant properties also contributed in the form of preventing the oxidation of P3HT [46] (see Fig. 9.5). Similar stabilizing effect has been reported for two other [60]PCBM-based organic photovoltaic systems, PTB7-Th:[60]PCBM and PffBT4T-2OD:[60]PCBM (see Fig. 9.6). Another interesting property of the

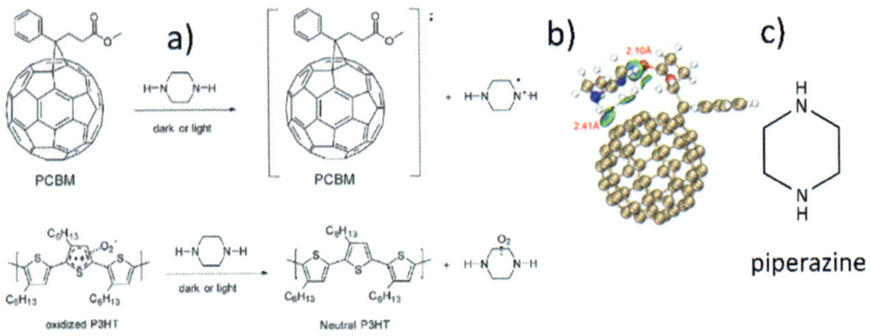

Fig. 9.5 (**a**) Proposed interactions of piperazine with [60]PCBM and P3HT. (**b**) Hydrogen bonds between the piperazine (H atom of N–H group) and [60]PCBM: via the pi aromatic system (distance from H of piperazine to closest C in pi system 2.41 Å) and via the carbonyl group on the ester side chain of PCBM (distance from H of piperazine to O of carbonyl group 2.10 Å). (**c**) Structure of the tested additive. Republished from Ref. [46], Copyright 2019, with permission from The Royal Society of Chemistry RSC. Republished from Ref. [47], Copyright 2020, with permission from The American Chemical Society ACS

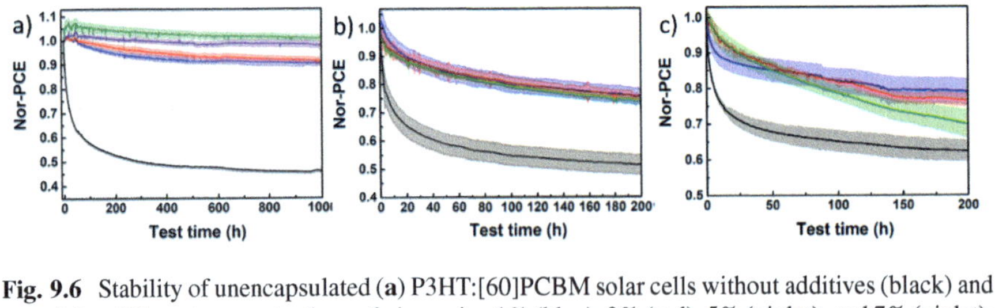

Fig. 9.6 Stability of unencapsulated (**a**) P3HT:[60]PCBM solar cells without additives (black) and containing different concentrations of piperazine 1% (blue), 3% (red), 5% (violet), and 7% (violet); (**b**) PTB7-Th:[60]PCBM solar cells without additives (black) and containing different concentrations of piperazine 0.05% (blue), 0.2% (red), and 0.5% (green); (**c**) PffBt4t-2OD:[60]PCBM solar cells without additives (black) and containing different concentrations of piperazine 0.1% (blue), 0.5% (red), and 3% (green). Degradation was performed under 1 Sun irradiation in nitrogen glove box. The values are normalized with respect to initial efficiency (PCE). Republished from Ref. [46], Copyright 2019, with permission from The Royal Society of Chemistry RSC

piperazine materials that proved beneficial for stability is their ability of forming hydrogen bonds with the active-layer materials [47], which can accelerate the electron transfer and transport, thus outcompeting the possible degradation reactions involving those charges [50]. By comparing the stabilization effects of seven different amine stabilizers (see Fig. 9.7) [47] as well as one polymeric amine stabilizer polyethyleneimine [48], a stabilizing effect was observed only using stabilizers containing N–H groups. This was explained by the formation of N–H... pi and N–H... O–C bonds between piperazine and fullerene, which brings the two closer together, promoting in this way the photoinduced electron transfer between P3HT and [60]PCBM, over the competing photoinduced energy transfer (which would otherwise result in the formation of fullerene triplet states, which could lead to dimerization of fullerene). The stabilizing effect of piperazine was reported also for semicrystalline PCE11:[70]PCBM and amorphous PBTZT-stat-BDTT-8:[70]PCBM systems, in a study where its effect was compared to that of two other stabilizers (see Fig. 9.8), chosen such that (a) their melting point is above the typical processing temperatures (100 °C), (b) their miscibility with fullerene is comparable to the miscibility of the fullerene with the polymer molecules, and (c) they have good stability and are small and symmetric in size [51]. The starting hypothesis was that the burn-in in these systems takes place because of morpho-

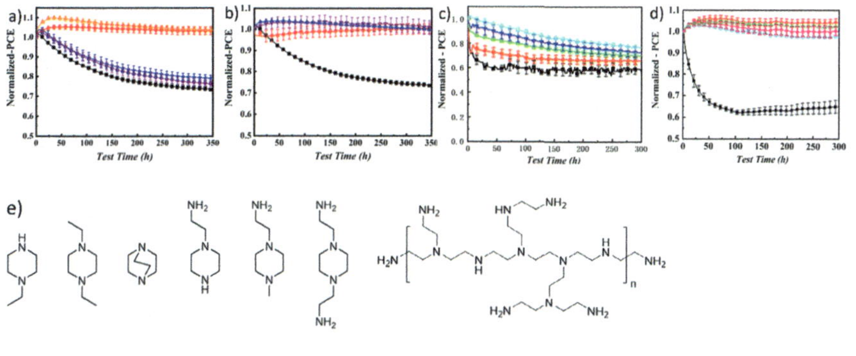

Fig. 9.7 Stability of unencapsulated P3HT:[60]PCBM solar cells without additives (black) and containing (**a**) piperazine (red), *N*-ethylpiperazine (*N*-EP) (orange), 1,4-bis (3-ethyl) piperazine (Bis-EP) (blue), and triethylenediamine (DABCO) (violet), and (**b**) *N*-aminoethylpiperazine (*N*-AEP) (red), 1-(2-aminoethyl)-4-methylpiperazine (M-AEP) (blue), and 1,4-bis (aminoethyl) piperazine (Bis-AEP) (violet). (**c**) Stability of unencapsulated PTB7-Th:[60]PCBM solar cells without additives (black) and containing different amounts of polyethyleneimine (PEI): 0.01% (red), 0.05% (green), 0.10% (blue), and 0.50% (turquoise). (**d**) Stability of unencapsulated P3HT:[60]PCBM solar cells without additives (black) and containing different amounts of polyethyleneimine: 0.05% (red), 0.10% (green), 0.20% (turquoise), and 0.30% (magenta). Degradation was performed under white LED light in nitrogen. The values are normalized with respect to initial efficiency (PCE). (**e**) Structures of tested additives. Republished from Ref. [47], Copyright 2020, with permission from The American Chemical Society ACS. Republished from Ref. [48], Copyright 2020, with permission from Wiley-VCH

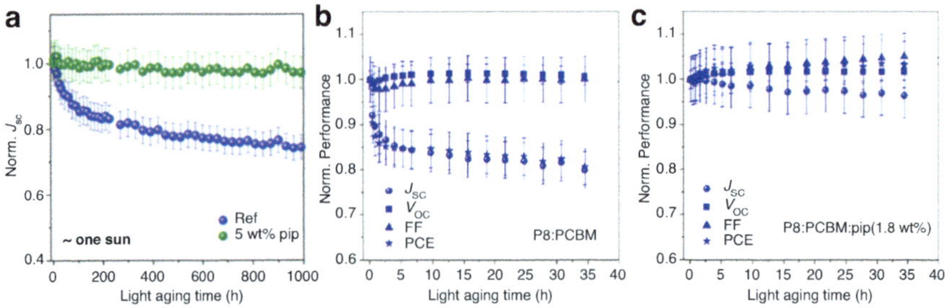

Fig. 9.8 Stability of (**a**) PCE11:[70]PCBM solar cells without additives and containing 5% of piperazine (pip); (**b**) PBTZT-stat-BDTT-8 (P8):PCBM without additives and (**c**) containing 1.8% of piperazine. Degradation is performed under white LED light (400–800 nm) irradiation in nitrogen. The values are normalized with respect to initial short-circuit current (J_{sc}), open-circuit voltage (V_{oc}), fill factor (FF), and efficiency (PCE). Republished from Ref. [51], Copyright 2019, with permission from The Royal Society of Chemistry RSC

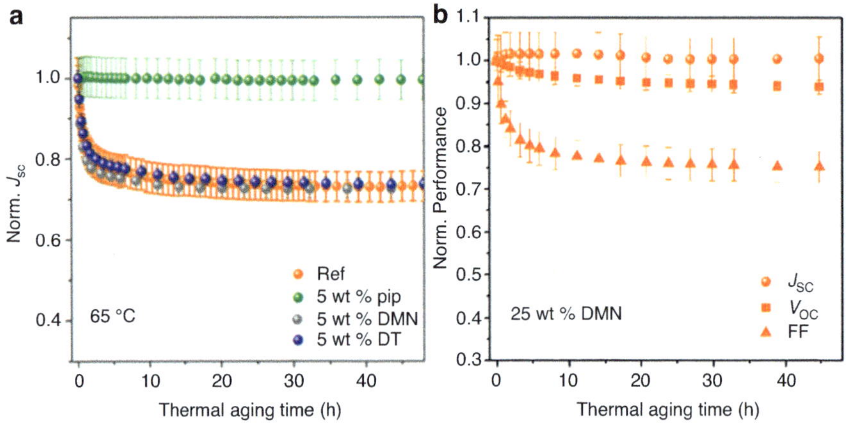

Fig. 9.9 Stability of PCE11:[70]PCBM solar cells without additives and containing 5% of (**a**) piperazine (pip), 2,6-dimethylnaphthalene (DMN), and 1,4-dithiane (DT), and (**b**) 25% of DMN. Degradation is performed under 65 °C temperature in nitrogen. The values are normalized with respect to initial short-circuit current (J_{sc}), open-circuit voltage (V_{oc}), fill factor (FF), and efficiency (PCE). Republished from Ref. [51], Copyright 2019, with permission from The Royal Society of Chemistry RSC

logical demixing of the polymer and the fullerene in the amorphous region. Of the three compounds, piperazine displayed the strongest stabilizing effect (see Fig. 9.9) already at a lower concentration, although a good effect was also observed using 2,6-dimethylnaphthalene at a higher concentration. To test the influence of hydrogen bonding on the stabilization, the stabilizing effect of piperazine was compared to another two stabilizers able to form hydrogen bonds (see Fig. 9.10), but with lower melting temperature (HALS molecule) and with poorer miscibility with the polymer (4,4′-biphenol). Here, despite the hydrogen bonding, the unstable microstructure

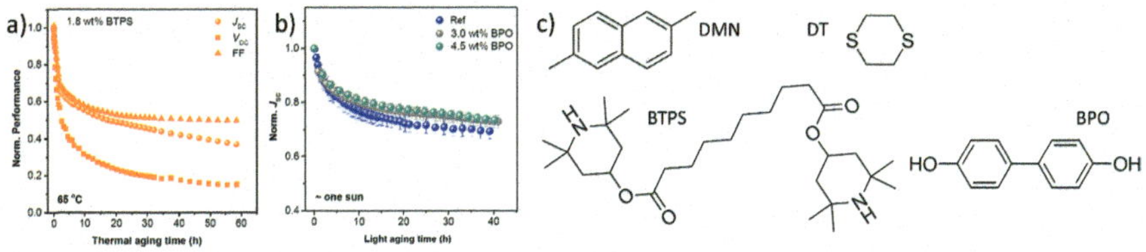

Fig. 9.10 Stability of PCE11:[70]PCBM solar cells without additives and containing (**a**) 1.8% of bis (2,2,6,6-tetramethyl-4-piperidyl) sebacate (BTPS), under 65 °C temperature in nitrogen, and (**b**) 3 and 4.5% of 4,4′-biphenol (BPO), under white LED light (400–800 nm) irradiation in nitrogen. The values are normalized with respect to short-circuit current (J_{sc}), open-circuit voltage (V_{oc}), and fill factor (FF). (**c**) Structures of tested additives. Republished from Ref. [51], Copyright 2019, with permission from The Royal Society of Chemistry RSC

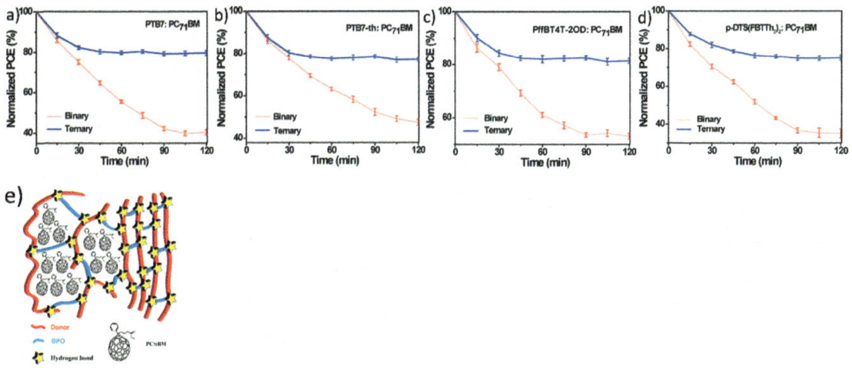

Fig. 9.11 Stability of (**a**) PTB7:[70]PCBM, (**b**) PTB7-Th:[70]PCBM, (**c**) PffBT4T-2OD:[70]PCBM, and (**d**) p-DTS(FBTTh$_2$)$_2$:[70]PCBM solar cells without additives and containing 3% of 4,4′-biphenol, under 1 Sun irradiation at 130 °C in nitrogen. The values are normalized with respect to initial efficiency (PCE). (**e**) Proposed hydrogen bond interactions between the BPO additive and the donor and [60]PCBM in the active layer. Republished from Ref. [52], Copyright 2016, with permission from Wiley-VCH

of HALS molecule and the low dispersion of biphenol among the polymer and fullerene molecules were found as the main culprits of the missing stabilization effect, and the ability of forming hydrogen bonds was attributed mainly to reducing the necessary stabilizer amount. The effect of hydrogen bonding using biphenol as a stabilizer was also investigated in a systematic study where its effect was compared in six different polymer:fullerene systems [52]. The stability of four systems containing F atoms, which can form hydrogen bonds (PTB7:[70]PCBM, PTB7-Th:[70]PCBM, PffBT4T-2OD:[70]PCBM, and p-DTS(FBTTh$_2$)$_2$:[70]PCBM), was strongly enhanced, as opposed to the other two systems (P3HT:[60]PCBM and PBDTTT-C-T:[70]PCBM) containing no atoms that could form hydrogen bonds, where no stabilization effect could be observed (see Fig. 9.11). The effect was explained by hydrogen bonds freezing the donor aggregated state and suppressing the excessive aggregation of [70]PCBM.

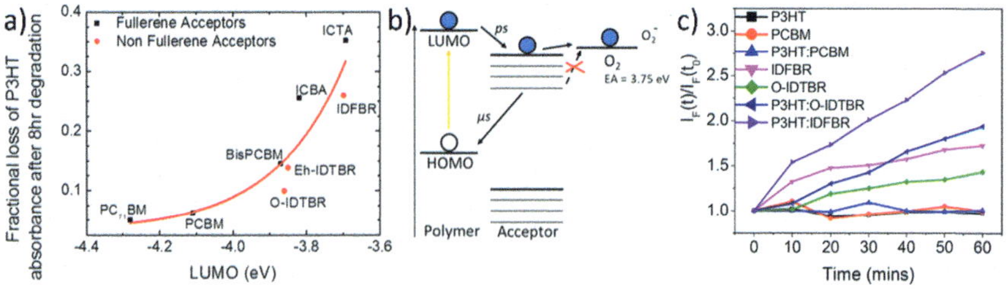

Fig. 9.12 (**a**) Fractional losses of the P3HT absorbance peaks in films blended with ICTA, ICBA, IDFBR, bis-PCBM, Eh-IDTBR, O-IDTBR, [60]PCBM, and [70]PCBM acceptor molecules, under 8 h of 1 Sun irradiation in dry air with RH <40% humidity, as a function of the measured LUMO level of the acceptors; (**b**) proposed degradation mechanism: the photodegradation of P3HT caused by the formation of superoxide (O_2^-) via electron transfer from the LUMO levels of the acceptors to molecular oxygen (O_2), which has an electron affinity (EA) of 3.75 eV; (**c**) fluorescence intensity increase of the hydroethidine (HE) probe, measuring superoxide generation, of neat and blend films at 610 nm (excited at 520 nm) as a function of time under 1 Sun irradiation in dry air with RH <40% humidity. The values are normalized with respect to initial fluorescence intensity. Republished from Ref. [69], Copyright 2019, with permission from The American Chemical Society ACS

Recently, the conventionally used fullerenes are being replaced by a wide variety of non-fullerene acceptor molecules, yielding breakthrough efficiencies of up to 18% [14, 53–57]. Due to their round shape, fullerenes are more easily movable within the active-layer blends, which results in strong phase segregations over the lifetime of the devices [21, 58, 59]. By replacing the fullerenes with small planar non-fullerene acceptor molecules, enhanced morphological (thermal) stability of the devices was reported for several systems [60–62]; however the photooxidative stability remains an issue [50, 63–68]. A study conducted on a series of non-fullerene- (O-IDTBR, Eh-IDTBR, IDFBR) and fullerene-based acceptors ([70]PCBM, [60]PCBM, bis-PCBM, ICBA) found that the degree of degradation strongly depends on the LUMO level of the acceptor molecule [69]. In accordance with the earlier study on the role of fullerenes' electron affinity on the formation of superoxide in the course of degradation of organic solar cells [34], it was found that, due to their shallow LUMO energies that lie above the electron affinity of molecular oxygen (3.75 eV), the non-fullerene acceptor systems suffered from photooxidative degradation to a larger extent than the fullerene systems (in which LUMO energies are more low lying, below 3.75 eV); see Fig. 9.12.

Since non-fullerene acceptors only recently started gaining attention due to the developments in the last few years yielding rapidly increasing power conversion efficiency records, there are only a few reports on additive-assisted stabilization in these systems. The most thorough study compared the effect of several classes of stabilizers, including TEMPO (hindered amine light stabilizer, HALS), two hydrogen donors, hydroperoxide decomposer, and two singlet/triplet/$^1O_2$ quenchers (see Fig. 9.13), on the stability of four non-fullerene acceptor-based systems (donor polymer J71 blended with acceptors ITIC, ITDIC, IT-4F, and O-IDTBR) [67]. In all of the tested systems, the strongest stabilization effect was reached using

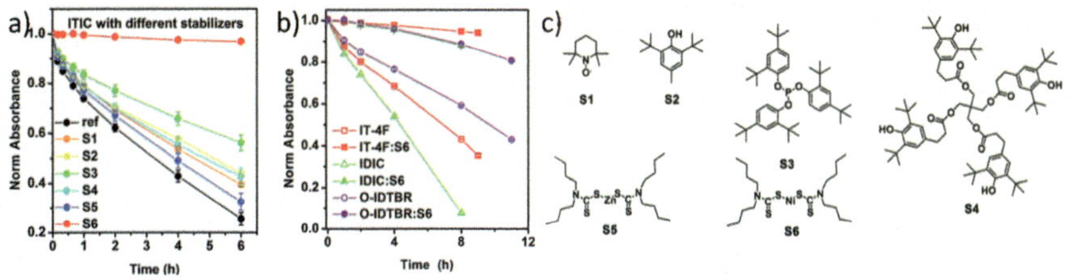

Fig. 9.13 (**a**) Stability of films of ITIC acceptor molecule without additives and containing 2% of 2,2,6,6-tetramethylpiperidinooxy (S1), 2,6-di-*t*-butyl-4-methylphenol (S2), tris (2,4-di-*ter*-butylphenyl) phosphite (S3), pentaerythritol tetrakis [3-(3,5-di-*tert*-butyl-4-hydroxyphenyl) propionate] (S4), zinc dibutyldithiocarbamate (S5), and nickel dibutyldithiocarbamate (S6); (**b**) stability of IT-4F, IDIC, and O-IDTBR acceptor molecules without additives and containing 2% of nickel dibutyldithiocarbamate. Degradation was performed under 1 Sun irradiation in air. The values are normalized with respect to initial absorbance. (**c**) Structures of tested additives. Republished from Ref. [67], Copyright 2019, with permission from The Royal Society of Chemistry RSC

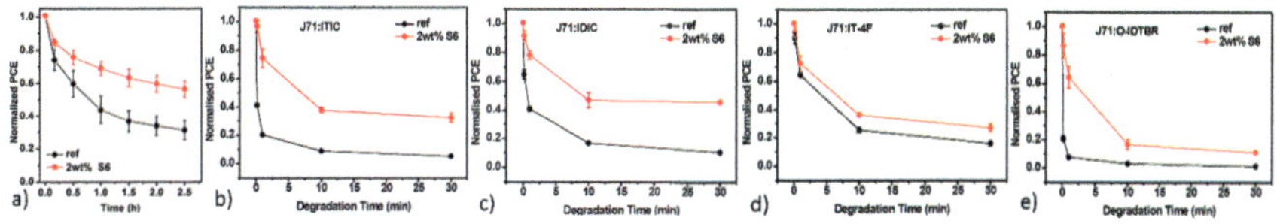

Fig. 9.14 Stability of unencapsulated (**a**) and (**b**) J71:ITIC, (**c**) J71:IDIC, (**d**) J71:IT-4F, and (**e**) J71:O-IDTBR solar cells without additives and containing 2% of nickel dibutyldithiocarbamate, under 1 Sun irradiation in dry air. In (**a**) degradation was performed on full devices, and in (**b–e**) on devices prior to deposition of back electrode layers. The values are normalized with respect to initial efficiency (PCE). Republished from Ref. [67], Copyright 2019, with permission from The Royal Society of Chemistry RSC

nickel dibutyldithiocarbamate (see Fig. 9.13), an additive previously reported as a successful polymer triplet-state quencher and radical scavenger in PTB7:[70]PCBM devices [70]. However, for this additive type, the stabilization comes somewhat at the expense of device performance, due to trap-assisted recombination effects (Fig. 9.14).

The effects of hydrogen bonding were also explored in non-fullerene-based cells, using two types of coumarin antioxidants, one that can easily form hydrogen bonds with the ITIC acceptor molecule in the blend, and the other which cannot form hydrogen bonds [50]. A good stabilization effect was achieved using the hydrogen-bonding coumarin additive with respect to thermal and photostability (both tested in nitrogen atmosphere) (see Fig. 9.15).

UV light is a common problem for both the non-fullerene and fullerene systems, as its high energy readily breaks the bonds of the active-layer molecules. A study reported enhanced device stability upon addition of benzophenone, a common UV-absorber stabilizer [68] (see Fig. 9.16).

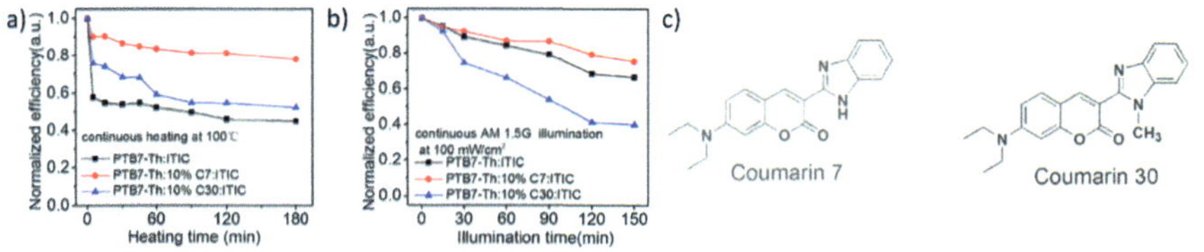

Fig. 9.15 Stability of unencapsulated PTB7-Th:ITIC solar cells without additives and containing 10% of coumarin 7 and coumarin 30, in nitrogen under (**a**) 100 °C temperature and (**b**) 1 Sun irradiation. The values are normalized with respect to initial efficiency (PCE). (**c**) Structures of tested additives. Republished from Ref. [50], Copyright 2018, with permission from The Royal Society of Chemistry RSC

Fig. 9.16 (**a**) Stability of FTAZ:ITIC-Th solar cells without additives and containing 7% of benzophenone, under 1 Sun irradiation in nitrogen. (**b**) Structure of the tested additive. Republished from Ref. [68], Copyright 2017, with permission from Wiley-VCH

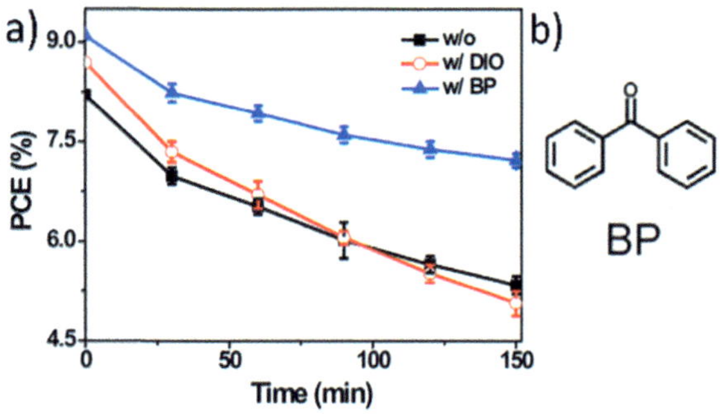

9.2 Perovskite Solar Cells

Perovskite solar cells today have efficiency records that resemble that of silicon (23.7% [71]). Their architecture resembles that of organic solar cells, with the main difference being that the active layer consists of a hybrid material of the $APbX_3$ type, where A is methylammonium ($CH_3NH_3^+$, MA), formamidinium ($CH_3(NH_2)_2^+$, FA), inorganic cations (Cs^+, Rb^+), or a mixture and X is a halide (Br, I, or a mixture), as shown in Fig. 9.17. Thanks to their strong absorption in the visible region and efficient charge generation and transport properties, as well as to their solution processability, they provide a highly attractive low-cost alternative to silicon technology. But, similar to organic materials, oxygen rapidly diffuses into the perovskite structure and occupies defect sites and grain boundaries. When exposed to light, photoexcited electrons created in the perovskite react with oxygen to form superoxide (O_2^-), which deprotonates the methylammonium cation ($CH_3NH_3^+$), leading to full structural decomposition of the perovskite structure into PbI_2, water, methylamine, and iodine [10]. The second point of instability of perovskite materials arises due to ionic migration [72]. In this process, which is intrinsic to perovskite materials, the perovskite material phase segregates [73] by migrating the constituting ions (e.g., I^-, Pb^{2+}, methylammonium cation, H^+). This also leaves behind vacancies in the perovskite lattice, which is the preferred

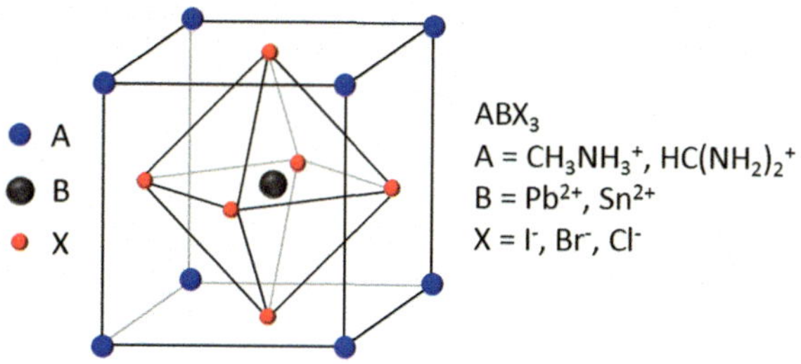

Fig. 9.17 General structure of perovskite as used in active layers of perovskite solar cells. Copyright unit.aist.go.jp

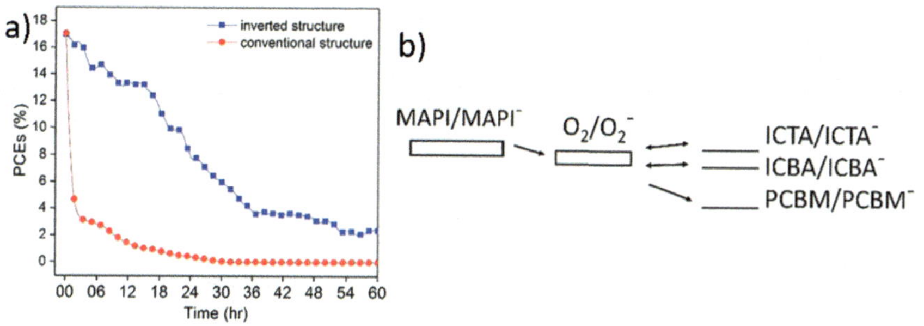

Fig. 9.18 (**a**) Stability of MA PbI_3 solar cells with inverted structure (ITO/PTAA/$CH_3NH_3PbI_3$/PCBM/BCP/Cu) and conventional structure (FTO/compact-TiO_2/meso-TiO_2/$CH_3NH_3PbI_3$/spiro-OMeTAD/Au) under 1 Sun irradiation in dry air. (**b**) Mechanism of electron transfer from superoxide to fullerene derivative acceptors. Republished from Ref. [74], Copyright 2018, with permission from The Royal Society of Chemistry RSC

site for formation of the photoinduced superoxide species [10], thus initiating the decomposition of perovskite structure and failure of solar cell devices.

Possible strategies of overcoming the superoxide-induced degradation pathways include employing (1) capping layers which suppress diffusion of oxygen into the perovskite layer, (2) perovskite-passivating layers which neutralize surface defects/voids which would otherwise present formation sites for superoxide, (3) tuned electron extraction layers to facilitate a faster removal of electrons generated in perovskite layer in order to reduce the rate of superoxide formation, and (4) scavengers which accept electrons from the superoxide [74]. Successful stabilization was reported by a combination of all of the above, by introduction of PCBM on top of perovskite layer, as shown in Fig. 9.18. Besides the ability to inhibit oxygen diffusion into the underlying perovskite layer and to passivate the ion vacancies on the perovskite surface [75], PCBM provides enhanced electron-selective transport (LUMO level is slightly below the conduction band of perovskites), thus inhibiting superoxide generation, along with a HOMO level deeper lying than the valence band of the perovskite layer, thus blocking exciton quenching and hole extraction [74]. This approach can lead to a stronger enhancement in the stabilization effect upon blending PCBM with piperazine, an aromatic amine hydrogen-donating antioxidant, as shown in Fig. 9.19. As reported previously, in case of organic solar cells, the performance and stabilization improvement was also attributed to the hydrogen bond

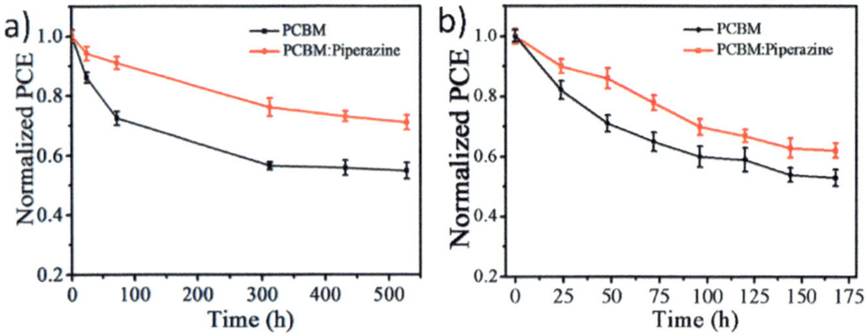

Fig. 9.19 Stability curves of unencapsulated devices without additives and containing 0.5% of piperazine, under storage in (**a**) nitrogen at room temperature, with alternating 10-h dark and 14-h 1000 lux irradiation by fluorescent lamp, (**b**) in air with RH 30–35% humidity. The values are normalized with respect to initial efficiency (PCE). Republished from Ref. [76], Copyright 2020, with permission from The Institute of Electrical and Electronics Engineers IEEE

between N–H of piperazine and O–C=O of butyric acid methyl ester on PCBM [76]. Another type of amine-based stabilizers, hindered amine light stabilizers (HALS), have also been reported to photooxidatively stabilize methylammonium lead iodide perovskite films, as manifested in reduced absorption losses in the visible range under 40 $mWcm^{-2}$ halogen bulb irradiation in air [77]. The antioxidant activity of these amines is based on their ability to form nitroxyl radicals, whose reaction rate with alkyl radicals is only slightly lower than that of alkyl radicals with oxygen. Upon addition of 10% of HALS salt, tetramethylpiperidinyl-4-ammonium iodide, into the MAI precursor, the optical absorption of the resulting films degraded to a lesser extent as compared to reference films without HALS salt (five times lower concentration of formed iodine, see Fig. 9.20). The stabilization mechanism is suggested to include inhibiting the oxidation of I^- to I_2 by deactivation of the intermediate radical species, preventing a cascade reaction between different iodine radicals formed, as well as deactivating superoxide formed during the oxidation, and the subsequent regeneration of the HALS. Another compound that includes stabilization via nitroxyl radicals is TEMPO (1-oxy-2,2,6,6-tetramethylpiperidine). The addition of TEMPO and several polymeric analogues of TEMPO, among them poly(1-oxy-2,2,6,6,-tetramethylpiperidine-4-yl methacrylate) (PTMA), to the perovskite precursor was also tested with respect to suppressing the degradation of organo-lead halide perovskite [78]. PTMA-containing perovskite films (0.1, 0.3, 0.6, 1.0 wt% concentrations tested, of which 0.3 wt% had an optimal effect on the efficiency) demonstrated a significantly reduced formation of superoxide radical anion and methylamine (the degradation intermediates of perovskite), as shown in Fig. 9.21. Besides antioxidant properties, this effect was assigned to an improved charge-transporting capability, improved perovskite film formation, as well as water-repellant properties.

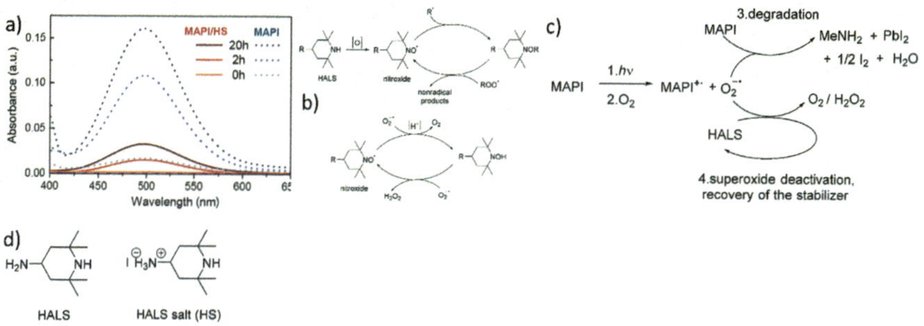

Fig. 9.20 (**a**) Absorption spectra of $MAPbI_3$ films in toluene solutions, without additives and containing 10% of HALS salt 2,2,6,6-tetramethylpiperidinyl-4-ammonium iodide (HS). 500 nm peak belongs to iodine as a degradation product. (**b**) Mechanism of stabilization via HALS—radical and superoxide deactivation. (**c**) Proposed mechanism of MAPI stabilization using HALS: (1) Excitation; (2) oxidation; (3) superoxide-induced degradation; or (4) (in the presence of HALS) deactivation of superoxide and recovery of HALS. (**d**) Structures of tested additives. Republished from Ref. [77], Copyright 2017, with permission from Wiley-VCH

Mechanistically similar to the aromatic amines, hindered phenols are highly efficient antioxidants, and as such they are the most widely used stabilizers for commodity plastics. The key reaction involves facilitating the formation of hydroperoxides by transfer of a hydrogen atom from the phenolic moiety to the peroxy radical, in which—instead of forming a radical in the stabilized compound—a stable phenoxyl radical (due to steric groups in 2- and/or 6-position) is formed. Also in perovskite devices, this type of stabilizers has demonstrated good stabilizing effects. Butylated hydroxytoluene (BHT) was reported to raise the stability of MAPI perovskite devices in several studies. Besides antioxidant capabilities, the ability of BHT's hydroxyl group to form hydrogen bonds with the iodine atoms of perovskite increased the cohesion of perovskite and contributed to slowing down of the crystal nucleation rate, thus forming larger grain size films, raising thus also the efficiency. Additionally, the hydrophobic nature of *tert*-butyl group and the hydrophilic nature of hydroxyl enhance the hydrophobicity of the perovskite layers, which suppresses the entry of oxygen and water in the perovskite layer [79], as shown in Fig. 9.22. Another study reported a similar stabilization effect also on MAPI-based perovskite devices (see Fig. 9.23), ascribing it to BHT-neutralizing reactive oxygen species formed in the intermediate degradation stages, thereby preventing iodine oxidation and deprotonation of ammonium ions, possibly also to inhibition of transformation of unstable Pb^+ into metallic Pb [80]. BHT was, alongside pentaerythritol tetrakis[3-(3,5-di-*tert*-butyl-4-hydroxyphenyl) propionate] (PTP), also tested in four cation-based perovskite devices, where they both demonstrated enhanced stability of devices as shown in Fig. 9.24, owing to a reduction of superoxide radical anion formation [81].

Although lead halide perovskite materials have attracted most attention in the photovoltaic community, due to their remarkably high efficiencies, their toxicity due to the lead content presents an obstacle to their worldwide industrial production and

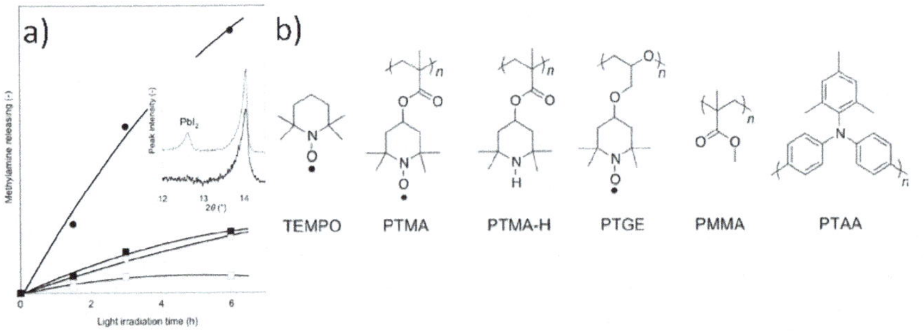

Fig. 9.21 (**a**) Methylamine gas concentration upon photooxidation of $MAPbI_3$ films without additives and containing 0.3% of PTMA, deposited on SnO_2 and m-TiO_2 transport layers, under 1 Sun irradiation in oxygen. Inset: XRD patterns of the $MAPbI_3$ films without additives and containing PTMA deposited on m-TiO_2. (**b**) Structures of tested additives. Republished from Ref. [78], Copyright 2019, with permission from Wiley-VCH

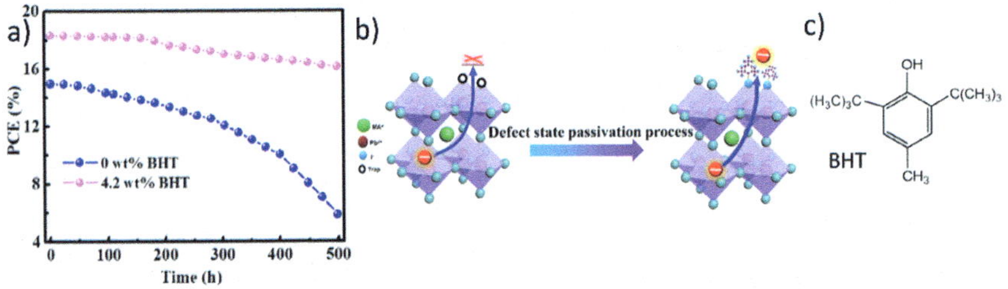

Fig. 9.22 (**a**) Stability of unencapsulated $MAPbI_3$ solar cells without additives and containing 4.2% of BHT, in air at room temperature. (**b**) Passivation of perovskite vacancies with BHT additive, in which superoxide formation is prevented. (**c**) Structure of the tested additive. Republished from Ref. [79], Copyright 2019, with permission from World Scientific Publishing

exploitation [82, 83]. Consequently, environmentally friendlier types of perovskite are being developed, the most promising candidates being the tin perovskites. With their optimal optical bandgap from 1.2 to 1.4 eV, small exciton binding energy (~18 meV), and high charge mobilities (~585 cm^2V^{-1} s^{-1}) [84–91], these materials have a great commercial potential. However, the metastable Sn^{2+} in the perovskite lattice can easily be oxidized to Sn^{4+} defects [92, 93], even in the presence of trace amounts of oxygen, which affects the crystallization of the perovskite and leads to the formation of pinholes and vacancies that act as recombination centers for charge carriers and cause the efficiency of these devices to be much lower than that of the lead halide analogues [85, 94–98]. To prevent this Sn oxidation process, different reducing compounds have been successfully introduced as sacrificial agents that reduce Sn^{4+} defects within the perovskite precursor [92, 95, 99–102]. Addition of excess tin halides, such as SnF_2 [85, 92], $SNCl_2$ [95], and SnI_2 [99], can also prevent Sn^{2+} oxidation by increasing the chemical potential of Sn and the Sn vacancies'

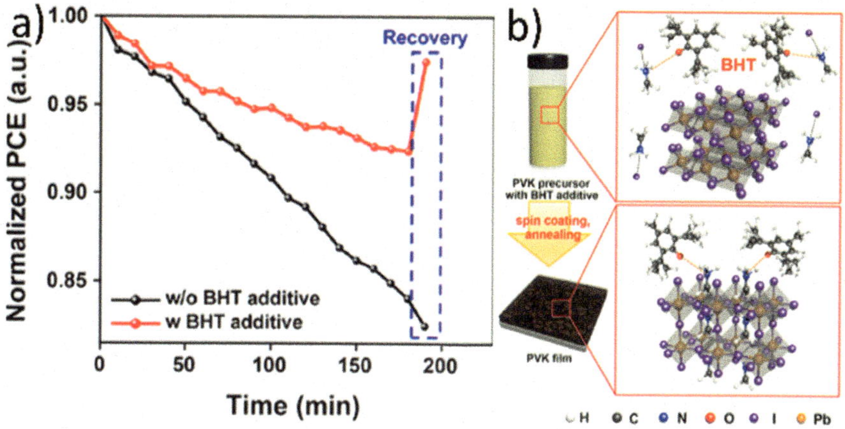

Fig. 9.23 (**a**) Stability of $MAPbI_3$ solar cells without additives and containing 0.02 M of di-BHT, under 1 Sun irradiation in dry air. The 95% recovery of the performance was observed upon a short storage period of 45 min in the dark in case of devices containing di-BHT; no such effect was observed for devices without the additive. The values are normalized with respect to initial efficiency (PCE). (**b**) Chemical interaction between the $MAPbI_3$ and the BHT additive in solution (top) and upon film formation (down). Republished from Ref. [80], Copyright 2019, with permission from The American Chemical Society ACS

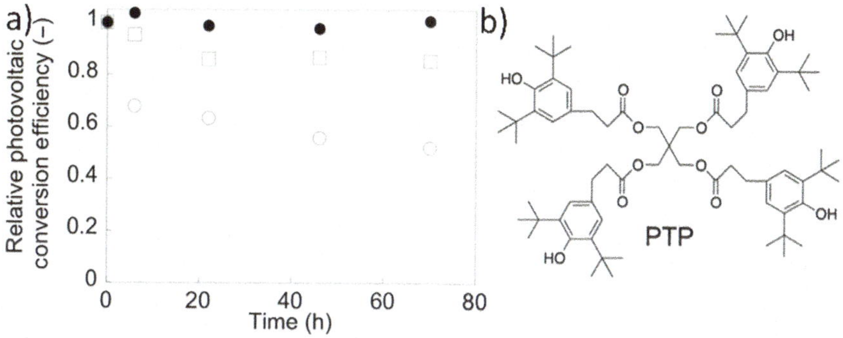

Fig. 9.24 (**a**) Stability of unencapsulated mixed-cation perovskite solar cells without additives and containing 0.1% of BHT and PTP, under 1 Sun irradiation in dry air. The values are normalized with respect to initial efficiency (PCE). (**b**) Structure of the tested additive. Republished from Ref. [81], Copyright 2020, with permission from The Materials Research Society MRS

formation of energy; however due to its inhomogeneous dispersion during the film formation process, it can lead to phase separation [92].

Hydrazine was found to have a stabilizing effect on different types of Sn-based perovskite absorbers, $MASnI_3$, $CsSnI_3$, and $CsSnBr_3$, due to its ability to act as oxygen scavengers, and also reduce metal salts and metal oxides to pure oxides, thereby increasing the performance of the devices [93]. The same group investigated the effect of the addition of piperazine, another diamine compound with

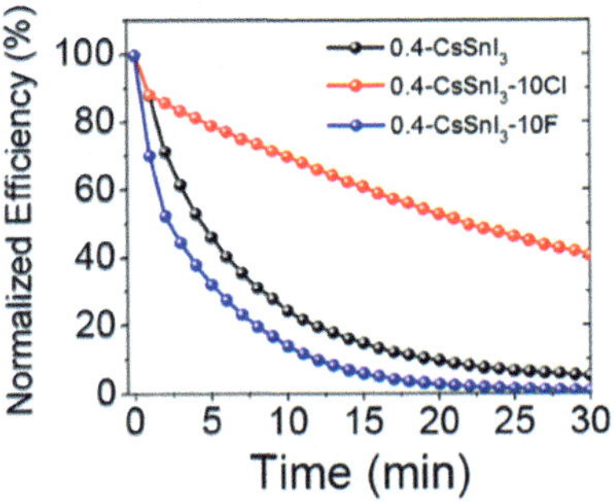

Fig. 9.25 Stability of unencapsulated 0.4-$CsSnI_3$, 0.4-$CsSnI_3$-10Cl, and 0.4-$CsSnI_3$-10F solar cells fabricated using 15 mol% piperazine, under 1 Sun irradiation in air with RH 40–50% humidity at room temperature. The values are normalized with respect to initial efficiency (PCE). Republished from Ref. [103], Copyright 2018, with permission from The American Chemical Society ACS

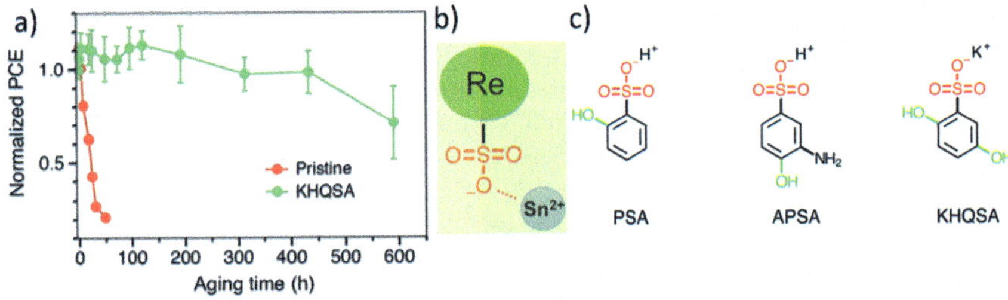

Fig. 9.26 (**a**) Stability of unencapsulated $FASnI_3$ solar cells without additives and containing 1.5% of KHQSA, in dark air with humidity RH 20%. The values are normalized with respect to initial efficiency (PCE). (**b**) Coordination interaction between the sulfonate group of KHQSA and Sn^{2+} ions in the perovskite (Re represents the rest of the KHQSA molecules). (**c**) Structure of the tested additive. Republished from Ref. [104], Copyright 2019, with permission from Wiley-VCH

a similar mechanism [103], to the precursor solution of tin-based halide perovskite films, which yielded an enhancement in device stability, as shown in Fig. 9.25. By comparing the aromatic pyrazine and the aliphatic piperazine, they could identify the diamine group (present only in piperazine) as the main stabilizing feature which enhances interaction with Sn-based halide perovskite, due to a greater electron donor tendency of the aliphatic amines as compared to the aromatic.

Greatly enhanced air stability was reported in a study comparing phenol-sulfonic acid, 2-aminophenol-4-sulfonic acid, and a salt derived from it, potassium salt of hydroquinone sulfonic acid (KHQSA) [104], as stabilizers of tin-based perovskite solar cells. The most stable performance of unencapsulated devices in dark air (RH 20%) was achieved using KHQSA (see Fig. 9.26), where the antioxidative activity is attributed to the existence of two OH groups, which can scavenge oxygen. Furthermore, the sulfonate group can additionally interact in solution with Sn^{2+} via electrostatic and coordination interactions, thus preventing the phase separation

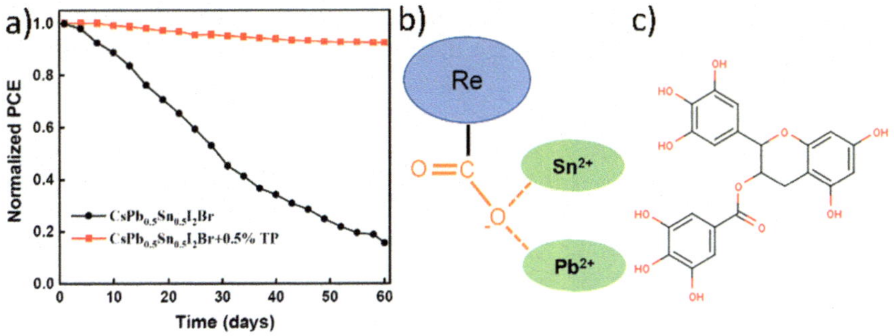

Fig. 9.27 (**a**) Stability of $CsPb_{0.5}Sn_{0.5}I_2Br$ solar cells without additives and containing 0.5% tea polyphenol (TP), under storage in nitrogen. The values are normalized with respect to initial efficiency (PCE). (**b**) Coordination interaction between the ester group of TP and Sn^{2+} and Pb^{2+} ions in the perovskite (Re represents the rest of the TP molecule). (**c**) The main molecular structure of the tested additive. Republished from Ref. [84], Copyright 2020, with permission from Wiley-VCH

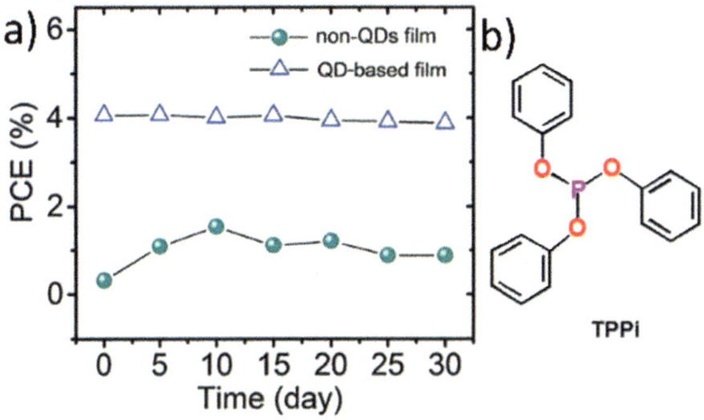

Fig. 9.28 (**a**) Stability of unencapsulated $CsSnI_3$ solar cells made of non-quantum dot (QD) films without additives, and quantum dots (QD) containing 4% of triphenyl phosphite, under 1 Sun irradiation in nitrogen. (**b**) Structure of the tested additive. Republished from Ref. [105], Copyright 2019, with permission from The Royal Society of Chemistry RSC

otherwise induced by $SnCl_2$. Similar mechanism, involving the antioxidant activity of the hydroxybenzene group, is also responsible for the good stabilizing effect of tea polyphenol upon storage of Pb/Sn binary metal halide perovskite solar cells in glove box [84], as shown in Fig. 9.27. Besides the phenols, organophosphorus compounds are also excellent antioxidants, and triphenyl phosphite was reported to increase the stability of $CsSnI_3$-based quantum dots [105]. The presence of antioxidant also raised the efficiency significantly, to over 5%, which was the highest performance of lead-free all-inorganic perovskite solar cells, and resulted in over 6 h of stabile performance upon exposure to 1 Sun irradiation in nitrogen (shown in Fig. 9.28).

Stabilizing additives have since decades been used to improve the durability of insulating plastics, and extend the lifetime of the products made thereof. The translation into the world of organic optoelectronic devices is not straightforward, as it requires additional consideration in order not to perturb the optimized morphology and not to introduce unwanted electronic traps into the active-layer system. Similarly to organic, perovskite devices also benefit from antioxidants, which help prevent the oxidation of constituent ions, prevent formation of superoxides, enhance the hydrophobicity, and at the same time provide smoother films with larger grain sizes, as jointly required for more stable and long-living devices. In this chapter we reviewed recent studies dealing with this problematic and gave an overview of tested additives along with their mechanisms. An overview of the tested compounds classified according to the stabilizing mechanisms is presented in Table 9.1.

Table 9.1 Overview of stabilizing additives used in organic and perovskite solar cells mentioned in this chapter

Stabilizing additive	Publications	Figure
Beta-carotene (BC)	[42, 43]	Figure 9.3, 9.4
Astaxanthin	[43]	Figure 9.4
Piperazine (pip)	[46–49, 51, 76, 103]	Figure 9.5, 9.6, 9.7, 9.8, 9.9, 9.19
N-ethylpiperazine (*N*-EP)	[47]	Figure 9.7
N-aminoethylpiperazine (*N*-AEP)	[47]	Figure 9.7
1,4-Bis(aminoethyl)piperazine (bis-AEP)	[47]	Figure 9.7
1-(2-Aminoethyl)-4-methylpiperazine (M-AEP)	[47]	Figure 9.7
1,4-Bis(3-ethyl)piperazine (Bis-EP)	[47]	Figure 9.7
Triethylenediamine (DABCO)	[47]	Figure 9.7
Polyethyleneimine (PEI)	[48]	Figure 9.7
2,6-Dimethylnaphthalene (DMN)	[51]	Figure 9.9
1,4-Dithiane (DT)	[51]	Figure 9.9
Bis(2,2,6,6-tetramethyl-4-piperidyl)sebacate (BTPS)	[51]	Figure 9.10
4,4′-Biphenol (BPO)	[51, 52]	Figure 9.10, 9.11
2,2,6,6-Tetramethylpiperidinooxy (TEMPO) (S1)	[67, 78]	Figure 9.13, 9.21
2,6-Di-*t*-butyl-4-methylphenol (S2)	[67]	Figure 9.13
Tris(2,4-di-*ter*-butylphenyl)phosphite (S3)	[67]	Figure 9.13
Pentaerythrityl tetrakis[3-(3,5-di-*tert*-butyl-4-hydroxyphenyl)propionate] (S4, PTP)	[67, 81]	Figure 9.13, 9.24
Zinc dibutyldithiocarbamate (S5)	[67]	Figure 9.13
Nickel dibutyldithiocarbamate (S6)	[67]	Figure 9.13, 9.14
Coumarin 7 (C7)	[50]	Figure 9.15
Coumarin 30 (C30)	[50]	Figure 9.15
Benzophenone (BP)	[68]	Figure 9.16
4-Amino-2,2,6,6-tetramethylpiperidine	[77]	Figure 9.20

(continued)

Table 9.1 (continued)

Stabilizing additive	Publications	Figure
2,2,6,6-Tetramethylpiperidinyl-4-ammonium iodide (HS)	[77]	Figure 9.20
Poly(1-oxy-2,2,6,6-tetramethylpiperidine-4-yl methacrylate) (PTMA)	[78]	Figure 9.21
Poly(1-oxy-2,2,6,6-tetramethylpiperidine-4-yl glycidyl ether) (PTGE)	[78]	Figure 9.21
Poly[bis(4-phenyl)(2,4,6-trimethylphenyl)amine] (PTAA)	[78]	Figure 9.21
Butylated hydroxytoluene (BHT)	[79–81]	Figure 9.22, 9.23, 9.24
Phenol-sulfonic acid (PSA)	[102]	Figure 9.26
2-Aminophenol-4-sulfonic acid (APSA)	[102]	Figure 9.26
Potassium salt of hydroquinone sulfonic acid (KHQSA)	[102]	Figure 9.26
Tea polyphenol (TP)	[84]	Figure 9.27
Triphenyl phosphite (TPPi)	[105]	Figure 9.28

References

1. H. Kang et al., Bulk-heterojunction organic solar cells: five core technologies for their commercialization. Adv. Mater. **28**(36), 7821–7861 (2016)
2. Y.S. Rim et al., Recent progress in materials and devices toward printable and flexible sensors. Adv. Mater. **28**(22), 4415–4440 (2016)
3. J. Jensen et al., Development and manufacture of polymer-based electrochromic devices. Adv. Funct. Mater. **25**(14), 2073–2090 (2015)
4. W.A. MacDonald et al., Latest advances in substrates for flexible electronics. J. Soc. Inf. Disp. **15**(12), 1075–1083 (2007)
5. H. Sirringhaus, 25th anniversary article: organic field-effect transistors: the path beyond amorphous silicon. Adv. Mater. **26**(9), 1319–1335 (2014)
6. K.-J. Baeg et al., Organic light detectors: photodiodes and phototransistors. Adv. Mater. **25**(31), 4267–4295 (2013)
7. W. Schnabel, *Polymers and Light* (Wiley-VCH, Weinheim, 2007)
8. M.A. Green et al., The emergence of perovskite solar cells. Nat. Photonics **8**(7), 506–514 (2014)
9. J. Pospisil et al., *Oxidation Inhibition in Organic Materials* (CRC Press, Boca Raton, 1990)
10. N. Aristidou et al., Fast oxygen diffusion and iodide defects mediate oxygen-induced degradation of perovskite solar cells. Nat. Commun. **8**, 15218 (2017)
11. H. Zweifel, *Stabilization of Polymeric Materials* (Springer, Berlin, 1998)
12. V. Turkovic et al., Long-term stabilization of organic solar cells using hindered phenols as additives. ACS Appl. Mater. Interfaces **6**(21), 18525–18537 (2014)
13. V. Turkovic et al., Chapter 12: Inhibiting photo-oxidative degradation in organic solar cells using stabilizing additives, in *Devices from Hybrid and Organic Materials*, ed. by V. Turkovic, M. Madsen, H. G. Rubahn, (World Scientific, Singapore, 2019), pp. 367–421. https://doi.org/10.1142/10993
14. Q. Liu et al., 18% efficiency organic solar cells. Sci. Bull. **65**(4), 272–275 (2020)
15. M. Jorgensen et al., Stability of polymer solar cells. Adv. Mater. **24**(5), 580–612 (2012)
16. M. Jorgensen et al., Stability/degradation of polymer solar cells. Sol. Energy Mater. Sol. Cells **92**(7), 686–714 (2008)

17. P. Cheng et al., Stability of organic solar cells: challenges and strategies. Chem. Soc. Rev. **45**(9), 2544–2582 (2016)
18. W.R. Mateker et al., Progress in understanding degradation mechanisms and improving stability in organic photovoltaics. Adv. Mater. **29**(10), 16 (2017)
19. S. Engmann et al., Direct correlation of the organic solar cell device performance to the in-depth distribution of highly ordered polymer domains in polymer/fullerene films. Adv. Energy Mater. **3**(11), 1463–1472 (2013)
20. M. Shin et al., Abrupt morphology change upon thermal annealing in poly(3-Hexylthiophene)/soluble fullerene blend films for polymer solar cells. Adv. Funct. Mater. **20**(5), 748–754 (2010)
21. V. Turkovic et al., Methods in determination of morphological degradation of polymer:fullerene solar cells. Synth. Met. **161**, 2534–2539 (2012)
22. T. Heumueller et al., Morphological and electrical control of fullerene dimerization determines organic photovoltaic stability. Energy Environ. Sci. **9**(1), 247–256 (2016)
23. J.L. Bolland, Kinetics of olefin oxidation. Q. Rev. Chem. Soc. **3**(1), 1–21 (1949)
24. S. Chambon et al., Photo- and thermal degradation of Mdmo-Ppv:Pcbm blends. Sol. Energy Mater. Sol. Cells **91**(5), 394–398 (2007)
25. M.O. Reese et al., Photoinduced degradation of polymer and polymer-fullerene active layers: experiment and theory. Adv. Funct. Mater. **20**(20), 3476–3483 (2010)
26. J.R. Morton et al., Addition of free radicals to C60. Acc. Chem. Res. **31**(2), 63–69 (1998)
27. A. Kahnt et al., Reaction of water-soluble fullerenes with O2 radical- and other reactive radical species. J. Mater. Chem. **20**(1), 83–89 (2010)
28. J.-J. Yin et al., The scavenging of reactive oxygen species and the potential for cell protection by functionalized fullerene materials. Biomaterials **30**(4), 611–621 (2009)
29. Y. Chen et al., Radical polymerization of styrene in the presence of C-60. J. Polym. Sci. A Polym. Chem. **37**(15), 2969–2975 (1999)
30. A. Distler et al., Effect of Pcbm on the photodegradation kinetics of polymers for organic photovoltaics. Chem. Mater. **24**(22), 4397–4405 (2012)
31. A. Sperlich et al., Reversible and irreversible interactions of poly(3-hexylthiophene) with oxygen studied by spin-sensitive methods. J. Phys. Chem. B **115**(46), 13513–13518 (2011)
32. Y.W. Soon et al., Correlating triplet yield, singlet oxygen generation and photochemical stability in polymer/fullerene blend films. Chem. Commun. **49**(13), 1291–1293 (2013)
33. Y. Yamakoshi et al., Active oxygen species generated from photoexcited fullerene (C-60) as potential medicines: O2- radical versus 1O2. J. Am. Chem. Soc. **125**(42), 12803–12809 (2003)
34. E.T. Hoke et al., The role of electron affinity in determining whether fullerenes catalyze or inhibit photooxidation of polymers for solar cells. Adv. Energy Mater. **2**(11), 1351–1357 (2012)
35. J.W. Arbogast et al., Photophysical properties of C60. J. Phys. Chem. **95**(1), 11–12 (1991)
36. D. Veldman et al., Compositional and electric field dependence of the dissociation of charge transfer excitons in alternating polyfluorene copolymer/fullerene blends. J. Am. Chem. Soc. **130**(24), 7721–7735 (2008)
37. D. Veldman et al., The energy of charge-transfer states in electron donor-acceptor blends: insight into the energy losses in organic solar cells. Adv. Funct. Mater. **19**(12), 1939–1948 (2009)
38. M. Liedtke et al., Triplet exciton generation in bulk-heterojunction solar cells based on endohedral fullerenes. J. Am. Chem. Soc. **133**(23), 9088–9094 (2011)
39. D. Di Nuzzo et al., Improved film morphology reduces charge carrier recombination into the triplet excited state in a small bandgap polymer-fullerene photovoltaic cell. Adv. Mater. **22**(38), 4321–4324 (2010)
40. T.A. Ford et al., Enhanced triplet exciton generation in polyfluorene blends. Phys. Rev. B **71**(12), 125212 (2005)
41. T. Offermans et al., Exciplex dynamics in a blend of pi-conjugated polymers with electron donating and accepting properties: Mdmo-Ppv and Pcnepv. Phys. Rev. B **72**(4), 045213 (2005)

42. V. Turkovic et al., Biomimetic approach to inhibition of photooxidation in organic solar cells using beta-carotene as an additive. ACS Appl. Mater. Interfaces **11**(44), 41570–41579 (2019)
43. M. Bregnhøj et al., Oxygen-dependent photophysics and photochemistry of prototypical compounds for organic photovoltaics: inhibiting degradation initiated by singlet oxygen at a molecular level. Methods Appl. Fluoresc. **8**(1), 014001 (2019)
44. M.O. Reese et al., Consensus stability testing protocols for organic photovoltaic materials and devices. Sol. Energy Mater. Sol. Cells **95**(5), 1253–1267 (2011)
45. Q. Burlingame et al., Photochemical origins of burn-in degradation in small molecular weight organic photovoltaic cells. Energy Environ. Sci. **8**(3), 1005–1010 (2015)
46. L.P. Yan et al., Simultaneous performance and stability improvement of polymer:fullerene solar cells by doping with piperazine. J. Mater. Chem. A **7**(12), 7099–7108 (2019)
47. Z. Li et al., The role of the hydrogen bond between piperazine and fullerene molecules in stabilizing polymer:fullerene solar cell performance. ACS Appl. Mater. Interfaces **12**(13), 15472–15481 (2020)
48. Z.R. Li et al., Organic amines as targeting stabilizer at the polymer/fullerene interface for polymer:Pc61bm solar cells. Energy Technol. **8**(12), 2000266 (2020)
49. L. Yan et al., External load-dependent degradation of P3ht:Pc61bm solar cells: behavior, mechanism, and method of suppression. J. Mater. Chem. A **5**(20), 10010–10020 (2017)
50. X. Kong et al., Hydrogen bond induced high performance ternary fullerene-free organic solar cells with increased current density and enhanced stability. J. Mater. Chem. C **6**(36), 9691–9702 (2018)
51. C.H. Zhang et al., A top-down strategy identifying molecular phase stabilizers to overcome microstructure instabilities in organic solar cells. Energy Environ. Sci. **12**(3), 1078–1087 (2019)
52. P. Cheng et al., Molecular lock: a versatile key to enhance efficiency and stability of organic solar cells. Adv. Mater. **28**(28), 5822–5829 (2016)
53. Z. Zheng et al., A highly efficient non-fullerene organic solar cell with a fill factor over 0.80 enabled by a fine-tuned hole-transporting layer. Adv. Mater. **30**(34), 1801801 (2018)
54. J. Yuan et al., Single-junction organic solar cell with over 15% efficiency using fused-ring acceptor with electron-deficient core. Joule **3**(4), 1140–1151 (2019)
55. Y. Cui et al., Over 16% efficiency organic photovoltaic cells enabled by a chlorinated acceptor with increased open-circuit voltages. Nat. Commun. **10**(1), 2515 (2019)
56. L. Meng et al., Organic and solution-processed tandem solar cells with 17.3% efficiency. Science **361**, 1094–1098 (2018)
57. Y. Lin et al., 17% efficient organic solar cells based on liquid exfoliated Ws2 as a replacement for Pedot:Pss. Adv. Mater. **31**(46), 1902965 (2019)
58. S. Bertho et al., Influence of thermal ageing on the stability of polymer bulk heterojunction solar cells. Sol. Energy Mater. Sol. Cells **91**(5), 385–389 (2007)
59. S. Bertho et al., Effect of temperature on the morphological and photovoltaic stability of bulk heterojunction polymer:fullerene solar cells. Sol. Energy Mater. Sol. Cells **92**(7), 753–760 (2008)
60. N. Gasparini et al., The physics of small molecule acceptors for efficient and stable bulk heterojunction solar cells. Adv. Energy Mater. **8**(12), 1703298 (2018)
61. Y.N. Zhang et al., Thermally stable all-polymer solar cells with high tolerance on blend ratios. Adv. Energy Mater. **8**(18), 1800029 (2018)
62. W.C. Zhao et al., Fullerene-free polymer solar cells with over 11% efficiency and excellent thermal stability. Adv. Mater. **28**(23), 4734–4739 (2016)
63. C. Wang et al., Effects of water vapor and oxygen on non-fullerene small molecule acceptors. J. Mater. Chem. C **7**(4), 879–886 (2019)
64. M.B. Upama et al., Photo-degradation of high efficiency fullerene-free polymer solar cells. Nanoscale **9**(47), 18788–18797 (2017)
65. N.Y. Doumon et al., Photostability of fullerene and non-fullerene polymer solar cells: the role of the acceptor. ACS Appl. Mater. Interfaces **11**(8), 8310–8318 (2019)

66. L. Zhang et al., Blade-cast nonfullerene organic solar cells in air with excellent morphology, efficiency, and stability. Adv. Mater. **30**(22), 1800343 (2018)
67. J. Guo et al., Suppressing photo-oxidation of non-fullerene acceptors and their blends in organic solar cells by exploring material design and employing friendly stabilizers. J. Mater. Chem. A **7**(43), 25088–25101 (2019)
68. M. Qin et al., Enhancing efficiency and stability of organic solar cells by Uv absorbent. Sol. RRL **1**(12), 1700148 (2017)
69. E.M. Speller et al., Toward improved environmental stability of polymer:fullerene and polymer:nonfullerene organic solar cells: a common energetic origin of light- and oxygen-induced degradation. ACS Energy Lett. **4**(4), 846–852 (2019)
70. M. Salvador et al., Suppressing photooxidation of conjugated polymers and their blends with fullerenes through nickel chelates. Energy Environ. Sci. **10**, 2005–2016 (2017)
71. M.A. Green et al., Solar cell efficiency tables (Version 53). Prog. Photovolt. **27**(1), 3–12 (2019)
72. A. Walsh et al., Taking control of ion transport in halide perovskite solar cells. ACS Energy Lett. **3**(8), 1983–1990 (2018)
73. E.T. Hoke et al., Reversible photo-induced trap formation in mixed-halide hybrid perovskites for photovoltaics. Chem. Sci. **6**(1), 613–617 (2015)
74. C.-T. Lin et al., Passivation against oxygen and light induced degradation by the Pcbm electron transport layer in planar perovskite solar cells. Sustain. Energy Fuels **2**(8), 1686–1692 (2018)
75. J. Xu et al., Perovskite-fullerene hybrid materials suppress hysteresis in planar diodes. Nat. Commun. **6**, 7081 (2015)
76. Y. He et al., Enhanced efficiency and stability of inverted planar perovskite solar cells with piperazine as an efficient dopant into Pcbm. IEEE J. Photovoltaics **10**(3), 811–817 (2020)
77. N. Marinova et al., Hindered amine light stabilizers increase the stability of methylammonium lead iodide perovskite against light and oxygen. ChemSusChem **10**(19), 3760–3764 (2017)
78. K. Suwa et al., Anti-oxidizing radical polymer-incorporated perovskite layers and their photovoltaic characteristics in solar cells. ChemSusChem **12**(23), 5207–5212 (2019)
79. L.F. Jin et al., Stable perovskite solar cells with improved hydrophobicity based on bht additive. Nano **14**(2), 1950022 (2019)
80. S. Kumar et al., Multifaceted role of a dibutylhydroxytoluene processing additive in enhancing the efficiency and stability of planar perovskite solar cells. ACS Appl. Mater. Interfaces **11**(42), 38828–38837 (2019)
81. K. Suwa et al., Phenolic antioxidant-incorporated durable perovskite layers and their application for a solar cell. MRS Commun. **10**(2), 312–316 (2020)
82. A. Babayigit et al., Toxicity of organometal halide perovskite solar cells. Nat. Mater. **15**(3), 247–251 (2016)
83. M. Grätzel, The light and shade of perovskite solar cells. Nat. Mater. **13**(9), 838–842 (2014)
84. H. Ban et al., Stabilization of inorganic Cspb0.5sn0.5i2br perovskite compounds by antioxidant tea polyphenol. Sol. RRL **4**(3), 1900457 (2020)
85. M.H. Kumar et al., Lead-free halide perovskite solar cells with high photocurrents realized through vacancy modulation. Adv. Mater. **26**(41), 7122 (2014)
86. N.K. Noel et al., Lead-free organic-inorganic tin halide perovskites for photovoltaic applications. Energy Environ. Sci. **7**(9), 3061–3068 (2014)
87. F. Hao et al., Solvent-mediated crystallization of Ch3nh3sni3 films for heterojunction depleted perovskite solar cells. J. Am. Chem. Soc. **137**(35), 11445–11452 (2015)
88. C.C. Stoumpos et al., Semiconducting tin and Lead iodide perovskites with organic cations: phase transitions, high mobilities, and near-infrared photoluminescent properties. Inorg. Chem. **52**(15), 9019–9038 (2013)
89. Z. Chen et al., Photoluminescence study of polycrystalline Cssni3 thin films: determination of exciton binding energy. J. Lumin. **132**(2), 345–349 (2012)
90. L.-y. Huang et al., Electronic band structure, phonons, and exciton binding energies of halide perovskites Cssncl3, Cssnbr3, and Cssni3. Phys. Rev. B **88**(16), 165203 (2013)

91. I. Chung et al., Cssni3: semiconductor or metal? High electrical conductivity and strong near-infrared photoluminescence from a single material. High hole mobility and phase-transitions. J. Am. Chem. Soc. **134**(20), 8579–8587 (2012)
92. S.J. Lee et al., Fabrication of efficient formamidinium tin iodide perovskite solar cells through Snf2-pyrazine complex. J. Am. Chem. Soc. **138**(12), 3974–3977 (2016)
93. T.-B. Song et al., Importance of reducing vapor atmosphere in the fabrication of tin-based perovskite solar cells. J. Am. Chem. Soc. **139**(2), 836–842 (2017)
94. W. Liao et al., Lead-free inverted planar formamidinium tin triiodide perovskite solar cells achieving power conversion efficiencies up to 6.22%. Adv. Mater. **28**(42), 9333 (2016)
95. K.P. Marshall et al., Enhanced stability and efficiency in hole-transport-layer-free Cssni3 perovskite photovoltaics. Nat. Energy **1**, 16178 (2016)
96. W. Ke et al., Efficient lead-free solar cells based on hollow {En}Masni(3) perovskites. J. Am. Chem. Soc. **139**(41), 14800–14806 (2017)
97. M. Konstantakou et al., A critical review on tin halide perovskite solar cells. J. Mater. Chem. A **5**(23), 11518–11549 (2017)
98. S. Gupta et al., How Snf2 impacts the material properties of lead-free tin perovskites. J. Phys. Chem. C **122**(25), 13926–13936 (2018)
99. T.-B. Song et al., Performance enhancement of lead-free tin-based perovskite solar cells with reducing atmosphere-assisted dispersible additive. ACS Energy Lett. **2**(4), 897–903 (2017)
100. W.Z. Li et al., Addictive-assisted construction of all-inorganic Cssnibr2 mesoscopic perovskite solar cells with superior thermal stability up to 473 K. J. Mater. Chem. A **4**(43), 17104–17110 (2016)
101. M.E. Kayesh et al., Enhanced photovoltaic performance of Fasni3-based perovskite solar cells with hydrazinium chloride coadditive. ACS Energy Lett. **3**(7), 1584–1589 (2018)
102. Q. Tai et al., Antioxidant grain passivation for air-stable tin-based perovskite solar cells. Angew. Chem. Int. Ed. **58**(3), 806–810 (2019)
103. T.B. Song et al., Piperazine suppresses self-doping in Cssni3 perovskite solar cells. ACS Appl. Energy Mater. **1**(8), 4221–4226 (2018)
104. Q. Tai et al., Antioxidant grain passivation for air-stable tin-based perovskite solar cells. Angew Chem Int Ed Engl **58**(3), 806–810 (2019)
105. Y.Y. Wang et al., Convenient preparation of Cssni3 quantum dots, excellent stability, and the highest performance of lead-free inorganic perovskite solar cells so far. J. Mater. Chem. A **7**(13), 7683–7690 (2019)

Part III
Implementation of Photochemical Upconversion in Solar Cells

Chapter 10
Optically Coupled Upconversion Solar Cells

Laszlo Frazer and Timothy W. Schmidt

Abstract Conventional solar cells do not use photons with energy less than the band gap. Upconversion uses sensitizers to enable capture of the remaining light. Since upconversion is a nonlinear process, efficient device design requires detailed knowledge of the coupling of sunlight to the sensitizer. This chapter describes the properties of the principal components of an upconversion-augmented solar cell.

Keywords Upconversion · Fusion · Photovoltaics · Optics · Energy

10.1 The Solar Spectrum

Blackbody Radiation

Solar energy systems start with the capture and storage of sunlight. The sun emits light because it is hot. Owing to the equipartition theorem, hot objects which are coupled to the electromagnetic field emit electromagnetic radiation. A blackbody is an object which is maximally coupled to the electromagnetic field. The spectrum of blackbody spectral irradiance is the Planck law,

$$\frac{8\pi hc^2}{\lambda^5}\frac{1}{e^{\frac{hc}{\lambda k_B T}}-1}, \tag{10.1}$$

L. Frazer
ARC Centre of Excellence in Exciton Science, School of Chemistry, Monash University, Clayton, VIC, Australia

T. W. Schmidt (✉)
ARC Centre of Excellence in Exciton Science, School of Chemistry, UNSW Sydney, Kensington, NSW, Australia
e-mail: timothy.schmidt@unsw.edu.au

J. S. Lissau, M. Madsen (eds.), *Emerging Strategies to Reduce Transmission and Thermalization Losses in Solar Cells*, https://doi.org/10.1007/978-3-030-70358-5_10

where h is Planck's constant, c is the speed of light, k_B is the Boltzmann constant, λ is the wavelength, and T is the temperature of the sun, which is often modeled with 5800 K. For the sun, the spectrum of this radiation is slightly modulated by atomic and molecular spectral lines [1].

Definition 1 Irradiance : Power per area.

Definition 2 Spectral Irradiance: Power per area per wavelength. Spectral irradiance is used to describe the power of light incident on a solar energy device when the spectrum of the light is important.

! Warning

Field installations of solar cells are illuminated fairly uniformly. However, most laboratory light sources do not provide uniform irradiance. Since upconversion is a nonlinear phenomenon, small irregularities in irradiance may cause large experimental errors.

Optical Properties of the Atmosphere

The spectrum of sunlight at ground level varies in both predictable and surprising ways [2]. This presents a challenge for solar cell engineers. The performance of solar energy devices varies with the solar spectrum. It is necessary to have a standard spectral irradiance into order to define the performance of devices. The accepted standard is AM1.5G. It is illustrated in Fig. 10.1. The performance of a device under standard conditions does not predict the performance of the device in a particular field installation.

Definition 3 AM1.5G: The standard solar spectral irradiance. This standard is an appropriate reference for the development of devices. "AM" indicates "Air Mass." "1.5" indicates that the path length of atmosphere's absorption is 1.5 times the mass of the atmosphere above sea level. The air mass is greater than 1 because sunlight usually does not travel perpendicularly to Earth's surface. As an example "AM0" would indicate the spectral irradiance in space near Earth's orbit. "G" indicates global; the irradiance includes scattered light as well as light which has come directly from the sun. The standard irradiance is $1\,\mathrm{kW\,m^{-2}}$.

The Earth's atmosphere contains molecules. The optically active vibrational modes of these molecules, such as the O–H stretching mode of water, absorb sunlight. The overtones of these modes also absorb light. As a result, there are a series of absorption lines in the near-infrared portion of the solar spectrum, as observed from Earth's surface. Sunlight is subject to Rayleigh (elastic) scattering. Rayleigh scattering is stronger at higher photon energies. As a result, blue and UV

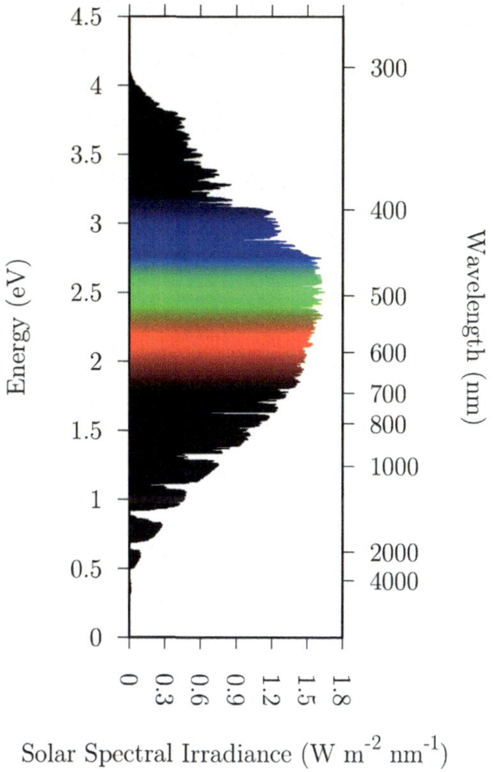

Fig. 10.1 The standard solar spectrum

sunlight are attenuated. Some of the scattered light eventually reaches the ground, but it comes from a different direction than the direct sunlight. Ozone molecules in the atmosphere attenuate some ultraviolet light [3].

As discussed in Chap. 2, the efficiency of upconversion increases with irradiance. Presently, upconversion materials only capture a limited portion of the solar spectrum. If the goal is to achieve the highest upconversion quantum yield, then it is best to capture the green-blue region of the spectrum. This is not usually an option in applications.

For optically coupled upconversion solar cells, the energy efficiency of converting a particular wavelength of light into current by way of upconversion cannot be greater than the efficiency of direct conversion; adding the extra step always results in energy loss. Therefore, as discussed in Chap. 3 and Ref. [4], in practical devices, only the red and infrared regions of the spectral irradiance are used via upconversion. Therefore, the details of AM1.5G in the infrared region are important to efficient device design.

Reflective Concentration

AM1.5G provides 1 kW m^{-2}. To achieve efficient upconversion, greater irradiance can be helpful. Concentration is an increase in irradiance created by a reduction in the area which is illuminated. A large area of mirrors can be used to collect sunlight into a smaller area, resulting in increased illumination. This concept was known in ancient Greece, but claims of its practical application by Archimedes are not supported by evidence [5]. A lens can also be used [6]. Historically, a lens concentrator played an important role in combustion research [7]. Solar concentrators continue to be used as thermal energy sources in industry and cooking.

Definition 4 Concentration: The brightness of sunlight, expressed in units of AM1.5G, which is the same as 1 kW m^{-2} when integrated over the entire spectrum. These units are informally called "suns" or "$\odot$".

! Warning

Illumination with 1 kW m^{-2} of monochromatic light is not equal to illumination with one sun of irradiance. For a given device, not all parts of the solar spectrum are used equally. A device designed to operate efficiently with a monochromatic light source may be inefficient in sunlight. When reporting illumination conditions, it is important to detail how an irradiance measurement was converted to suns.

Reflective solar concentrators are expensive. Lens-based concentrators are even more costly. The Earth rotates relative to the sun. As a result, the location of sunlight reflected off a ground-mounted mirror changes over time. Typically, the illuminated device is built to be the same size as the region of concentrated illumination. In order to utilize reflective concentration on a photovoltaic device, a mechanical tracking system is required to ensure the concentrated light reaches the device[8]. There are also fundamental efficiency limits [9, 10].

Luminescent Concentration

Luminescent concentration[11, 12] achieves an increase in irradiance by a different mechanism from reflective concentration. Where reflective concentration uses ray optics, luminescent concentration uses both ray optics and quantum mechanics.

The typical luminescent concentrator consists of a slab of material. Owing to the index of refraction n of the material, the slab is a waveguide. Light inside the concentrator reaches the surface at an angle θ away from perpendicular. If $\theta > \arcsin(1/n)$, then all of the light is reflected. When this happens repeatedly at both the top and bottom of the slab, the light is guided to the edge of the slab. The aim of a luminescent concentrator is to achieve an irradiance at the edge of the slab which is greater than the irradiance of the sun.

The material of the slab is designed to absorb as much of the solar spectrum as possible. Instead of converting the light to current, the slab material emits light at a different wavelength. Typically the emission is in a random direction.

Luminescent solar concentration runs into two difficulties. First, the random direction of emission often does not meet the requirement $\theta > \arcsin(1/n)$ for total reflection, defeating the waveguiding function. Second, it is challenging to create a material which absorbs sunlight but does not absorb its own emission. Even a low level of absorption can ruin a solar concentrator if the optical path that light takes from the center of a concentrator to the edge is meters long.

The concept of a luminescent solar concentrator is that it can cheaply collect light from a large area and deliver it to a small, expensive device. Thus far, cheap photovoltaics have outcompeted that design. For applications where upconversion is necessary, the usefulness of luminescent solar concentrators may be increased. Since the efficiency of upconversion is nonlinear, luminescent solar concentrators that produce a high irradiance but poor energy efficiency may, in the future, enable efficient upconversion. In addition, luminescent solar concentrators convert the broad spectrum of the sun into a relatively narrow emission spectrum. This greatly simplifies the design of the absorption spectrum of upconversion materials.

10.2 Attenuation of Sunlight by Solar Cells

There are two geometric strategies available for assembling optically coupled upconversion solar cells. The upconversion device can be on the sun side of the solar cell, or it can be on the shadow side of the solar cell. Generally, the shadow side is selected. In addition to absorbing the light that a solar cell cannot capture, upconversion materials are able to absorb some of the light that the solar cell can use. Since the upconversion device does not convert with perfect efficiency, it is best to have the solar cell capture all the light it can before upconversion is used. Therefore, the solar cell is located closer to the sun.

Tauc Model of Semiconductor Absorption

To understand what light is available for upconversion after sunlight has passed through a solar cell, it is helpful to know the absorption spectrum of semiconductors. As described in Chap. 1, semiconductors have a band gap. The Tauc model [13, 14] describes the absorption of semiconductors in the energy range just above the band gap. It can be derived from the Fermi Golden Rule.

The Tauc model predicts that the proportion of spectral irradiance at wavelength λ remaining after propagation distance x is

$$T(\lambda, x) = e^{-\alpha(\lambda)x}, \quad (10.2)$$

where α is defined differently depending on the category of the band gap transition b [14]:

$$\left(\frac{\alpha hc}{\lambda}\right)^b \propto \frac{hc}{\lambda} - E_g \text{where} \begin{cases} b = 2 & \text{direct allowed} \\ b = \frac{2}{3} & \text{direct forbidden} \\ b = \frac{1}{2} & \text{indirect allowed} \\ b = \frac{1}{3} & \text{indirect forbidden} \end{cases}. \quad (10.3)$$

The Tauc model does not account for all semiconductor absorption properties. It assumes that there is only one valence band and one conduction band. Ultimately, there are always additional bands which can absorb higher-energy photons. It does not include absorption caused by exciton states, vibrational (phonon) states, or defects. It is inappropriate for solar cells which are not ordered semiconductors, such as dye solar cells. However, the Tauc model is often adequate.

! Warning

According to the Tauc model, the band gap of a semiconductor cannot be accurately determined by fitting a line to an absorption spectrum.

Experimental Determination

While semiconductor absorption is the mechanism for light capture in solar cells, a variety of other things can reduce the light available for upconversion. The solar cell is partially reflective (Fig. 10.2). There are metal contacts on the solar cell, which may not be homogenous. Solar cells also have protective coatings. The rear of commercial solar cells has a protective backsheet which is opaque [15]. To create upconversion-enhanced solar cells, it is necessary to place the upconversion material between the solar cell and the backsheet.

To accurately determine the light available for upconversion, it is necessary to obtain the associated solar cell and measure its light transmission properties. The angle of incidence of sunlight is variable because the Earth rotates. The reflectivity of the solar cell depends on that angle of incidence. Detailed measurements should account for the angular distributions of sunlight and reflectivity.

To understand the utilization of upconverted light by a solar cell, the absorption spectrum of the solar cell is insufficient. It is necessary to know the external quantum efficiency spectrum of the solar cell under the conditions where it is illuminated by upconversion (typically from the rear). Since the upconversion can be designed to illuminate the solar cell at the wavelength where it is most efficient, the conversion

Fig. 10.2 Demonstration of the reflectivity of a solar cell

of upconverted light to current can be much more efficient than the conversion of sunlight to current.

10.3 The Beer-Lambert Law

Exponential Decay

The propagation of light through an upconversion material can be predicted with the Beer-Lambert law [16]:

$$T(\lambda, x) = 10^{-x(\epsilon_s(\lambda)C_s+\epsilon_e(\lambda)C_e)}. \tag{10.4}$$

T is the proportion of spectral irradiance at wavelength λ remaining after propagation distance x, $\epsilon(\lambda)$ is the molar absorption coefficient, and C is the chemical concentration. The subscript s indicates sensitizer properties, and the subscript e indicates emitter properties.

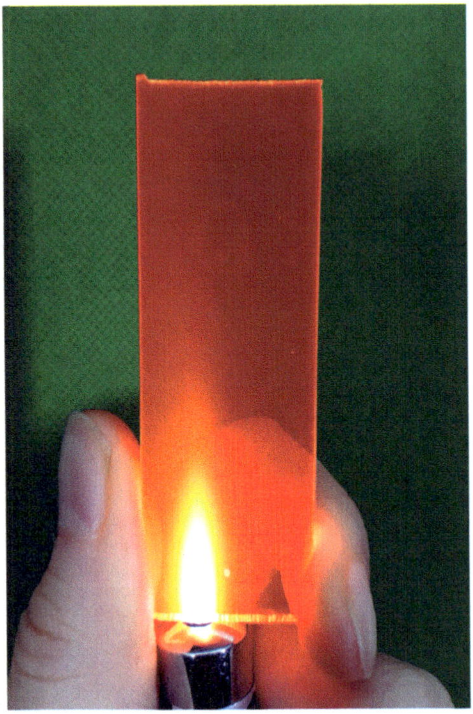

Fig. 10.3 A laser pointer illuminating a uniform, luminescent material. The decay in the luminescence brightness across the material is caused by the Beer-Lambert law

Definition 5 Molar absorption coefficient: $\epsilon(\lambda) = \frac{\log_{10}\left(\frac{I_0}{I_1}\right)}{LC}$ describes the ability of a molecule to capture light. I_0 is the power of the light with wavelength λ illuminating a sample. I_1 is the power transmitted through the sample. L is the length of the path the light takes through the sample. C is the chemical concentration of the molecule in the sample. Molar absorption coefficient is among the most important properties of sensitizers and emitters.

The Beer-Lambert law indicates that light is not absorbed at the surface. Instead, absorption is gradual. The more concentrated the sensitizer, the greater the absorption over a path length. In the Beer-Lambert theory, there is always some light left after passing through a material (Fig. 10.3).

Importance to Efficiency

If the excitation rate of sensitizer is k_ϕ at the illuminated surface, then the excitation rate in the interior is $k_\phi T(x)$. For the moment, we will consider monochromatic illumination as a simplification. As discussed in Chap. 2, the intensity of upconversion is proportional to

$$k_{\phi} T(x) \text{ at high intensities or} \tag{10.5}$$

$$\left(k_{\phi} T(x)\right)^2 \text{ at low intensities.} \tag{10.6}$$

! Warning

The transition between expressions 10.5 and 10.6 is gradual. Unlike a laser, upconversion does not have an abrupt threshold at which it turns on. Demonstration of the linear behavior of Expression 10.5 is necessary but not sufficient to achieve an efficient device.

Since $T(x)$ decays across a device, the intensity of upconversion also decays. In Regime (10.6), the *efficiency of upconversion also decays* [17]. For any upconversion device which obeys the Beer-Lambert law, a portion of the incident light is absorbed in a region where the device is inefficient. In practice, this is a large portion of the light. Our conclusion is that good device designs have elements which do not obey the Beer-Lambert law.

In addition to attenuating the incident light, the Beer-Lambert law also attenuates the upconversion produced by the emitter. In this case, the Beer-Lambert law distance x is not measured from the surface of the device to the point of absorption, but instead is measured from the interior location where the light is emitted to the point where it is absorbed. Unlike sunlight, the upconversion emission propagates in a random direction. The energy captured when emission is absorbed can be recycled as new fluorescence or upconversion, but the recycling is not perfectly efficient. To achieve efficient optical coupling, it is important to select the molar absorption coefficient so that it is small at the wavelength where the emission occurs.

10.4 Scattering

To work around the problems caused by the Beer-Lambert law, the incident light can be scattered. The goal of scattering is to increase the path length traversed by light without increasing the volume of the material. This has the potential to boost efficiency by increasing k_{ϕ} while reducing material cost.

We will assume a device that has planar geometry. The theoretically simplest form of scattering is reflection. Placing a mirror [18] on the opposite side of the upconversion material from the illumination doubles the path length-to-volume ratio.

Random scattering is also an option. For example, Rayleigh or Mie scattering mechanisms distribute light in all directions [19]. Small particles can be placed inside the upconversion material. If these particles have a different index of refraction but no absorption, they will scatter the incident light. A portion of the scattered light will have reduced path length because scattering can occur near the surface and in the direction of the surface. Most of the light will have increased path length.

Lambertian scattering [20] performs better than isotropic random scattering. It is achieved by placing a white material in the shadow of the upconversion device [18, 21]. Since no scattering occurs until all the sensitizer has been traversed, the path length-to-volume ratio of light with a Lambertian scatterer is always better than with no scatterer. If the illumination is normal to the device surface, then the Lambertian scatterer always increases path length compared to a mirror.

Definition 6 Lambertian: Scattered light intensity proportional to the cosine of the angle measured from the surface normal. Also known as "diffuse."

The ideal scattering material will optimize path length-to-volume ratio by scattering light into the plane of the device. The geometry of this scattering would be similar to Wood's anomaly [22]. Creating such an engineered scattering material is difficult because the illumination is broadband and the angle of incidence is variable.

Scattering is also important to the coupling of upconversion emission to a solar cell. The emission is in a random direction. Scattering can redirect upconversion which is headed away from a solar cell so that it heads toward the solar cell. This can increase performance by up to a factor of 2.

Acknowledgments We gratefully acknowledge the support of the ARC Centre of Excellence in Exciton Science (CE170100026). We also thank Anum Nisar for technical assistance.

References

1. J. Fontenla, O.R. White, P.A. Fox, E.H. Avrett, R.L. Kurucz, Astrophys. J. **518**(1), 480 (1999)
2. R.R. Cordero, G. Seckmeyer, A. Damiani, S. Riechelmann, J. Rayas, F. Labbe, D. Laroze, Photochem. Photobiol. Sci. **13**(1), 70 (2014)
3. T.K. Van Heuklon, Solar Energy **22**(1), 63 (1979)
4. M.J. Tayebjee, D.R. McCamey, T.W. Schmidt, J. Phys. Chem. Lett. **6**(12), 2367 (2015)
5. T.W. Africa, Classical World **68**(5), 305 (1975)
6. E. Fennig, G. Schmidt, D.T. Moore, *Optics for Solar Energy* (Optical Society of America, 2017), pp. RW3B–4
7. C.E. Perrin, Br. J. Hist. Sci. **22**(1), 3 (1989)
8. H. Mousazadeh, A. Keyhani, A. Javadi, H. Mobli, K. Abrinia, A. Sharifi, Renew. Sustain. Energy Rev. **13**(8), 1800 (2009)
9. J.M. Gordon, D. Feuermann, Appl. Opt. **44**(12), 2327 (2005)
10. R. Winston, J.M. Gordon, Opt. Lett. **30**(19), 2617 (2005)
11. W.G. Van Sark, K.W. Barnham, L.H. Slooff, A.J. Chatten, A. Büchtemann, A. Meyer, S.J. McCormack, R. Koole, D.J. Farrell, R. Bose, et al., Opt. Express **16**(26), 21773 (2008)
12. A. Hermann, Solar Energy **29**(4), 323 (1982)
13. J. Tauc, Mater. Res. Bull. **3**(1), 37 (1968)
14. B.D. Viezbicke, S. Patel, B.E. Davis, D.P. Birnie III, Physica Status Solidi (b) **252**(8), 1700 (2015)
15. M.T. Zarmai, N. Ekere, C. Oduoza, E.H. Amalu, Appl. Energy **154**, 173 (2015)

16. D. Swinehart, J. Chem. Educ. **39**(7), 333 (1962)
17. L. Frazer, J.K. Gallaher, T. Schmidt, ACS Energy Lett. **2**(6), 1346 (2017)
18. Y.Y. Cheng, B. Fückel, R.W. MacQueen, T. Khoury, R.G. Clady, T.F. Schulze, N. Ekins-Daukes, M.J. Crossley, B. Stannowski, K. Lips, et al., Energy Environ. Sci. **5**(5), 6953 (2012)
19. T. Wriedt, in *The Mie Theory* (Springer, 2012), pp. 53–71
20. C. Battaglia, M. Boccard, F.J. Haug, C. Ballif, J. Appl. Phys. **112**(9), 094504 (2012)
21. T.F. Schulze, Y.Y. Cheng, B. Fückel, R.W. MacQueen, A. Danos, N.J. Davis, M.J. Tayebjee, T. Khoury, R.G. Clady, N. Ekins-Daukes, et al., Aust. J. Chem. **65**(5), 480 (2012)
22. D. Maystre, in *Plasmonics* (Springer, Berlin, 2012), pp. 39–83

Chapter 11
Electronically Coupled TTA-UC Solar Cells

Yan Zhou and Kenneth Hanson

Abstract Optical and electronic coupling architectures are two distinct strategies for harnessing photon upconversion via triplet-triplet annihilation (TTA-UC) in a solar cell. The former combines a standard solar cell with an UC film (Chap. 10), while the latter integrates TTA-UC directly into the solar cell. In this chapter we will review the various strategies for integrating TTA-UC into dye-sensitized and layered heterojunction solar cells. These strategies include heterogeneous sensitization, multilayers, metal-organic frameworks, co-deposition, and more. We describe these architectures, note advantages and disadvantages of each, and summarize progress in the field to date. We also discuss the efficiency-limiting factors of current devices and the prospects for integrated TTA-UC solar cells.

Keywords Upconversion · Triplet-triplet annihilation · Photocurrent · Charge separation · Electronically coupled

11.1 Introduction

Chapter 10 described the efforts in harnessing TTA-UC for solar energy conversion using an optical coupling strategy. In an optically coupled scheme, an UC solution or film is used in conjunction with a standard solar cell (Fig. 11.1a). Sub-bandgap, low-energy light that is transmitted through the solar cell is absorbed by the sensitizer in the UC solution/film. Following energy transfer to the annihilator (a.k.a. acceptor or emitter) and TTA, a higher energy photon is emitted and the upconverted light can then be harnessed by the solar cell.

An alternative strategy to harness TTA-UC is with an electronically coupled scheme (a.k.a. an integrated TTA-UC solar cell) where, rather than relying on the generation of upconverted photons, these devices charge separate the upconverted

Y. Zhou · K. Hanson (✉)
Department of Chemistry and Biochemistry, Florida State University, Tallahassee, FL, USA
e-mail: yz15b@my.fsu.edu; hanson@chem.fsu.edu

J. S. Lissau, M. Madsen (eds.), *Emerging Strategies to Reduce Transmission and Thermalization Losses in Solar Cells*, https://doi.org/10.1007/978-3-030-70358-5_11

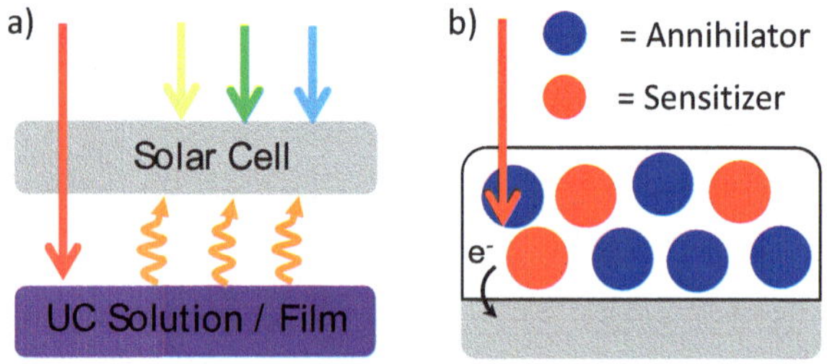

Fig. 11.1 (**a**) Optically and (**b**) electronically coupled TTA-UC solar cell schemes

state prior to emission (Fig. 11.1b). In this scheme, TTA-UC is incorporated directly into the solar cell and is intrinsic to the photocurrent generation mechanism. Despite being a recent development (ca. 2015), the number of reports and architectures for directly harnessing TTA-UC in an electronically coupled scheme has greatly expanded over the past several years. Herein we summarize the efforts focused on integrating TTA-UC into dye-sensitized (Sect. 11.2) and layered heterojunction (Sect. 11.3) solar cells.

11.2 Dye-Sensitized TTA-UC Solar Cells

Introduction

Charge separation and photocurrent generation using dye-metal oxide interfaces have been known for over 50 years [1] but interest dramatically increased in 1991 with the first reported use of high-surface-area, mesoporous metal oxides by Gratzel and coworkers [2]. The latter report laid the foundation to modern dye-sensitized solar cells (DSSCs), and more recently also served as an effective scaffolding for facilitating and directly harnessing TTA-UC in a solar cell. Given that a majority of integrated TTA-UC solar cells rely on a dye-metal oxide interface, it is worth briefly introducing the basic architecture and working mechanism of DSSCs as depicted in Fig. 11.2 [3–5].

A typical DSSC has a catalyst-coated, conductive glass anode and a cathode composed of a transparent conducting substrate (e.g., FTO or ITO glass) coated with nanocrystalline semiconducting metal oxide (usually TiO_2) that is functionalized with a monolayer of dye(s). Closing the circuit between the two electrodes is a solid state or solution electrolyte containing a redox mediator (Fig. 11.2a) [6]. The relative energetics and dynamics of a typical DSSC under operational conditions are depicted in Fig. 11.2b with productive processes in black and nonproductive in red. After excitation (k_{ex}), the dye molecule has sufficient excited-state potential to inject an electron (k_{inj}) into the conduction band (CB) of the semiconductor [7]. The injected electron can then enter the external circuit to do useful work before moving to the cathode to reduce the mediator in the electrolyte solution (k_{red}). Concurrently, reduced mediator in the electrolyte transfers an electron to the previously oxidized

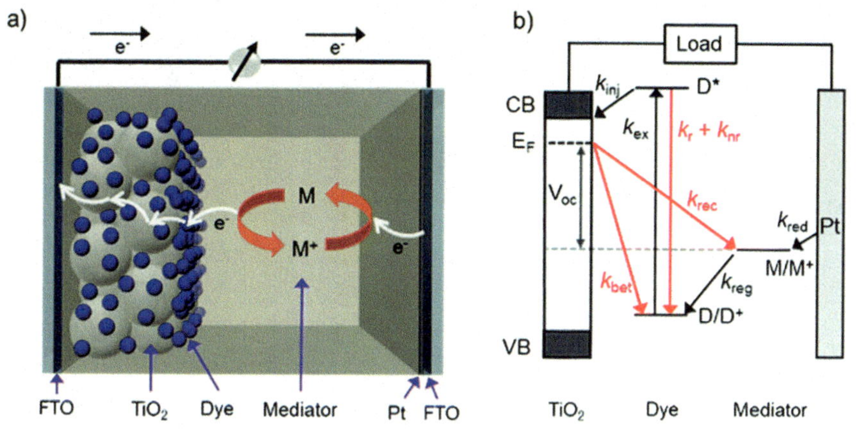

Fig. 11.2 The (**a**) components and (**b**) relative energetics and dynamic events for a standard DSSC. Advantageous and loss events are depicted with black and red arrows, respectively (E_F = Fermi level, k_{ex} = excitation, k_{inj} = electron injection, k_r = radiative decay, k_{nr} = non-radiative decay, k_{BET} = back electron transfer, k_{reg} = dye regeneration, k_{rec} = recombination, and k_{red} = mediator reduction rate constants). Reprinted from [6]. Copyright 2018 Royal Society of Chemistry

dye effectively regenerating the dye and closing the circuit (k_{reg}) [8–10]. In addition to the productive processes, a number of nonproductive loss pathways can occur including radiative (k_r) decay, non-radiative (k_{nr}) decay, and back electron transfer from the CB to the oxidized dye (k_{bet}) and/or mediator (k_{rec}) [8, 10].

For dye-sensitized TTA-UC solar cells, the general device structure and mechanism are similar to a conventional DSSCs but there are several additional steps (e.g., triplet energy transfer, TTA) as well as new nonproductive loss pathways (e.g., back energy transfer, O_2, or mediator quenching). Also, whereas DSSCs typically rely on one dye, TTA-UC solar cells require both sensitizer and annihilator molecules. Below we will recount the various strategies for combining sensitizers and annihilators in TTA-UC DSSCs (Fig. 11.3) including heterogeneous sensitization (section "Heterogeneous Sensitization"), metal ion-linked multilayers (section "Metal Ion-Linked Multilayers"), surface-supported metal-organic frameworks (section "Surface-Supported Metal-Organic Frameworks"), co-deposition (section "Co-deposition"), and organic-inorganic hybrid multilayer (section "Organic-Inorganic Hybrid Multilayer") architectures.

TTA-UC at Dye-Metal Oxide Interfaces

In 2011, Morandeira, Lissau, and coworkers were the first to demonstrate that metal oxide surfaces were an effective substrate for combining sensitizer and annihilator molecules and facilitating TTA-UC emission [11]. In their seminal

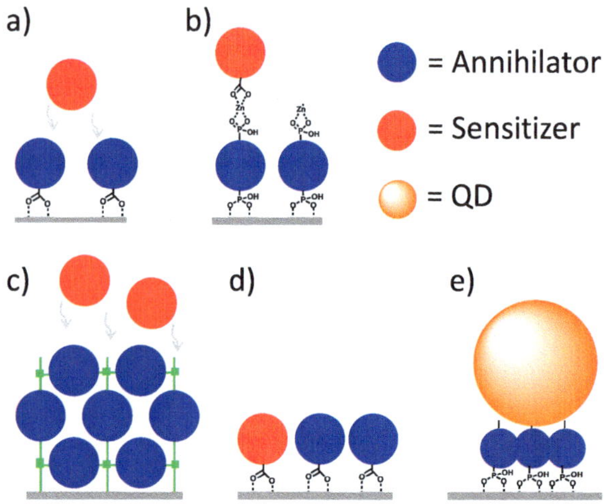

Fig. 11.3 Schematic depiction of (**a**) heterogeneous sensitization, (**b**) metal ion-linked multilayers, (**c**) surface-supported metal-organic frameworks, (**d**) co-deposition, and (**e**) organic-inorganic hybrid multilayer architectures for facilitating TTA-UC on a metal oxide substrate

report, they physiosorbed platinum(II) octaethylporphyrin sensitizer and 9,10-diphenylanthracene on a mesoporous ZrO_2 substrate. Excitation of the porphyrin was followed by cross-surface sensitizer to annihilator energy transfer, annihilator-to-annihilator energy transfer, and finally TTA-UC emission. Intensity dependence measurements showed quadratic dependence that is symptomatic of TTA-UC [12–15] at intensities as low as 5 mW/cm^2. While a definitive proof of concept, this scheme suffered from low surface coverages and limited stability of the physiosorbed UC pairs on the surface.

In a second iteration, Lissau et al. introduced a heterogeneous TTA-UC scheme where the annihilator molecule was anchored to the surface via carboxylate binding groups and the sensitizer was dissolved in the external solution (Fig. 11.3a) [16, 17]. This strategy increased TTA-UC emission efficiencies by several orders of magnitude (0.04%) and a quadratic to linear intensity dependence was observed with a strong annihilation limit onset of ~8 mW/cm^2. Chemically linking the annihilator molecule to the surface increased the stability and surface coverage. Unfortunately, non-radiative sensitizer decay and/or nonproductive sensitizer-sensitizer TTA were competitive with diffusion-limited sensitizer to annihilator triplet energy transfer (TET) [17].

The third scheme introduced by Morandeira and coworkers was to functionalize both the annihilator and sensitizer molecules with carboxylic acid anchor groups and chemically link both to the ZrO_2 surface (Fig. 11.3d) [18]. The proximity of the sensitizer and annihilator molecules increased TET yields but at the cost of lower annihilator loading and presumably reduced cross-surface triplet percolation as well as concentration-dependent sensitizer-sensitizer aggregation resulting in a $2 \times 10^{-3}\%$ TTA-UC emission quantum yield. Nonetheless, the films exhibited remarkable stability with TTA-UC emission observed even 1 year after sample preparation.

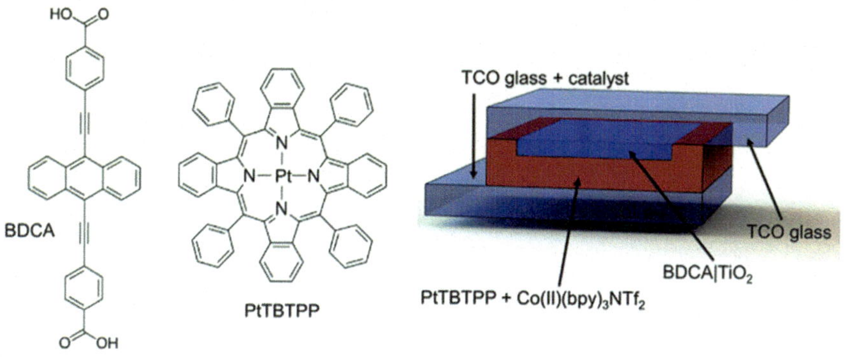

Fig. 11.4 Molecular structures for BDCA and PtTBTPP and device architecture of the heterogenous TTA-UC DSSC. Reprinted with modification from [20]. Copyright 2015 Royal Society of Chemistry

In the above examples, ZrO_2 was used as a photo- and electrochemically inert substrate where excited-state dynamics and TTA-UC emission could be observed without the concern of electron transfer quenching. To facilitate charge separation of the upconverted state, attention was then shifted towards binding sensitizer and/or annihilators to mesoporous TiO_2 whose CB is ~1 eV lower than that of ZrO_2 [19]. The lower CB of TiO_2 enables excited-state electron transfer from the upconverted state of the annihilator molecule as the first step towards photocurrent generation. All dye-sensitized TTA-UC solar cells to date rely on TiO_2 substrate but the key difference is how the sensitizer and annihilator molecules are combined at the interface (Fig. 11.3) including heterogeneous, metal ion-linked multilayer metal-organic framework, co-deposition, and organic-inorganic hybrid multilayer architectures.

Heterogeneous Sensitization

In 2015, Nattestad and coworkers published the first example of an integrated TTA-UC solar cell [20]. Their device architecture, depicted in Fig. 11.4, utilized the heterogenous sensitization scheme (Fig. 11.3a) with a Pt cathode, a photoanode consisting of a monolayer of 4,4′-(anthracene-9,10-diylbis(ethyne-2,1-diyl)) dibenzoic acid (BDCA) on mesoporous TiO_2, and an electrolyte solution containing platinum(II) tetrabenzotetraphenylporphyrin (PtTBTPP) sensitizer and $Co^{II/III}$tris-bipyridine ($Co(bpy)_3{}^{3+/2+}$) as the redox mediator.

Under selective excitation of the sensitizer with 560–640 nm irradiation at 1 Sun intensities, the device exhibited a short-circuit photocurrent density (J_{sc}) of 0.1 $\mu A\ cm^{-2}$ that was not observed in the absence of the sensitizer molecule. Direct excitation of the BDCA annihilator-only device showed a linear J_{sc} response with

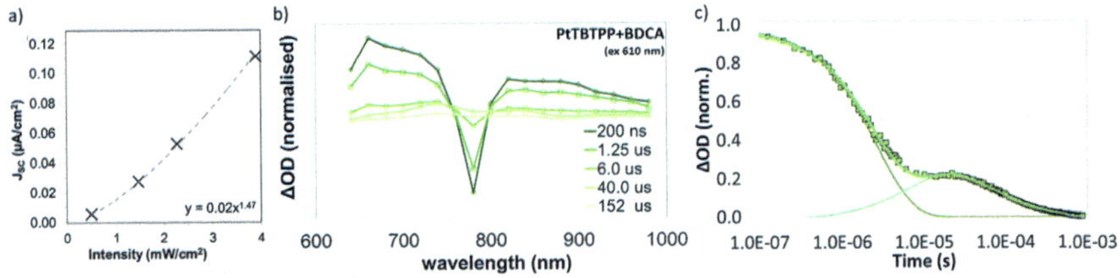

Fig. 11.5 (**a**) Superlinear current density response of the heterogenous sensitized TTA-UC solar cell ($\lambda_{ex} = 560$–640 nm). (**b**) Transient absorption spectrum of TiO_2-BDCA+PtTPTBP ($\lambda_{ex} = 610$ nm). (**c**) Normalized ΔOD for TiO_2-BDCA + PtTPTBP at 720 nm. Reprinted from [20]. Copyright 2015 Royal Society of Chemistry

respect to excitation intensity that was consistent with a one-photon absorption and one-electron injection mechanism. However, upon PtTBTPP excitation, the photocurrent response was superlinear (slope = 1.47) suggesting a nonstandard, multiphoton photocurrent generation process (Fig. 11.5a).

To gain further insights into the mechanism, the authors performed transient absorption measurements on the photoanodes and their results are shown in Fig. 11.5b, c. After selectively exciting PtTBTPP, early spectra closely resembled the triplet excited state of the sensitizer molecule that was then followed by the formation of spectral features of the BDCA cation.

Based on their results, the proposed photocurrent generation mechanism was (1) excitation of PtTBTPP followed by rapid intersystem crossing to the triplet excited state, (2) PtTBTPP diffusion to the metal oxide surface, (3) PtTBTPP-to-BDCA triplet energy transfer, (4) cross-surface triplet energy migration, (5) TTA, (6) electron transfer from the singlet excited state of BDCA to TiO_2, and finally (7) regeneration of the BDCA cation by the $Co(bpy)_3{}^{2+}$ redox mediator. Although the generated photocurrent was relatively low (0.1 μA cm^{-2}) compared to traditional DSSCs (>10 mA cm^{-2}) [3], this was the first example of directly extracting charge and generating photocurrent from an upconverted state.

Metal Ion-Linked Multilayers

An alternative scaffolding for facilitating TTA-UC is with metal ion-linked multilayers (Fig. 11.3b) that are prepared through a stepwise soaking procedure resulting in alternating layers of molecules with metal ion-binding groups (e.g., COOH, PO_3H_2) and metal ions (typically Zn^{II} or Zr^{IV}) on an inorganic surface. These multilayer films were first introduced in 1988 by Mallouk and coworkers using planar substrates [21, 22]. The molecule and substrate scope were subsequently expanded by the work of Katz [23, 24], Haga [25, 26], and others [27, 28]. In 2012 Hanson et al. published the first example of multilayer assembly on a mesoporous, nanocrystalline metal oxide films [29]. The high surface area and

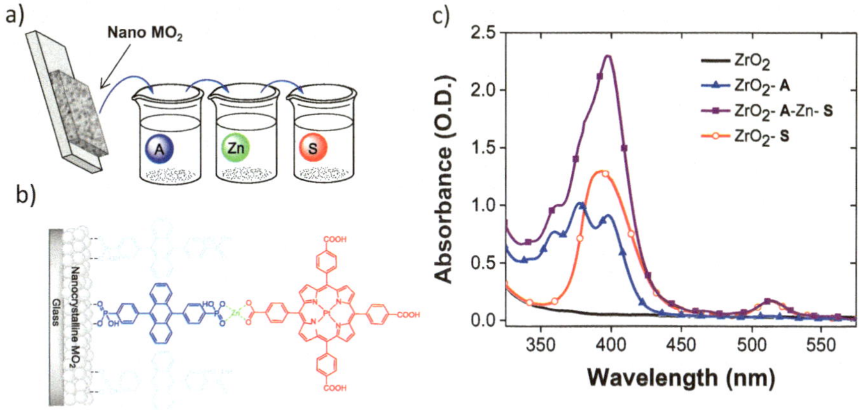

Fig. 11.6 Preparation (**a**) and structure (**b**) of self-assembled bilayer on a nanocrystalline metal oxide substrate. (**c**) UV–vis absorption of the annihilator (blue), bilayer (purple), and sensitizer (red) films on ZrO_2 (black). Adopted in part with permission from [35]. Copyright 2015 American Chemical Society

micrometer thickness of these films significantly increase their overall absorption cross section and were quickly incorporated into solar energy conversion devices [30–33]. Subsequently, metal ion-linked multilayers proved to be an effective strategy to assemble sensitizer and annihilator molecules on a metal oxide surface and facilitate both TTA-UC and charge separation of the upconverted state [34]. In this section we will recount the progress using metal ion-linked multilayers for integrated TTA-UC solar cells including the prototype, molecular structure-performance relationships, and moving beyond sensitizer-annihilator bilayers.

The Metal Ion-Linked Multilayer Prototype

The first TTA-UC multilayer was reported in 2015 and was composed of 4,4′-(anthracene-9,10-diyl)bis(4,1-phenylene)phosphonic acid (**A**) and Pt(II)meso-tetra(4-carboxyphenyl)porphine (**S**) as annihilators and sensitizers (Fig. 11.6b), respectively [35], which are derivatives of the well-studied DPA and PtOEP UC pair [36, 37]. The metal ion-linked multilayer was prepared by soaking the mesoporous metal oxide (ZrO_2 or TiO_2) in the solution of **A**, then Zn^{2+}-linking ion, and finally **S** (Fig. 11.6a). As shown in Fig. 11.6c, the absorption of the bilayer (ZrO_2-**A**-Zn-**S**) is effectively the sum of the two monolayers. In the absence of Zn^{2+} treatment, nominal absorption from **S** was observed, indicating that there was no co-deposition of **A** and **S** but instead they generated a summative, bilayer structure.

Upon the excitation with 532 nm light, upconverted blue emission (380–520 nm) was observed from ZrO_2-**A**-Zn-**S** with a quantum yield of 0.0023 ± 0.0003. The emission intensity exhibited the symptomatic quadratic (slope = 2) to linear (slope = 1) dependence on incident light intensity, lending further support for the TTA-UC mechanism [12–15].

With the same bilayer but on TiO_2, the authors then generated a sandwich, electrochemical cell composed of TiO_2-**A**-Zn-**S** working electrode, platinum counter electrode, and $TBAClO_4$ in MeCN as electrolyte. A transient photocurrent peak at 0.31 mA/cm^2 was detected, under 2 Sun irradiation (2 × AM 1.5) passed through a 495 nm long-pass filter, to selectively excite the Q-band of **S**. Despite having the same total absorption, this photocurrent was three times the sum of its parts (TiO_2-A = 0.008 mA/cm^2 and TiO_2-PtP = 0.1 mA/cm^2). It also showed quadratic-to-linear-like photocurrent to excitation intensity relationship, indicating that the self-assembled bilayer film is an effective strategy to facilitate both TTA-UC and the subsequent charge separation of the upconverted state. However, without a redox mediator, only transient photocurrent was observed.

To realize sustained photocurrent and a fully functioned solar cell, a $Co(bpy)_3^{3+/2+}$ redox mediator was then introduced to the MeCN electrolyte solution [38]. The electronic transitions and energetics in the device are summarized in Fig. 11.7a. Under 2 Sun (2 × AM 1.5) irradiation, with a 495 nm long-pass filter, TiO_2-**A**-Zn-**S** exhibits a fivefold increase in sustained photocurrent density (J_{sc} = 0.028 ± 0.002 mA/cm^2) relative to that of TiO_2-**S** (0.005 ± 0.002 mA/cm^2) and TiO_2-**A** (0.002 ± 0.002 mA/cm^2) devices (Fig. 11.7b). Again, support for a TTA-UC photocurrent generation mechanism was provided by the photocurrent dependence on excitation intensity (Fig. 11.7c). Whereas the **S**-only devices, and direct excitation of **A** in the bilayer, showed a linear dependence that is expected for a single-photon excitation-electron injection process [39], excitation of **S** in the bilayer device resulted in quadratic (slope = 2) to linear (slope = 1) intensity-dependent behavior that is indicative of a TTA-UC mechanism. Collectively, these results demonstrated that the bilayer films can effectively generate photocurrent by two different mechanisms: (1) direct excitation and electron injection from **A** and

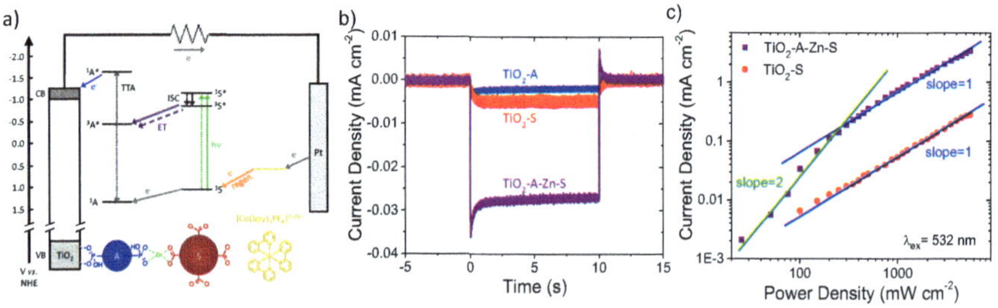

Fig. 11.7 (**a**) Electronic transitions and energetics for TiO_2, **A**, **S**, and $[Co(bpy)_3]^{3+/2+}$ (vs. NHE) in the TTA-UC solar cell. (**b**) Amperometric *i*–*t* curves (V = 0 V; light on at 0 s, light off at 10 s) for solar cells containing TiO_2-**A**, TiO_2-**S**, and TiO_2-**A**-Zn-**S** photoanodes under two equivalent AM1.5 solar irradiation. (**c**) Photocurrent density from TiO_2-**A**-Zn-**S** and TiO_2-**S** devices with respect to 532 nm excitation intensity. Reproduced in part from [38]. Copyright 2016 American Chemical Society

(2) low-energy light absorption by **S** followed by **S**-to-**A** triplet energy transfer, TTA, and electron injection from the upconverted state on **A**.

Molecular Structure-Performance Relationships

Expanding on the prototypical metal ion-linked bilayer devices, subsequent work was to explore the role of the structure of the annihilator molecule and redox mediator on TTA-UC and photocurrent generation. In TTA-UC solutions, molecules are free to rotate and collide in random orientations. However, the geometrically fixed nature of the multilayer assembly means that the molecular structure can have an increasingly important influence on energy and electron transfer events. To probe the role of the structure, Zhou et al. generated a series of anthracene annihilator molecules where the chromophore core was functionalized with 9,10- or 2,6 phenyl, methyl or directly bound phosphonic acid function groups (**A** and **A2–4** in Fig. 11.8) [40, 41].

Annihilator molecules **A** and **A2–5** readily formed monolayer and bilayers on ZrO_2 with center-to-center distances on the order of 10 Å. Of all the annihilators, the **A3** bilayer exhibited the fastest and most efficient sensitizer to annihilator triplet energy transfer (60%) presumably due, at least in part, to the lack of phenyl/methyl groups and shorter distance in between the anthracene core and the phosphonate metal ion-binding group. Interestingly, despite having three orders of magnitude lower overlap integral ($J = 2.4 \times 10^{-8}$ cm^3 M^{-1}), relative to **A** and **A2–4**, **A5** exhibited the fastest and most efficient annihilator-to-sensitizer back energy transfer. Presumably **A5** and **S** in the ZrO_2-**A5**-Zn-**S** bilayer have a significantly more favorable orientation for this back energy transfer event.

Upon excitation with 532 nm light, all five bilayer films (ZrO_2-**Ax**-Zn-**S**) facilitated green-to-blue upconverted emission, with a quantum yield as high as 0.0023 for the parent **A**, followed by the order of **A2** > **A4** > **A3** > **A5**. Quadratic to linear intensity dependence was observed from which the cross-surface triplet

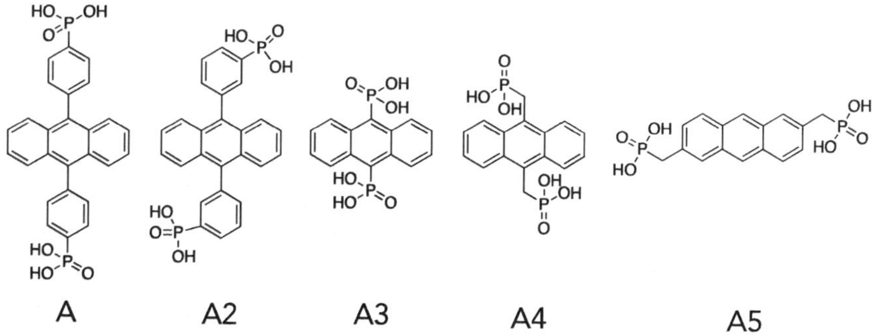

Fig. 11.8 Structures of annihilator molecules **A** and **A2–5**. Reproduced in part from [40]. Copyright 2018 Royal Society of Chemistry

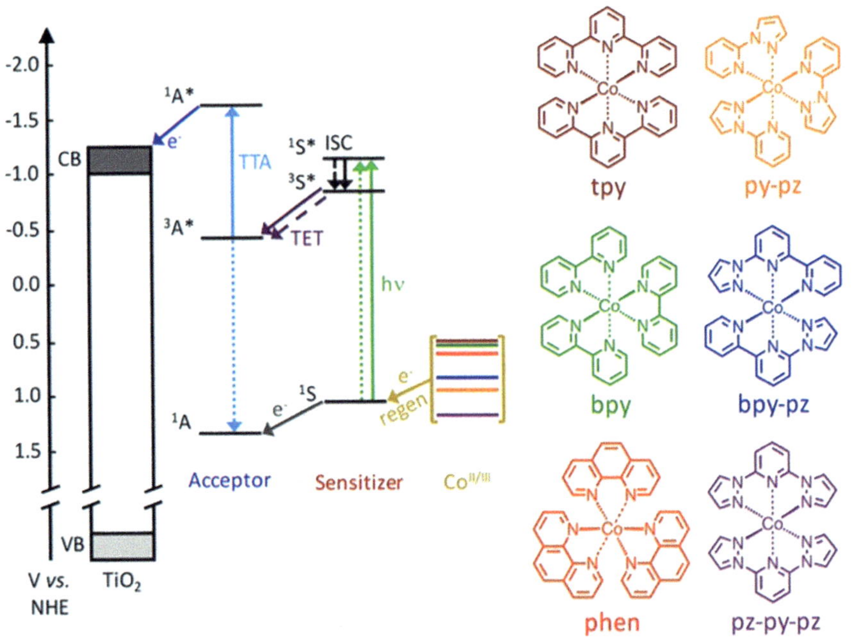

Fig. 11.9 Electronic transitions and energetics for TiO_2, **A**, **S**, and mediators (vs. NHE) alongside the structure of the $Co^{III/II}$ redox mediators. Reprinted from [43]. Copyright 2018 American Chemical Society

exciton diffusion constant (D) was calculated [13, 42]. The origin of the two orders of magnitude higher D (2×10^{-7} cm^2s^{-1}) for **A5** compared to the other annihilators is still unclear, but the authors suggested that lower steric hindrance and/or forced relative orientation may play a role.

In addition to the bilayer, the nature of the redox mediator can strongly influence the TTA-UC solar cell performance. To explore their influence, Hill et al. incorporated a series of cobalt polypyridine-based redox shuttles (Fig. 11.9) into the TiO_2-**A**-Zn-**S** DSSC [43]. The six redox mediators have similar size, to mitigate the potential difference in diffusion rate, but a range of redox potentials from 0.50 to 1.20 V vs. NHE [44, 45].

It was found that variation in the redox mediator had a dramatic effect on the J_{sc} and maximum TTA efficiency onset threshold (I_{th} value). The highest J_{sc} due to TTA-UC under 1 Sun was 0.158 mA/cm^2 using **phen**. In contrast the I_{th} value decreased with increasing oxidation potential with the lowest being 0.8 mW/cm^2 for **pz-py-pz** which is well below ambient solar intensities and on par with the lowest I_{th} values reported to date. Time-resolved spectroscopic measurements indicate that the wide range of J_{sc} and I_{th} values was due to a balance between k_{regen} and quenching of dyes' triplet excited state by Co^{III} and Co^{II} mediators via electron transfer and energy transfer/paramagnetic quenching, respectively. The results indicated that careful selection of redox mediator plays a pivotal role in the efficiency of the device

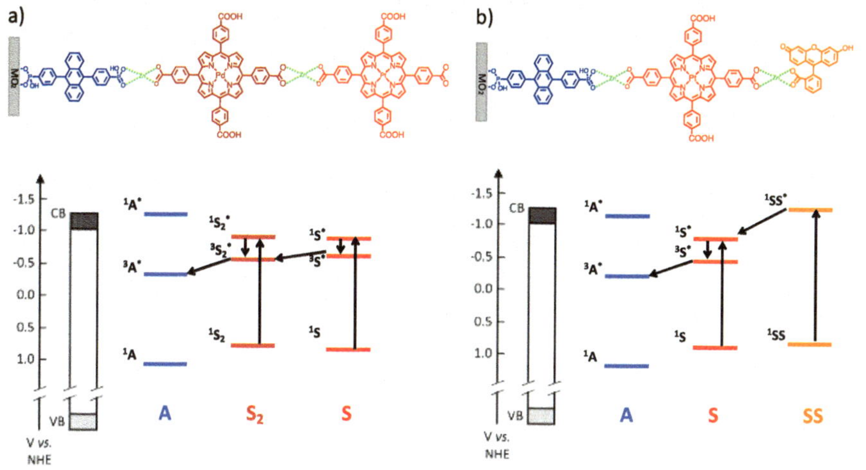

Fig. 11.10 (**a**) Dual-triplet sensitizer and (**b**) singlet sensitization-enhanced TTA-UC self-assembled trilayers on nanocrystalline metal oxide substrates, with energetics for photoanode and selected electronic transitions

with the goal being to inhibit excited-state quenching while still maximizing the dye regeneration.

Beyond Bilayers

In any solar energy conversion scheme, intense broadband absorption is necessary to maximize light-harvesting efficiency. Unfortunately, unlike inorganic crystalline materials, molecular sensitizers typically exhibit relatively narrow absorption features. In DSSCs, co-deposition of two chromophores with complementary absorption spectra is one strategy to broaden absorption but due to the limited surface area at a given film thickness, it decreases the total absorbance of each chromophore. Fortunately, with the appropriately placed metal ion-binding groups, the second dye in the metal ion-linked bilayer can also serve as the support for a third molecular layer, giving all the advantages of complementary absorption while still retaining full monolayer coverage of each dye [33]. For TTA-UC, this increased light absorption can help lower the light intensity necessary to reach the strong annihilation limit and increase overall, external TTA-UC efficiency. To date there have been two reports of three-component, metal ion-linked trilayers for TTA-UC and the molecular structure and energetic scheme for each are depicted in Fig. 11.10. The post-sensitizer-to-annihilator energy transfer events are omitted for clarity.

The first example of a TTA-UC trilayer was to incorporate a second sensitizer Pd(II) meso-tetra(4-carboxyphenyl)porphine (**S**$_2$ in Fig. 11.10a) [46] into the prototype bilayer generating a TiO_2-**A**-Zn-**S**$_2$-Zn-**S** film. The two sensitizers were selected for their complementary absorption features and organized such that it was

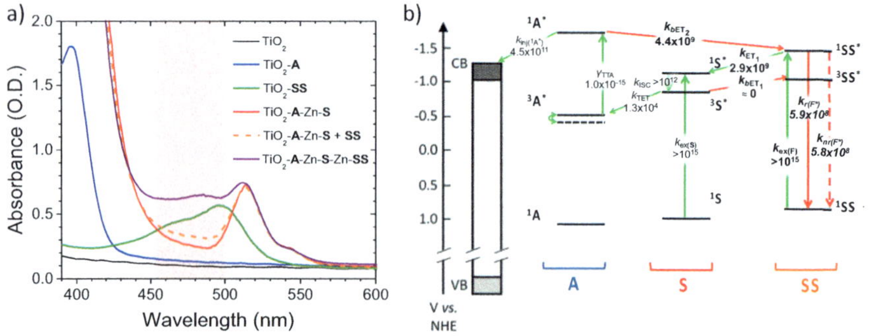

Fig. 11.11 (**a**) Absorption spectra for the trilayer, bilayer, and constituent dyes on TiO_2 with the **S**-**A** transparency window are highlighted in pink, and (**b**) productive (green) and nonproductive (red) dynamic events that occur in the TiO_2-**A**-**S**-**SS** trilayer with their associated experimentally determined rate constants (all units in s^{-1} except γ_{TTA} which is in $cm^3\ s^{-1}$). Reprinted with modification from [47]. Copyright 2019 American Chemical Society

energetically favorable for a triplet energy transfer cascade from **S** to $\mathbf{S_2}$ to **A**. This interlayer, directional energy transfer was supported by both steady-state and time-resolved spectroscopic measurements. The cooperative absorption and triplet energy pooling resulted in a TTA-UC photocurrent of 0.074 mA/cm^2 and an I_{th} value <1 mW/cm^2. Incident photon-to-current efficiency (IPCE) measurements further confirmed that both **S** and $\mathbf{S_2}$ were contributing to the photocurrent generation mechanism. The trilayer did have improved performance but the less-than-unity energy transfer efficiency and the fact that the absorption of **S** and $\mathbf{S_2}$ was not sufficiently separated limited the improvements in the overall device performance.

In a second iteration of the trilayer, Zhou et al. incorporated fluorescein as a singlet sensitizer (**SS**), to generate a TiO_2-**A**-Zn-**S**-Zn-**SS** assembly (Fig. 11.10b) [47]. This scheme was inspired by the report from Monguzzi and coworkers that the incorporation of a singlet sensitizer could increase TTA-UC emission quantum yield by a factor of 1.2 [48].

In the prototype bilayer there is a large absorption gap from ~450 to 500 nm (Fig. 11.11a). The addition of fluorescein (**SS**) in the trilayer increased absorption in this transmission window and resulted in nearly a twofold increase in J_{sc} for **A**-**S**-**SS** trilayer relative to **A**-**S** bilayer device. The IPCE closely resembles the absorptance spectra but with a threefold enhancement in IPCE from 440 to 530 nm for the **A**-**S**-**SS** trilayer relative to **A**-**S** bilayer device. The proposed mechanism and associated rate constants for the singlet sensitized TTA-UC solar cell are depicted in Fig. 11.11b with select productive and nonproductive processes highlighted in green and red, respectively. In this device, the **SS** harvests photons in the **S**-**A** transmission window, initiating directional **SS**-to-**S** singlet energy transfer and **S**-to-**A** triplet energy transfer, followed by TTA, and electron injection from $^1\mathbf{A}^*$. The resulting photocurrent density of 0.315 mA cm^{-2} under 1 Sun irradiation is, as far as we know, the highest photocurrent yet generated by a TTA-UC solar cell.

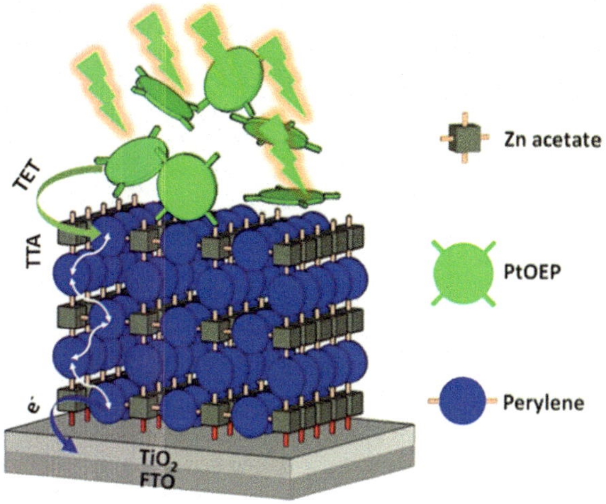

Fig. 11.12 Schematic illustration of a SURMOF, TTA-UC photoanode. Reprinted from [50]. Copyright 2018 American Chemical Society

It is important to note that the devices described above still only harvest visible light and absorption must be shifted into the near IR for this to be a viable solar energy conversion scheme. Nonetheless, given their performance it is clear that dual-triplet sensitizers and singlet sensitization-enhanced TTA-UC in self-assembled trilayers are promising strategies for enhancing broadband light absorption and improving the performance of integrated TTA-UC solar cell.

Surface-Supported Metal-Organic Frameworks

A third category of dye-sensitized TTA-UC devices is with surface-supported metal-organic frameworks (SURMOFs). SURMOFs are epitaxially, in a layer-by-layer fashion, grown metal-organic frameworks on an inorganic substrate [49]. In contrast to the amorphous, metal ion-linked multilayers described above, SURMOFs are crystalline structures. In 2018, Sun and coworkers [50] were the first to incorporate SURMOFs into a TTA-UC solar cell (Fig. 11.12). The device consisted of a Zn-perylene SURMOF [51] grown off a mesoporous TiO_2 substrate, and PtOEP as the sensitizer in a heterogenous sensitization scheme.

Under 530 nm excitation of PtOEP, the SURMOF film exhibited emission from 400–525 nm consistent with upconverted emission from the singlet state of perylene. Electrochemical cells were then prepared with TiO_2-SURMOF + PtOEP, TiO_2-SURMOF, and TiO_2 + PtOEP as working electrodes; Ag/$AgNO_3$ as reference electrode; and platinum wire as counter electrode with 0.01 μM $[Co(bpy)_3]^{2+/3+}$ in MeCN as the electrolyte solution. Under 1 Sun irradiation (AM 1.5) passing through a 530 nm long-pass filter, photocurrents of ~2 μA/cm^2 for TiO_2-SURMOF + PtOEP were generated which was ten times larger than the TiO_2-SURMOF or TiO_2 + PtOEP control samples (Fig. 11.13a). Intensity dependence showed a

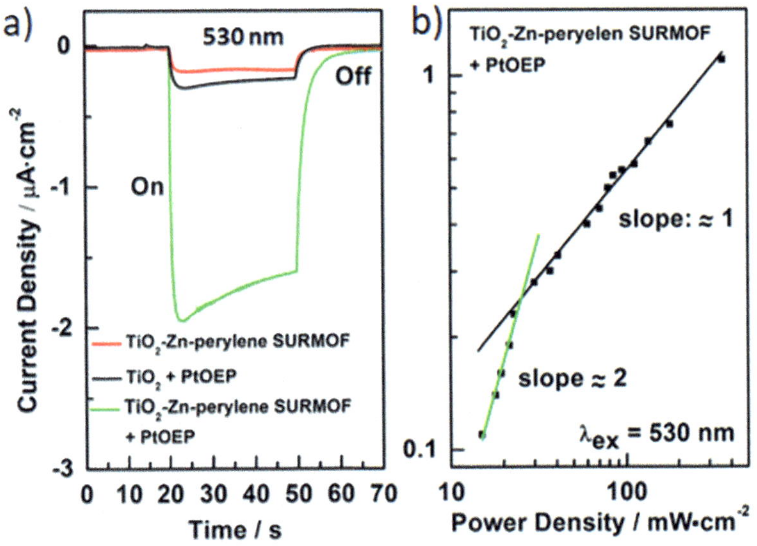

Fig. 11.13 (**a**) Amperometric *i–t* curves for solar cell containing TiO_2 + PtEOP, TiO_2-SURMOF, and TiO_2-SURMOF + PtOEP photoanodes under AM1.5 solar irradiation passing through a 530 nm long-pass filter. (**b**) Photocurrent density from solar cell containing TiO_2 SURMOF + PtOEP photoanode with respect to 530 nm excitation intensity. Reprinted from [50]. Copyright 2018 American Chemical Society

quadratic to linear behavior, which confirms that the photocurrent was due to TTA-UC (Fig. 11.13b).

While the TTA-UC photocurrent was relatively low, this report was the first to demonstrate that SURMOFs are effective scaffolding to facilitate TTA-UC and charge separate the upconverted state. Given the ordered nature of SURMOF and the high density of annihilator molecules, this may be an ideal scaffold for facilitating fast triplet energy migration [52, 53]. However, unless annihilator molecules have the appropriate orientation and spatial separation, the TTA event itself may be hindered. Additionally, following upconversion, the upconverted singlet excited state must migrate to the metal oxide surface for charge separation prior to relaxation which can be challenging with random exciton walk in a continuous solid. Meeting all of these criteria will likely require a fine balance of structure and thickness but the degree of tunability of SURMOFs makes this an intriguing avenue of pursuit.

Co-deposition

Harkening back to the early work of Morandeira, Nagata and coworkers introduced a co-deposited, integrated TTA-UC solar cell, where both Pt(II)-porphyrin sensitizers

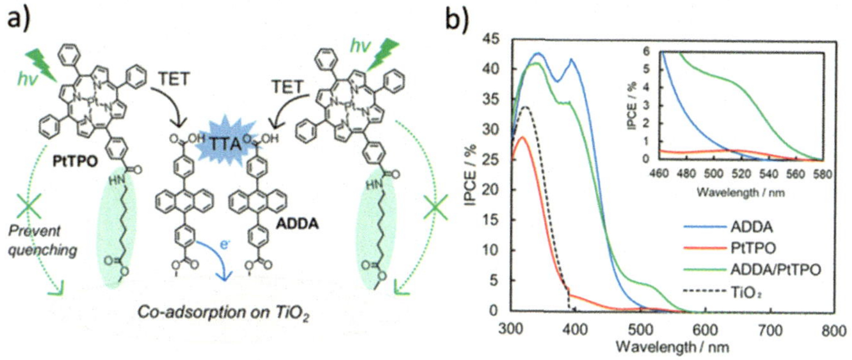

Fig. 11.14 (**a**) Schematic representation of TiO_2 electrode with co-deposited TTA-UC molecules. (**b**) Incident photon-to-current conversion efficiency (IPCE) spectra of solar cells based on TiO_2 (black), TiO_2-ADDA (blue), TiO_2-PtTPO (red), and TiO_2-ADDA/PtTPO (green). Reprinted in part from [54]. Copyright 2019 American Chemical Society

(PtTPO) and diphenylanthracene annihilators (ADDA) were bound to mesoporous TiO_2 via carboxylate linker (Fig. 11.14a) [54].

The TiO_2 photoanode was fabricated by immersing TiO_2 in a solution containing both ADDA and PtTPO as the annihilator and sensitizer molecules, respectively. The anodes were then incorporated into a sandwich cell using I^-/I_3^- as the redox mediator. Under 1 Sun (AM 1.5) irradiation, the ADDA/PtTPO co-depositing device generated higher J_{sc} than the PtTPO- or ADDA-only device. Moreover, the incident photon-to-current conversion efficiency (IPCE) was enhanced from 500 to 550 nm in ADDA/PtTPO device (Fig. 11.14b), while the PtTPO device showed very weak IPCE at that range. For the intensity-dependent measurements, a slope change from 1.5 to 1 was observed in ADDA/PtTPO device while only linear dependence was detected in ADDA-only device. Through loading ratio studies, the authors realized a maximum upconverted photocurrent of 0.036 mA/cm^2 with a 10:1 ADDA:PtTPO ratio. Like the previous UC systems, co-deposition reduces total absorption and may hinder cross-surface migration and lower overall performance, but the ease of synthesis and loading makes this an appealing strategy for realizing a TTA-UC DSSC.

Organic-Inorganic Hybrid Multilayer

The integrated TTA-UC solar cells discussed to this point have relied on platinum and/or palladium porphyrins as the sensitizer. While successful, it is important to acknowledge that these porphyrin sensitizers require multistep syntheses and suffer from inherent energy losses due to the >300 meV S_1-T_1 gap [36, 55]. Quantum dots (QDs) have emerged as an alternative triplet sensitizer [56, 57] for TTA-UC [58–61]

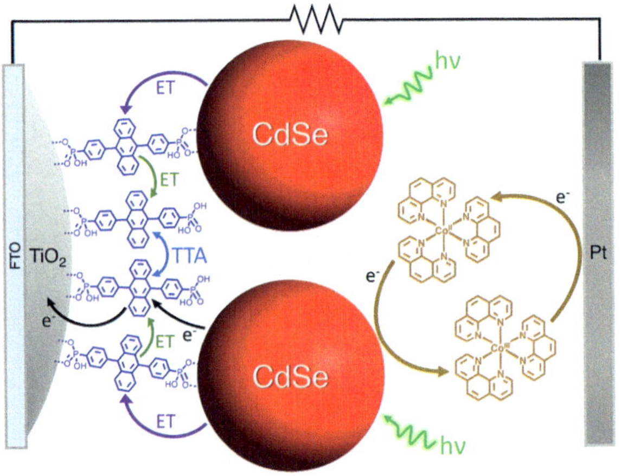

Fig. 11.15 (**a**) Integrated TTA-UC solar cell architecture with dynamic events and molecular structures. Reprinted from [66]. Copyright 2019 American Chemical Society

and have even been shown to facilitate TTA-UC and charge separation at a metal oxide interface [62]. QDs are appealing as sensitizers because they are relatively easy to synthesize, contain more earth abundant elements, have small to no S_1-T_1 gap (<15 meV) [63, 64], and have readily tunable energetics [65].

In 2019, Beery and coworkers reported the first QD (CdSe)-sensitized, integrated TTA-UC solar cell [66]. Their device architecture can be seen in Fig. 11.15 and was composed of a TiO_2-**A**-**QD** inorganic-organic-inorganic multilayer, photoanode, a platinum counter electrode, and $[Co(bpy)_3]^{3+/2+}$ as the mediator in the electrolyte solution.

Like the metal ion-linked multilayers, the TiO_2-**A**-**QD** was prepared bay a stepwise soaking procedure. The maximum **QD**-to-**A** loading ratio of ~1:15 was notably lower than the previously reported TiO_2-**A**-Zn-**S** films A:S of 3:1 at similar **A** surface coverage presumably to the larger size of the **QDs**. Despite having the same absorption cross section as the TiO_2-**A** and TiO_2-**B**-**QD** control samples, under 1 Sun irradiation passed through a 455 nm long-pass filter, the J_{sc} for the TiO_2-**A**-**QD** device was ~1.4 times the sum of its parts (Fig. 11.16a). Although the **QD** is capable of injecting directly into the TiO_2, the quadratic to linear behavior in the intensity-dependent measurements (Fig. 11.16b) indicated that TTA-UC was the dominant photocurrent generation mechanism with an I_{th} value (0.9 mW cm^{-2}) below solar flux and on par with the previously reported molecular sensitized systems.

Despite having similar absorption as the porphyrin-sensitized devices, the photocurrent for the TiO_2-**A**-**QD** DSSC only reached a maximum of 29 μA cm^{-2}. Interestingly, time-resolved spectroscopic measurements showed that the low photocurrent was not due to issues with TTA-UC but instead because of a combination of excited-state quenching of the CdSe **QD** by the $[Co(phen)_3]^{3+/2+}$ mediator and slow regeneration of the oxidized **QD** ($k_{regen} < 1 \times 10^4$ s^{-1}) [67]. Presumably these issues can be resolved with known troubleshooting strategies including increasing **QD**-to-**A** energy transfer rates via increased coupling as well as sur-

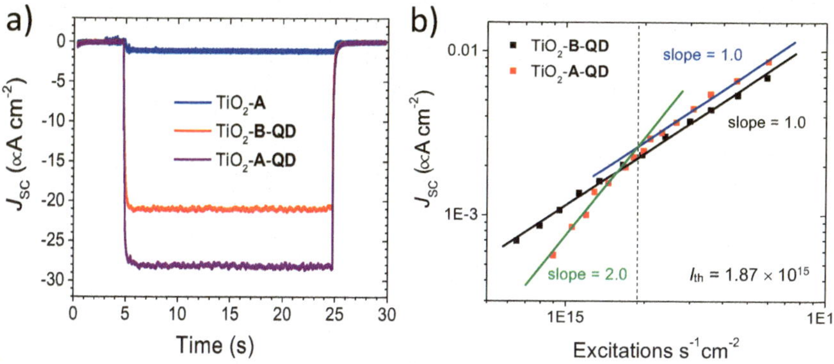

Fig. 11.16 (**a**) Amperometric *i–t* curves under AM1.5 solar irradiation passed through a 455 nm long-pass filter ($V = 0$ V; light on at 5 s, light off at 25 s) for TiO_2-**A**, TiO_2-**B**-**QD**, and TiO_2-**A**-**QD** devices; (**b**) J_{sc} with respect to excitation rate ($\lambda_{ex} = 532$ nm). Reprinted from [66]. Copyright 2019 American Chemical Society

factant modification to increase the regeneration rates. Additionally, we anticipate significantly improved performance with increased film porosity and greater **QD** surface loadings. Furthermore, in terms of bathochromically shifting TTA-UC, QDs are an ideal candidate because of their well-studied absorption and tunability even out the near-IR region [68, 69].

11.3 Layered Heterojunction TTA-UC Solar Cells

A majority of integrated TTA-UC solar cells rely on a dye-sensitized architecture but in 2017, Rand and coworkers showed that a heterojunction organic photovoltaic (OPV) was also an effective scaffold for directly harnessing TTA-UC [70]. As depicted in Fig. 11.17a the TTA-UC layer was composed of a platinum (II) tetraphenyltetrabenzoporphyrin (PtTPBP) sensitizer doped at 5 wt% into a 50 nm layer of α-sexithiophene (α-6T) annihilator.

Transient absorption measurements showed a spectral evolution consistent with triplet energy transfer from PtTPBP to α-6T. Upon direct excitation of α-6T at 405 nm or PtTPBP at 633 nm, the film exhibited fluorescent emission from α-6T with linear and quadratic intensity dependence, respectively (Fig. 11.18a). These observations were consistent with a TTA-UC emission mechanism in the doped film.

For photocurrent measurements, the author generated two similar devices, an intermediate-band solar cell (IBSC) and a triplet-sensitized solar cell (TSSC) that were to serve as the TTA-UC and non-TTA solar cells, respectively (Fig. 11.17a). Whereas the former incorporated a diindenoperylene (DIP) that can only charge separate the singlet excited state of α-6T, the latter contained C60 which has

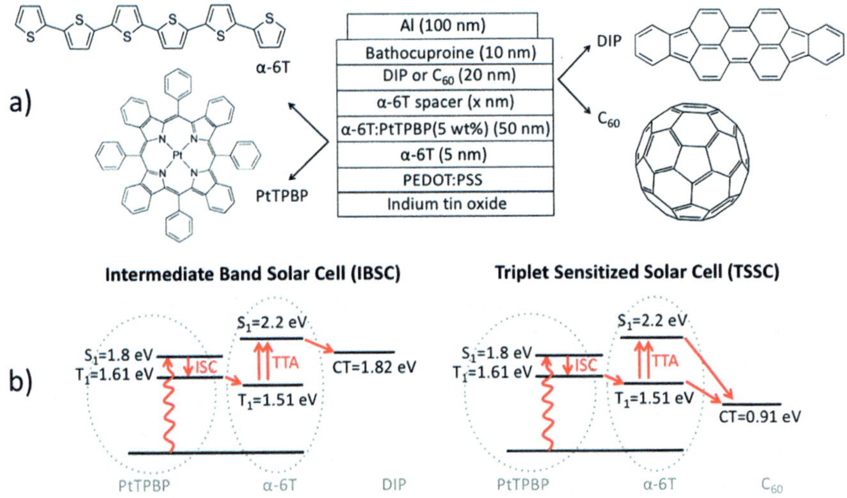

Fig. 11.17 (**a**) Device structure of the intermediate-band solar cell (IBSC) and triplet-sensitized solar cell (TSSC). Molecules shown are α-sexithiophene (α-6T), diindenoperylene (DIP), platinum(II) tetraphenyltetrabenzoporphyrin (PtTPBP), and fullerene C60. (**b**) Energy-level diagrams of the IBSC and TSSC. Reprinted from [70]. Copyright 2017 Royal Society of Chemistry

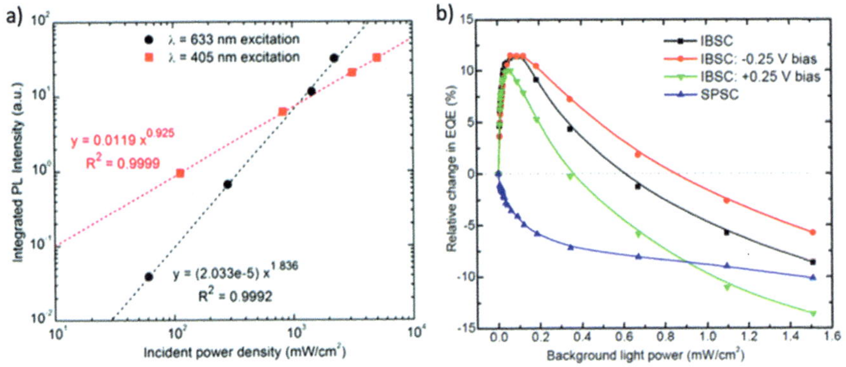

Fig. 11.18 (**a**) Integrated emission intensity from 500 to 600 nm vs. incident power density for directly exciting PtTPBP (black) and α-6T (red). (**b**) Percentage change in 615 nm EQE response of the IBSC with a 20 nm α-6T spacer and TSSC with a 25 nm α-6T spacer as a function of background white light intensity. Reprinted in part from [70]. Copyright 2017 Royal Society of Chemistry

sufficiently low energy to charge separate both the singlet and the triplet states of α-6T (Fig. 11.17b).

Traditional photocurrent measurement with respect to excitation intensity was inconclusive, likely due to competitive intensity-dependent exciton-polaron quenching events, but the authors offered several pieces of indirect evidence that the

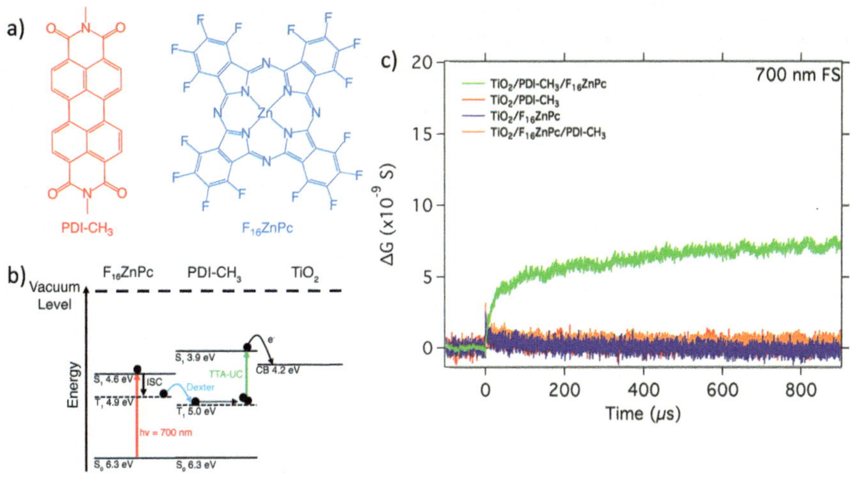

Fig. 11.19 (**a**) Molecular structures of PDI-CH_3 and F_{16}ZnPc. (**b**) The energy-level diagram for the trilayer films. (**c**) Photoconductivity transients of the bilayer, trilayer, and inverted trilayer films under 700 nm excitation. Reprinted with modification from [71]. Copyright 2019 American Chemical Society

IBSC cell was generating photocurrent via a TTA-UC mechanism. The first was by varying the thickness of the undoped α-6T layer (× nm in Fig. 11.17a). The IBSC and TSSC devices exhibited a positive and negative trend, respectively, in J_{sc} with α-6T layer thickness. The contrasting behavior was notable and was attributed to an increased likelihood of TTA-UC in the IBSC device with increasing volume of α-6T. Similar contrasting behavior was observed in the external quantum efficiency at 615 nm with respect to different intensity background white light bias (Fig. 11.18b). Whereas the SPSC device simply decreased, the IBSC showed an applied potential-dependent increase and then decreased with increasing light bias. Collectively, the emission, transient absorption, and device measurements point to a unique photocurrent generation mechanism that is most likely TTA-UC. The upconverted photocurrent was relatively low (13 μA cm^{-2}) due in part to low sensitizer absorption, annihilator-to-sensitizer back energy transfer, and exciton-polaron quenching. But they offer several suggestions to remedy these shortcomings including independent sensitizer and annihilator layers and improved charge/exciton transport layers, among others.

Recently, Felter et al. introduced a layered TTA-UC architecture that combines aspects of both the OPV and DSSC solar cells [71]. On a planar polycrystalline TiO_2 surface, the authors thermally evaporated a pristine layer of perylenetetracarboxylic acid diimide (PDI-CH_3 in Fig. 11.19a) followed by fluorinated zinc phthalocyanine (F_{16}ZnPc) creating the TiO_2/PDI-CH_3/F_{16}ZnPc trilayer. In this film, F_{16}ZnPc, PDI-CH_3, and TiO_2 are energetically appropriate to act as the sensitizer, annihilator, and electron acceptor, respectively (Fig. 11.19b).

With microwave photoconductance measurements under 700 nm excitation (Fig. 11.19c), the authors showed drastically more injected electrons from the trilayer film than the bilayer or inverter trilayer. Transient absorption measurements confirmed that excitation of F_{16}ZnPc is followed by sub-nanosecond triplet energy transfer to PDI-CH_3. Thus, the proposed mechanism is excitation of the F_{16}ZnPc layer, TET to the PDI-CH_3 layer, triplet migration, and TTA on a microsecond timescale, followed by electron injection into TiO_2 from the upconverted state of PDI-CH_3. It is important to note that this trilayer is not a closed circuit and that photocurrent was not measured. However, one can readily envision depositing a counter electrode on top of the F_{16}ZnPc layer to realize a fully functional, TTA-UC solar cell.

11.4 Comparing Optical and Electronic Coupling Schemes

The progress to date harnessing TTA-UC in both optically and electronically coupled solar cells is summarized in Fig. 11.20 with each denoted by gray and red circles, respectively. This graph is an updated version of the one originally introduced by Schmidt and coworkers [72, 73]. Before discussing the performance trends, there are several important things to note about this graph. First is that Schmidt and coworkers define photocurrent density (J), not open-circuit voltage (V_{oc}) or power conversion efficiency (η or PCE), as the critical metric for tracking the progression of harnessing TTA-UC in a solar cell. The rationale behind this decision is that V_{oc} and η are primarily dictated by the high-bandgap components of the solar cell and that the photocurrent contribution from TTA-UC is a supplement that helps to increase the device J_{sc}.

Second is that to account for variations in excitation intensity (i.e., >1 Sun), all photocurrent is scaled to 1 Sun intensity conditions, that is, the measured J_{sc} (mA cm^{-2}) divided by the square of the solar equivalent intensities used to irradiate the solar cell (J_{UC} in mA cm^{-2} sun^{-2}) [72–74]. The assumption here is that J_{sc} is

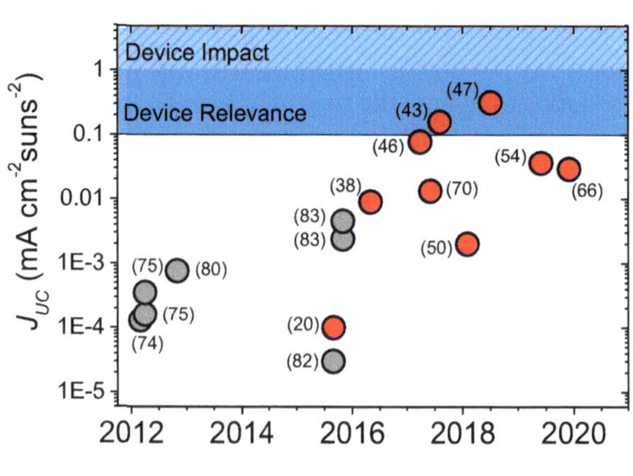

Fig. 11.20 Progression in photocurrent contribution from TTA-UC (J_{UC}) with respect to time. Reference numbers are in parentheses

measured in the quadratic regime and by dividing by the equivalent Suns squared extrapolates to 1 Sun conditions. For TTA-UC solar cells operating in the linear regime, this extrapolation will underestimate their performance. With this in mind, it is strongly recommended that, in addition to the solar concentrator measurements, any reports of TTA-UC solar cells should also perform the measurement and report the outcome under 1 Sun conditions.

Finally, Schmidt and coworkers defined a device relevance threshold of 0.1 mA cm^{-2} which was described as the minimum J_{UC} that is a clearly distinguishable contribution to a traditional solar cell performance [72]. Later, Dilbeck et al. added a device impact threshold of 1 mA cm^{-2}, which is the point where TTA-UC makes notable contributions to the devices, increasing power conversion efficiencies by 1 percentage point or more [34].

As can be seen in Fig. 11.20, in less than 10 years, the collective upconversion community has increased TTA-UC photocurrent enhancements by more than three orders of magnitude [74, 75]. Despite this progress there is still much room for improvement. Regardless of optically or electronically coupled strategies, engineering advances including oxygen exclusion/scavenging [70, 76–79], improved light management, optimized thicknesses, etc. are expected to significantly increase already available systems [80, 81]. Additionally, new sensitizers and/or annihilator molecules with increased extinction coefficients [82, 83], longer triplet excited-state lifetimes, and increased annihilation rates and that can surpass [84, 85] the "spin statistical limit" [86–89] will only help to improve J_{UC} across all devices.

Looking at Fig. 11.20 and J_{UC} alone, one might be tempted to suggest that electronically coupled schemes are the more promising strategy for harnessing TTA-UC. However, the J_{UC} is only part of the story because upconversion is a supplement to the wide-bandgap, high-energy solar energy conversion. With that in mind, the application's advantage almost certainly goes to the optically coupled strategy because the TTA-UC filter/reflector can readily be added to commercially available solar cells without concerns of disrupting or optimizing the solar cell performance. However, there are still some disadvantages. First is that due to the isotropic nature of UC emission, back reflectors or waveguides are required to maximize the number of upconverted photons that are sent back to solar cells [80]. Second, a high sensitizer concentration is desirable for maximizing the triplet excited-state densities for efficient TTA-UC, but the sensitizer concentration must be kept relatively low to limit parasitic self-absorption or back energy transfer from the UC state to the sensitizer. Because of these and other limitations, optically coupled TTA-UC solar cells are yet to cross the device relevance threshold. Nonetheless, system analysis by Frazer et al. indicates that with the appropriate reflectors, improved annihilation rates/efficiencies, and increased sensitizer concentrations/absorbance, upconverted photocurrent enhancements of 1 mA/cm^2 or more are achievable [72].

In contrast, electronically coupled TTA-UC solar cells have already passed the device relevance threshold and are on the verge of device impact. In fact, for the highest J_{UC} device to date [47], it is likely that with further engineering (i.e., $TiCl_4$ treatments, mediator concentration optimization, co-surfactants, thickness optimization) [90, 91] a $J_{UC} > 1$ mA/cm^2 could be achieved. However, there are two important caveats that must be given with these J_{UC} performance numbers. The first is that most of the integrated UC solar cells are simply upconverting green light (<550 nm) with only a few reports extending to absorption of 650 nm [20, 70] with the most bathochromic being ~800 nm [71]. These devices are encouraging but still well short of the 800–1000 nm sensitizer absorption window necessary to improve record dye-sensitized, OPV, and perovskite solar cells [34]. As compared to the optically coupled where most operate in the near-IR spectral region, the use of green-to-blue upconversion pairs has resulted in artificially high J_{UC} for the electronically coupled devices because (1) there are more solar photons in the visible than the near IR and (2) due to the energy gap law [92, 93], triplet excited-state lifetimes and the resulting upconversion efficiencies decrease with lower energy sensitizers. That is not to say long-lived sensitizers and/or faster TTA rates cannot give efficient TTA-UC in the near IR but only that to date, TTA-UC efficiencies for the near IR are notably lower than in the visible.

The second, and perhaps more important, caveat is that a majority of integrated TTA-UC solar cells are exclusively focused on enhancing upconverted photocurrent and are not particularly good at harnessing the high-energy portion of the spectrum. These devices could be used in tandem with a higher energy-absorbing solar cells but then they are not really utilizing TTA-UC to circumvent the Shockley-Queisser limit [94] and it would most likely be more efficient and cost effective to simply couple a traditional near-IR solar cell instead of TTA-UC. Thus, to realize their full potential in an integrated solar cell, both TTA-UC and direct, high-energy solar energy conversion must be integrated into a single-junction device.

For both dye-sensitized and layered heterojunction devices, one can envision a scenario where the annihilator molecule acts as both the annihilator for TTA-UC and the high-energy absorber (i.e., direct excitation to the singlet state followed by charge separation). In fact, in all the integrated solar cells mentioned above, this latter mechanism is active under broadband excitation. However, for this strategy to be efficient, the annihilator molecule must perform on par with the highest efficiency dyes used in world record solar cells and there must be no competitive high-energy absorption by the sensitizer molecule. Finding sensitizers, annihilators, and architectures that meet these criteria will be difficult, if not impossible.

However, for both integrated architectures, it may be possible to generate a layered structure that is still a single-junction solar cell, but high-energy photons are absorbed before hitting the TTA-UC layer. For example, in the proposed layered TiO_2 anode shown in Fig. 11.21 [34], the high-energy-absorbing layer and TTA-UC layer are spatially separated on the bottom and top of the nanocrystalline TiO_2 film. As the incident light enters from the bottom of the anode, there is no competing high-energy photon absorption from TTA-UC portion. Layered DSSCs with complementary chromophores of this type have been demonstrated [6, 95], and there is no obvious reason that this strategy could not be extended to TTA-UC solar cells.

Rough calculations indicate that if near-IR-absorbing TTA-UC with moderate efficiency can be coupled with world record DSSC or perovskite materials [96, 97], then >1 mA cm^{-2} and more than 2 percentage point efficiency enhancements could

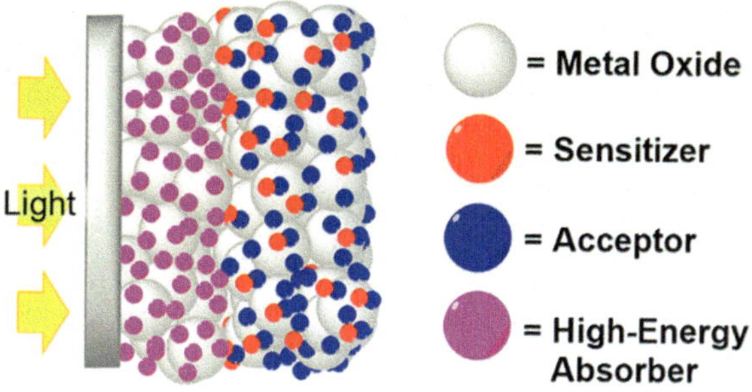

Fig. 11.21 Layered TiO_2 anode with the high-energy absorber (purple sphere) and the TTA-UC molecules (red and blue spheres) spatially separated. Reprinted in part from [34]. Copyright 2018 American Chemical Society

be achieved [34]. Because similar enhancements have been calculated for optically coupled TTA-UC solar cells [72], we are optimistic about the future of harnessing TTA-UC for more efficient solar energy conversion.

11.5 Conclusion

In this chapter we have summarized the efforts to date to directly harness TTA-UC in dye-sensitized and layered heterojunction solar cells. The dye-sensitized schemes rely on combining sensitizer and acceptor/annihilator molecules on a metal oxide substrate and undergo a similar mechanism (i.e., sensitizer excitation, sensitizer-to-annihilator energy transfer, TTA, and electron transfer from the upconverted state to TiO_2). The primary difference lies in how the molecules are assembled at the surface including co-deposition, metal ion-linked multilayers, surface-supported metal-organic frameworks, and organic-inorganic layered architectures. Considerable progress has been with dye-sensitized devices demonstrating photocurrents >0.3 mA/cm^2 under one Sun intensities. However, much improvement still needs to be made including (1) understanding and controlling the geometry of molecules at the dye-semiconductor interface to maximize the efficiency of energy transfer and TTA events, (2) improving the redox mediator to increase photocurrent and minimize excited-state quenching, (3) broadening and shifting the sensitizer absorption into the near IR, and (4) coupling the TTA-UC scheme to a high-energy solar energy conversion scheme.

Layered heterojunctions are an alternative means of harnessing TTA-UC using a sensitizer-doped annihilator layer or two independent sensitizer and annihilator layers. In both cases sensitizer excitation, sensitizer-to-annihilator energy transfer, TTA, and through-film exciton diffusion are followed by charge separation at the heterojunction. In comparison to the dye-sensitized scheme, these devices have the distinct advantage of being entirely solid state, they do not require synthesis of dyes with surface-binding groups, and they use well-known chemical vapor deposition techniques to control doping and film formation. However, intrinsic to these devices is the need for exciton migration which can be problematic for long-lived triplet states that can undergo exciton-polaron quenching which is partially responsible for their relatively low performance to date. This issue may be addressable with the appropriate charge/exciton transport layers and/or a gradient for directional energy migration. Also, like the dye-sensitized devices, the heterojunction devices must still harness high-energy light and shift their absorption into the near IR.

Finally, we close with a discussion comparing optically and electronically coupled TTA-UC solar cell schemes. In terms of immediate application, optically coupled strategies have the distinct advantage of simply adding TTA-UC to well-established solar cells but still have relatively low photocurrent contributions. On the other hand, dye-sensitized devices have the highest upconverted photocurrent reported to date but with the disclaimer that they are only currently operating in the visible regime and are yet to be coupled to a high-efficiency, high-energy solar energy conversion schemes. Nonetheless, both strategies will benefit from progress generating new upconversion pairs, improved oxygen exclusion/scavenging, engineering advances, etc. Regardless of the strategy, system analyses indicate that with these and other advances, TTA-UC photocurrent enhancements of >1 mA/cm^2 and 1 or 2 percentage point efficiency enhancements are achievable. While there is still much progress to be made, we are optimistic and excited to see what improvements and new architectures lie ahead on the journey towards a viable TTA-UC solar energy conversion device that pushes the Shockley-Queisser limit.

References

1. S. Namba, Y. Hishiki, Color sensitization of zinc oxide with cyanine Dyes1. J. Phys. Chem. **69**(3), 774–779 (1965)
2. B. O'Regan, M. Grätzel, A low-cost, high-efficiency solar cell based on dye-sensitized colloidal TiO2 films. Nature **353**(6346), 737–740 (1991)
3. A. Hagfeldt, G. Boschloo, L. Sun, L. Kloo, H. Pettersson, Dye-sensitized solar cells. Chem. Rev. **110**(11), 6595–6663 (2010)
4. A. Hagfeldt, Brief overview of dye-sensitized solar cells. Ambio **41**(Suppl 2), 151–155 (2012)
5. J. Gong, K. Sumathy, Q. Qiao, Z. Zhou, Review on dye-sensitized solar cells (DSSCs): advanced techniques and research trends. Renew. Sust. Energ. Rev. **68**, 234–246 (2017)
6. J.C. Wang, S.P. Hill, T. Dilbeck, O.O. Ogunsolu, T. Banerjee, K. Hanson, Multimolecular assemblies on high surface area metal oxides and their role in interfacial energy and electron transfer. Chem. Soc. Rev. **47**(1), 104–148 (2018)
7. M.K. Nazeeruddin, R. Humphry-Baker, P. Liska, M. Grätzel, Investigation of sensitizer adsorption and the influence of protons on current and voltage of a dye-sensitized nanocrystalline TiO2 solar cell. J. Phys. Chem. B **107**(34), 8981–8987 (2003)
8. J. Albero, P. Atienzar, A. Corma, H. Garcia, Efficiency records in mesoscopic dye-sensitized solar cells. Chem. Rec. **15**(4), 803–828 (2015)
9. V.S. Manthou, E.K. Pefkianakis, P. Falaras, G.C. Vougioukalakis, Co-adsorbents: a key component in efficient and robust dye-sensitized solar cells. ChemSusChem **8**(4), 588–599 (2015)
10. Z. Zhang, S.M. Zakeeruddin, B.C. O'Regan, R. Humphry-Baker, M. Grätzel, Influence of 4-guanidinobutyric acid as coadsorbent in reducing recombination in dye-sensitized solar cells. J. Phys. Chem. B **109**(46), 21818–21824 (2005)

11. J.S. Lissau, J.M. Gardner, A. Morandeira, Photon upconversion on dye-sensitized nanostructured ZrO2 films. J. Phys. Chem. C **115**(46), 23226–23232 (2011)
12. A. Haefele, J. Blumhoff, R.S. Khnayzer, F.N. Castellano, Getting to the (square) root of the problem: how to make noncoherent pumped upconversion linear. J. Phys. Chem. Lett. **3**(3), 299–303 (2012)
13. A. Monguzzi, J. Mezyk, F. Scotognella, R. Tubino, F. Meinardi, Upconversion-induced fluorescence in multicomponent systems: steady-state excitation power threshold. Phys. Rev. B **78**(19), 195112 (2008)
14. Y.Y. Cheng, T. Khoury, R.G.C.R. Clady, M.J.Y. Tayebjee, N.J. Ekins-Daukes, M.J. Crossley, T.W. Schmidt, On the efficiency limit of triplet-triplet annihilation for photochemical upconversion. Phys. Chem. Chem. Phys. **12**(1), 66–71 (2010)
15. J.E. Auckett, Y.Y. Chen, T. Khoury, R.G. Clady, N. Ekins-Daukes, M.J. Crossley, T.W. Schmidt, Efficient up-conversion by triplet-triplet annihilation. J. Phys.: Conf. Ser. **185**, 012002 (2009). IOP Publishing
16. J.S. Lissau, D. Nauroozi, M.-P. Santoni, S. Ott, J.M. Gardner, A. Morandeira, Anchoring energy acceptors to nanostructured ZrO2 enhances photon upconversion by sensitized triplet–triplet annihilation under simulated solar flux. J. Phys. Chem. C **117**(28), 14493–14501 (2013)
17. J.S. Lissau, D. Nauroozi, M.-P. Santoni, T. Edvinsson, S. Ott, J.M. Gardner, A. Morandeira, What limits photon upconversion on mesoporous thin films sensitized by solution-phase absorbers? J. Phys. Chem. C **119**(9), 4550–4564 (2015)
18. J.S. Lissau, D. Nauroozi, M.-P. Santoni, S. Ott, J.M. Gardner, A. Morandeira, Photon upconversion from chemically bound triplet sensitizers and emitters on mesoporous ZrO2: implications for solar energy conversion. J. Phys. Chem. C **119**(46), 25792–25806 (2015)
19. J.M. Giaimuccio, J.G. Rowley, G.J. Meyer, D. Wang, E. Galoppini, Heavy atom effects on anthracene-rigid-rod excited states anchored to metal oxide nanoparticles. Chem. Phys. **339**(1), 146–153 (2007)
20. C. Simpson, T.M. Clarke, R.W. MacQueen, Y.Y. Cheng, A.J. Trevitt, A.J. Mozer, P. Wagner, T.W. Schmidt, A. Nattestad, An intermediate band dye-sensitised solar cell using triplet-triplet annihilation. Phys. Chem. Chem. Phys. **17**(38), 24826–24830 (2015)
21. H. Lee, L.J. Kepley, H.G. Hong, S. Akhter, T.E. Mallouk, Adsorption of ordered zirconium phosphonate multilayer films on silicon and gold surfaces. J. Phys. Chem. **92**(9), 2597–2601 (1988)
22. H. Lee, L.J. Kepley, H.G. Hong, T.E. Mallouk, Inorganic analogs of Langmuir-Blodgett films: adsorption of ordered zirconium 1,10-decanebisphosphonate multilayers on silicon surfaces. J. Am. Chem. Soc. **110**(2), 618–620 (1988)
23. H.E. Katz, Multilayer deposition of novel organophosphonates with zirconium(IV). Chem. Mater. **6**(12), 2227–2232 (1994)
24. H.E. Katz, G. Scheller, T.M. Putvinski, M.L. Schilling, W.L. Wilson, C.E.D. Chidsey, Polar orientation of dyes in robust multilayers by zirconium phosphate-phosphonate interlayers. Science **254**(5037), 1485–1487 (1991)
25. T. Ishida, K.-I. Terada, K. Hasegawa, H. Kuwahata, K. Kusama, R. Sato, M. Nakano, Y. Naitoh, M.-A. Haga, Self-assembled monolayer and multilayer formation using redox-active Ru complex with phosphonic acids on silicon oxide surface. Appl. Surf. Sci. **255**(21), 8824–8830 (2009)
26. T. Keiichi, K. Katsuaki, H. Jiro, H. Masa-aki, Electric conduction properties of self-assembled monolayer films of Ru complexes with disulfide/phosphonate anchors in a Au–(molecular ensemble)–(Au nanoparticle) junction. Chem. Lett. **38**(5), 416–417 (2009)
27. L.A. Vermeulen, J.L. Snover, L.S. Sapochak, M.E. Thompson, Efficient photoinduced charge separation in layered zirconium viologen phosphonate compounds. J. Am. Chem. Soc. **115**(25), 11767–11774 (1993)
28. S.B. Ungashe, W.L. Wilson, H.E. Katz, G.R. Scheller, T.M. Putvinski, Synthesis, self-assembly, and photophysical dynamics of stacked layers of porphyrin and viologen phosphonates. J. Am. Chem. Soc. **114**(22), 8717–8719 (1992)

29. K. Hanson, D.A. Torelli, A.K. Vannucci, M.K. Brennaman, H. Luo, L. Alibabaei, W. Song, D.L. Ashford, M.R. Norris, C.R.K. Glasson, J.J. Concepcion, T.J. Meyer, Self-assembled bilayer films of ruthenium(II)/polypyridyl complexes through layer-by-layer deposition on nanostructured metal oxides. Angew. Chem. Int. Ed. **51**(51), 12782–12785 (2012)
30. O.O. Ogunsolu, I.A. Murphy, J.C. Wang, A. Das, K. Hanson, Energy and electron transfer cascade in self-assembled bilayer dye-sensitized solar cells. ACS Appl. Mater. Interfaces **8**(42), 28633–28640 (2016)
31. O.O. Ogunsolu, J.C. Wang, K. Hanson, Inhibiting interfacial recombination events in dye-sensitized solar cells using self-assembled bilayers. ACS Appl. Mater. Interfaces **7**(50), 27730–27734 (2015)
32. X. Ding, Y. Gao, L. Zhang, Z. Yu, J. Liu, L. Sun, Visible light-driven water splitting in photoelectrochemical cells with supramolecular catalysts on photoanodes. ACS Catal. **4**(7), 2347–2350 (2014)
33. A.J. Robb, E.S. Knorr, N. Watson, K. Hanson, Metal ion linked multilayers on mesoporous substrates: energy/electron transfer, photon upconversion, and more. J. Photochem. Photobiol. A Chem. **390**, 112291 (2020)
34. T. Dilbeck, K. Hanson, Molecular photon upconversion solar cells using multilayer assemblies: progress and prospects. J. Phys. Chem. Lett. **9**(19), 5810–5821 (2018)
35. S.P. Hill, T. Banerjee, T. Dilbeck, K. Hanson, Photon upconversion and photocurrent generation via self-assembly at organic–inorganic interfaces. J. Phys. Chem. Lett. **6**(22), 4510–4517 (2015)
36. T.N. Singh-Rachford, F.N. Castellano, Photon upconversion based on sensitized triplet–triplet annihilation. Coord. Chem. Rev. **254**(21), 2560–2573 (2010)
37. D.V. Kozlov, F.N. Castellano, Anti-Stokes delayed fluorescence from metal–organic bichromophores. Chem. Commun. **24**, 2860–2861 (2004)
38. S.P. Hill, T. Dilbeck, E. Baduell, K. Hanson, Integrated photon upconversion solar cell via molecular self-assembled bilayers. ACS Energy Lett. **1**(1), 3–8 (2016)
39. M.K. Nazeeruddin, A. Kay, I. Rodicio, R. Humphry-Baker, E. Mueller, P. Liska, N. Vlachopoulos, M. Graetzel, Conversion of light to electricity by cis-X2bis(2,2′-bipyridyl-4,4′-dicarboxylate)ruthenium(II) charge-transfer sensitizers (X = Cl-, Br-, I-, CN-, and SCN-) on nanocrystalline titanium dioxide electrodes. J. Am. Chem. Soc. **115**(14), 6382–6390 (1993)
40. Y. Zhou, S. Ayad, C. Ruchlin, V. Posey, S.P. Hill, Q. Wu, K. Hanson, Examining the role of acceptor molecule structure in self-assembled bilayers: surface loading, stability, energy transfer, and upconverted emission. Phys. Chem. Chem. Phys. **20**(31), 20513–20524 (2018)
41. Y. Zhou, S.P. Hill, K. Hanson, Influence of meta- and para-phosphonated diphenylanthracene on photon upconversion in self-assembled bilayers. J. Photonics Energy **8**(11), 1–11 (2018)
42. T. Ogawa, N. Yanai, A. Monguzzi, N. Kimizuka, Highly efficient photon upconversion in self-assembled light-harvesting molecular systems. Sci. Rep. **5**, 10882 (2015)
43. S.P. Hill, K. Hanson, Harnessing molecular photon upconversion in a solar cell at sub-solar irradiance: role of the redox mediator. J. Am. Chem. Soc. **139**(32), 10988–10991 (2017)
44. S.A. Sapp, C.M. Elliott, C. Contado, S. Caramori, C.A. Bignozzi, Substituted polypyridine complexes of cobalt(II/III) as efficient electron-transfer mediators in dye-sensitized solar cells. J. Am. Chem. Soc. **124**(37), 11215–11222 (2002)
45. E. Mosconi, J.-H. Yum, F. Kessler, C.J. Gómez García, C. Zuccaccia, A. Cinti, M.K. Nazeeruddin, M. Grätzel, F. De Angelis, Cobalt electrolyte/dye interactions in dye-sensitized solar cells: a combined computational and experimental study. J. Am. Chem. Soc. **134**(47), 19438–19453 (2012)
46. T. Dilbeck, S.P. Hill, K. Hanson, Harnessing molecular photon upconversion at sub-solar irradiance using dual sensitized self-assembled trilayers. J. Mater. Chem. A **5**(23), 11652–11660 (2017)
47. Y. Zhou, C. Ruchlin, A.J. Robb, K. Hanson, Singlet sensitization-enhanced upconversion solar cells via self-assembled trilayers. ACS Energy Lett. **4**(6), 1458–1463 (2019)
48. J. Pedrini, A. Monguzzi, F. Meinardi, Cascade sensitization of triplet–triplet annihilation based photon upconversion at sub-solar irradiance. Phys. Chem. Chem. Phys. **20**(15), 9745–9750 (2018)

49. H. Gliemann, C. Wöll, Epitaxially grown metal-organic frameworks. Mater. Today **15**(3), 110–116 (2012)
50. S. Ahmad, J. Liu, C. Gong, J. Zhao, L. Sun, Photon up-conversion via epitaxial surface-supported metal–organic framework thin films with enhanced photocurrent. ACS Appl. Energy Mater. **1**(2), 249–253 (2018)
51. Z. Wang, A. Knebel, S. Grosjean, D. Wagner, S. Bräse, C. Wöll, J. Caro, L. Heinke, Tunable molecular separation by nanoporous membranes. Nat. Commun. **7**, 13872–13872 (2016)
52. J. Lin, X. Hu, P. Zhang, A. Van Rynbach, D.N. Beratan, C.A. Kent, B.P. Mehl, J.M. Papanikolas, T.J. Meyer, W. Lin, S.S. Skourtis, M. Constantinou, Triplet excitation energy dynamics in metal–organic frameworks. J. Phys. Chem. C **117**(43), 22250–22259 (2013)
53. C.A. Kent, B.P. Mehl, L. Ma, J.M. Papanikolas, T.J. Meyer, W. Lin, Energy transfer dynamics in metal−organic frameworks. J. Am. Chem. Soc. **132**(37), 12767–12769 (2010)
54. T. Morifuji, Y. Takekuma, M. Nagata, Integrated photon upconversion dye-sensitized solar cell by co-adsorption with derivative of Pt–porphyrin and anthracene on mesoporous TiO2. ACS Omega **4**(6), 11271–11275 (2019)
55. J.C. Goldschmidt, S. Fischer, Upconversion for photovoltaics—a review of materials, devices and concepts for performance enhancement. Adv. Opt. Mater. **3**(4), 510–535 (2015)
56. C. Mongin, S. Garakyaraghi, N. Razgoniaeva, M. Zamkov, F.N. Castellano, Direct observation of triplet energy transfer from semiconductor nanocrystals. Science **351**(6271), 369 (2016)
57. Z. Huang, X. Li, M. Mahboub, K.M. Hanson, V.M. Nichols, H. Le, M.L. Tang, C.J. Bardeen, Hybrid molecule–nanocrystal photon upconversion across the visible and near-infrared. Nano Lett. **15**(8), 5552–5557 (2015)
58. Z. Huang, X. Li, B.D. Yip, J.M. Rubalcava, C.J. Bardeen, M.L. Tang, Nanocrystal size and quantum yield in the upconversion of green to violet light with CdSe and anthracene derivatives. Chem. Mater. **27**(21), 7503–7507 (2015)
59. M. Mahboub, Z. Huang, M.L. Tang, Efficient infrared-to-visible upconversion with subsolar irradiance. Nano Lett. **16**(11), 7169–7175 (2016)
60. A. Ronchi, P. Brazzo, M. Sassi, L. Beverina, J. Pedrini, F. Meinardi, A. Monguzzi, Triplet–triplet annihilation based photon up-conversion in hybrid molecule–semiconductor nanocrystal systems. Phys. Chem. Chem. Phys. **21**(23), 12353–12359 (2019)
61. K. Okumura, K. Mase, N. Yanai, N. Kimizuka, Employing core-shell quantum dots as triplet sensitizers for photon upconversion. Chem. Eur. J. **22**(23), 7721–7726 (2016)
62. B. Shan, T.-T. Li, M.K. Brennaman, A. Nayak, L. Wu, T.J. Meyer, Charge transfer from upconverting nanocrystals to semiconducting electrodes: optimizing thermodynamic outputs by electronic energy transfer. J. Am. Chem. Soc. **141**(1), 463–471 (2019)
63. C. Mongin, P. Moroz, M. Zamkov, F.N. Castellano, Thermally activated delayed photoluminescence from pyrenyl-functionalized CdSe quantum dots. Nat. Chem. **10**(2), 225–230 (2018)
64. S. Garakyaraghi, F.N. Castellano, Nanocrystals for triplet sensitization: molecular behavior from quantum-confined materials. Inorg. Chem. **57**(5), 2351–2359 (2018)
65. R. Rossetti, J.L. Ellison, J.M. Gibson, L.E. Brus, Size effects in the excited electronic states of small colloidal CdS crystallites. J. Chem. Phys. **80**(9), 4464–4469 (1984)
66. D. Beery, J.P. Wheeler, A. Arcidiacono, K. Hanson, CdSe quantum dot sensitized molecular photon upconversion solar cells. ACS Appl. Energy Mater. **3**(1), 29–37 (2020)
67. Y. Xie, J. Baillargeon, T.W. Hamann, Kinetics of regeneration and recombination reactions in dye-sensitized solar cells employing cobalt redox shuttles. J. Phys. Chem. C **119**(50), 28155–28166 (2015)
68. S. Chinnathambi, N. Shirahata, Recent advances on fluorescent biomarkers of near-infrared quantum dots for in vitro and in vivo imaging. Sci. Technol. Adv. Mater. **20**(1), 337–355 (2019)
69. J.M. Pietryga, Y.-S. Park, J. Lim, A.F. Fidler, W.K. Bae, S. Brovelli, V.I. Klimov, Spectroscopic and device aspects of nanocrystal quantum dots. Chem. Rev. **116**(18), 10513–10622 (2016)
70. Y.L. Lin, M. Koch, A.N. Brigeman, D.M.E. Freeman, L. Zhao, H. Bronstein, N.C. Giebink, G.D. Scholes, B.P. Rand, Enhanced sub-bandgap efficiency of a solid-state organic intermediate band solar cell using triplet-triplet annihilation. Energy Environ. Sci. **10**(6), 1465–1475 (2017)

71. K.M. Felter, M.C. Fravventura, E. Koster, R.D. Abellon, T.J. Savenije, F.C. Grozema, Solid-state infrared upconversion in perylene diimides followed by direct electron injection. ACS Energy Lett. **5**(1), 124–129 (2020)
72. L. Frazer, J.K. Gallaher, T.W. Schmidt, Optimizing the efficiency of solar photon upconversion. ACS Energy Lett. **2**(6), 1346–1354 (2017)
73. Y.Y. Cheng, B. Fückel, T. Schulze, R. MacQueen, M. Tayebjee, A. Danos, T. Khoury, R.G. Clady, N. Ekins-Daukes, M. Crossley, B. Stannowski, K. Lips, T. Schmidt, Improving the light-harvesting of second generation solar cells with photochemical upconversion. SPIE **8477** (2012)
74. Y.Y. Cheng, B. Fuckel, R.W. MacQueen, T. Khoury, R.G.C.R. Clady, T.F. Schulze, N.J. Ekins-Daukes, M.J. Crossley, B. Stannowski, K. Lips, T.W. Schmidt, Improving the light-harvesting of amorphous silicon solar cells with photochemical upconversion. Energy Environ. Sci. **5**(5), 6953–6959 (2012)
75. T.F. Schulze, J. Czolk, Y.-Y. Cheng, B. Fückel, R.W. MacQueen, T. Khoury, M.J. Crossley, B. Stannowski, K. Lips, U. Lemmer, A. Colsmann, T.W. Schmidt, Efficiency enhancement of organic and thin-film silicon solar cells with photochemical upconversion. J. Phys. Chem. C **116**(43), 22794–22801 (2012)
76. A.L. Hagstrom, F. Deng, J.-H. Kim, Enhanced triplet–triplet annihilation upconversion in dual-sensitizer systems: translating broadband light absorption to practical solid-state materials. ACS Photonics **4**(1), 127–137 (2017)
77. D. Dzebo, K. Moth-Poulsen, B. Albinsson, Robust triplet–triplet annihilation photon upconversion by efficient oxygen scavenging. Photochem. Photobiol. Sci. **16**(8), 1327–1334 (2017)
78. C. Li, C. Koenigsmann, F. Deng, A. Hagstrom, C.A. Schmuttenmaer, J.-H. Kim, Photocurrent enhancement from solid-state triplet–triplet annihilation upconversion of low-intensity, low-energy photons. ACS Photonics **3**(5), 784–790 (2016)
79. D. Di, L. Yang, J.M. Richter, L. Meraldi, R.M. Altamimi, A.Y. Alyamani, D. Credgington, K.P. Musselman, J.L. MacManus-Driscoll, R.H. Friend, Efficient triplet exciton fusion in molecularly doped polymer light-emitting diodes. Adv. Mater. **29**(13), 1605987 (2017)
80. T.F. Schulze, Y.Y. Cheng, B. Fückel, R.W. MacQueen, A. Danos, N.J.L.K. Davis, M.J.Y. Tayebjee, T. Khoury, R.G.C.R. Clady, N.J. Ekins-Daukes, M.J. Crossley, B. Stannowski, K. Lips, T.W. Schmidt, Photochemical upconversion enhanced solar cells: effect of a back reflector. Aust. J. Chem. **65**(5), 480–485 (2012)
81. T. Schulze, Y.Y. Cheng, T. Khoury, M. Crossley, B. Stannowski, K. Lips, T. Schmidt, Micro-optical design of photochemical upconverters for thin-film solar cells. J. Photonics Energy **3**(1), 034598 (2013)
82. A. Monguzzi, S.M. Borisov, J. Pedrini, I. Klimant, M. Salvalaggio, P. Biagini, F. Melchiorre, C. Lelii, F. Meinardi, Efficient broadband triplet–triplet annihilation-assisted photon upconversion at subsolar irradiance in fully organic systems. Adv. Funct. Mater. **25**(35), 5617–5624 (2015)
83. Y.Y. Cheng, A. Nattestad, T.F. Schulze, R.W. MacQueen, B. Fuckel, K. Lips, G.G. Wallace, T. Khoury, M.J. Crossley, T.W. Schmidt, Increased upconversion performance for thin film solar cells: a trimolecular composition. Chem. Sci. **7**(1), 559–568 (2016)
84. S. Hoseinkhani, R. Tubino, F. Meinardi, A. Monguzzi, Achieving the photon up-conversion thermodynamic yield upper limit by sensitized triplet–triplet annihilation. Phys. Chem. Chem. Phys. **17**(6), 4020–4024 (2015)
85. M.C. DeRosa, R.J. Crutchley, Photosensitized singlet oxygen and its applications. Coord. Chem. Rev. **233–234**, 351–371 (2002)
86. J.L. Charlton, R. Dabestani, J. Saltiel, Role of triplet-triplet annihilation in anthracene dimerization. J. Am. Chem. Soc. **105**(11), 3473–3476 (1983)
87. J. Saltiel, B.W. Atwater, Spin-Statistical Factors in Diffusion-Controlled Reactions. In *Advances in Photochemistry* (eds D. H. Volman, G. S. Hammond and K. Gollnick). (1988)
88. A.J. McLean, T.G. Truscott, Faraday communications. Efficiency of triplet-photosensitised singlet oxygen generation in benzene. J. Chem. Soc. Faraday Trans. **86**(14), 2671–2672 (1990)
89. S.M. Bachilo, R.B. Weisman, Determination of triplet quantum yields from triplet−triplet annihilation fluorescence. Chem. A Eur. J. **104**(33), 7711–7714 (2000)

90. B.M. Klahr, T.W. Hamann, Performance enhancement and limitations of cobalt bipyridyl redox shuttles in dye-sensitized solar cells. J. Phys. Chem. C **113**(31), 14040–14045 (2009)
91. P.M. Sommeling, B.C. O'Regan, R.R. Haswell, H.J.P. Smit, N.J. Bakker, J.J.T. Smits, J.M. Kroon, J.A.M. van Roosmalen, Influence of a TiCl4 post-treatment on nanocrystalline TiO2 films in dye-sensitized solar cells. J. Phys. Chem. B **110**(39), 19191–19197 (2006)
92. R. Englman, J. Jortner, The energy gap law for radiationless transitions in large molecules. Mol. Phys. **18**(2), 145–164 (1970)
93. K.F. Freed, J. Jortner, Multiphonon processes in the nonradiative decay of large molecules. J. Chem. Phys. **52**(12), 6272–6291 (1970)
94. W. Shockley, H.J. Queisser, Detailed balance limit of efficiency of p-n junction solar cells. J. Appl. Phys. **32**(3), 510–519 (1961)
95. S.K. Balasingam, M. Lee, M.G. Kang, Y. Jun, Improvement of dye-sensitized solar cells toward the broader light harvesting of the solar spectrum. Chem. Commun. **49**(15), 1471–1487 (2013)
96. K. Kakiage, Y. Aoyama, T. Yano, K. Oya, J.-I. Fujisawa, M. Hanaya, Highly-efficient dye-sensitized solar cells with collaborative sensitization by silyl-anchor and carboxy-anchor dyes. Chem. Commun. **51**(88), 15894–15897 (2015)
97. W.S. Yang, B.-W. Park, E.H. Jung, N.J. Jeon, Y.C. Kim, D.U. Lee, S.S. Shin, J. Seo, E.K. Kim, J.H. Noh, S.I. Seok, Iodide management in formamidinium-lead-halide-based perovskite layers for efficient solar cells. Science **356**(6345), 1376–1379 (2017)

Part IV
Addressing Transmission Losses: Sequential Absorption in Rare Earth Ions

Chapter 12
Rare-Earth Ion-Based Photon Up-Conversion for Transmission-Loss Reduction in Solar Cells

Hai-Qiao Wang, Andres Osvet, Miroslaw Batentschuk, and Christoph J. Brabec

Abstract Photon up-conversion (UC) describes an anti-Stokes emission process, in which a luminophor emits one higher energy photon after being excited by multiple low-energy photons, among which rare-earth (RE) ion-doped materials present promising UC properties due to unique electron configuration. RE UC materials have been widely studied in solar cells with the purpose to reduce transmission losses, i.e., achieve wide/full solar spectral harvesting and high-power conversion efficiency, by converting unutilized sub-bandgap photons into sensitive resonant photons. This chapter exclusively focuses on RE-doped UC materials and their applications in solar cells. The RE-based UC photophysics, UC enhancement, and applications in solar cells will be reviewed and briefly discussed.

Keywords Photon up-conversion · Rare-earth ion · Solar cell · Transmission-loss

H.-Q. Wang (✉)
School of Chemical and Biological Engineering, NingboTech University, Ningbo, China
e-mail: whqbme@gmail.com

A. Osvet (✉) · M. Batentschuk
Department of Materials Science and Engineering, Institute of Materials for Electronics and Energy Technology (i-MEET), Friedrich-Alexander Universität Erlangen-Nürnberg (FAU), Erlangen, Germany
e-mail: andres.osvet@fau.de

C. J. Brabec
Department of Materials Science and Engineering, Institute of Materials for Electronics and Energy Technology (i-MEET), Friedrich-Alexander Universität Erlangen-Nürnberg (FAU), Erlangen, Germany

Helmholtz-Institute Erlangen-Nürnberg (HI ERN), Research Center Julich (FZJ), Erlangen, Germany

J. S. Lissau, M. Madsen (eds.), *Emerging Strategies to Reduce Transmission and Thermalization Losses in Solar Cells*, https://doi.org/10.1007/978-3-030-70358-5_12

12.1 Introduction

In photovoltaic solar cells, several energy losses are related to device electronic properties, including contact voltage loss, recombination loss, and junction loss [1, 2]. Besides, there are two more primary loss mechanisms linked to photon harvesting, that is, thermalization loss of above-bandgap photons and transmission loss of sub-bandgap photons (Fig. 12.1). Both of the two losses arise from the intrinsic properties of the semiconductor material [1]. In thermalization loss, above-bandgap photons lose part of their energy in the form of heat by non-radiative relaxation after being absorbed, which is discussed in another chapter of the book. The transmission loss is caused by the fact that incident sub-bandgap photons cannot be absorbed by semiconductor and cannot generate charges due to low energy. For example, about 20% of the solar energy is lost owing to the transition losses in a crystalline-silicon (c-Si) solar cell. All these fundamental losses directly lead to an energy conversion efficiency limit of ca. 30% for single-junction solar cells under non-concentrated AM1.5 illumination [2].

Among the above loss mechanisms, the sub-bandgap energy loss is one of the main. Over 50% of the solar irradiation photons on the earth are located in the infrared (IR) range (UV ~5%, visible ~43%, IR ~52%) [5], which generally cannot be harvested by most existing solar cells. With a bandgap of approx. 2 eV (~650 nm), solar cells absorb around 24.3% of solar photons. Even with a bandgap of 1.1 eV (~1100 nm), only 63% of the solar photons can be utilized. Most current thin-film photovoltaics like dye-sensitized solar cells, organic solar cells, and perovskite solar cells are sensitive to solar photons only in visible range, generally with wavelength shorter than 800 nm. They usually suffer from this common sub-bandgap loss. Especially, most available organic polymers and small-molecule semiconductors exhibit energy bandgaps ≥2 eV, with only a few low-bandgap candidates available [6, 7]. Hence there is a clear mismatch between the absorption spectrum of active

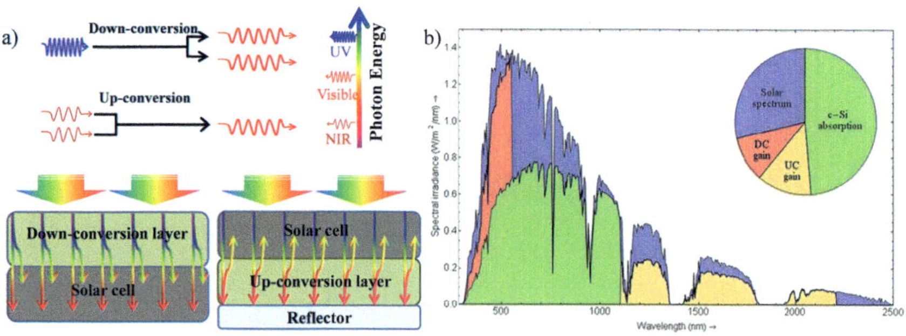

Fig. 12.1 Schematic illustration of photon down-conversion and up-conversion concept, and their incorporations in devices to reduce thermalization and transmission energy losses in solar cells (left, redraw from [3]). Potential efficiency gain of down- and two-photon up-conversion for a c-Si solar cell (right, reproduced from [4] with permission from the Royal Society of Chemistry)

materials and the terrestrial solar spectrum [8]. As such, it is clear that the loss of sub-bandgap photons is one of the major obstacles on the way of improving efficiency, and exploitation of the NIR/IR photons of solar radiation is essential to go beyond the current efficiency limitation of single-junction solar cells [2].

To reduce the sub-bandgap losses in photovoltaic solar cells, different strategies including tandem or multi-bandgap fabrication [9, 10], design of new absorbers [6, 11], and photon up-conversion (UC) [3, 12–14] have been developed to exploit the sub-bandgap solar photons. For the first two strategies, we refer the readers to previously published reports [6, 10, 15]. The photon UC can convert nonresonant sub-bandgap photons into the resonant region of solar cells, and can then be harvested by the solar cell contributing to current density. Particularly, photon UC provides a solution to circumvent the transmission loss without changing the existing designs and merits of solar cells (Fig. 12.1).

Photon UC describes a nonlinear optical process in which UC materials emit one high-energy photon by absorbing two or more lower energy photons. Contrary to the Stokes emission process, UC refers to anti-Stokes emission. Since first proposed in 1959 for developing infrared detectors [16], photon UC has attracted great attention due to its promising potential in the applications of optical detector, solid-state laser, bioanalysis, display technology, and photovoltaics [17]. Different approaches including rare-earth (RE) ion-based UC [18] and triplet-triplet annihilation-based UC [19] can achieve photon UC, among which RE ion-based UC is achieved by doping trivalent lanthanide (Ln^{3+}) ions (mostly erbium and ytterbium) in a solid-state matrix [18], making use of their discrete and long-lifetime atomic states that can facilitate stepwise absorption of multiple photons, followed by UC photon emission. The triplet-triplet-annihilation (TTA) UC is discussed in other chapters of this book.

RE ion-based UC is not only a smart concept to exploit the IR photons and use them at a higher thermodynamic potential. Further, this strategy is the only one which does not require specific engineering of semiconductor/absorber and solar cell device, but can be applied to any type of solar cells including silicon, thin film, organic/hybrid, as well as perovskite solar cells [3, 20]. RE up-converters are preferentially placed at the rear (or both sides) of a solar cell as an external layer, where it takes up the sub-bandgap photons and re-emits visible photons resonant to solar cell, hence improving photon harvest. Alternatively, nanosized up-converters can also be incorporated inside a solar cell as "internal layer" to construct electrode or comprised in active layer, where they play multifunctional roles. Different theoretical predictions show that promising increase of the Shockley-Queisser efficiency limit for solar cells can be expected [21, 22], by using ideal up-converters under non-concentrated solar irradiation (AM 1.5).

Although theory predicts a significant performance increase when Ln-based UC is coupled to a solar cell, so far no absolute increase in the performance of UC-coupled solar cells was recorded under normal one sun (AM 1.5) irradiation. Up to now, the reported improvement of the overall solar cell efficiency remains rather limited even under intense laser beam irradiation [3, 22]. This is mainly attributed to two limitations. First, the UC efficiency of current up-converters is still far from perfect. Most absorbed sub-bandgap photons cannot be effectively converted into resonant photons. Especially the UC efficiency is extremely low when excited under AM1.5 solar irradiation. For instance, for a laser irradiation close to one sun, the reported highest UC quantum yield (QY) is ~0.6% in alpha-($NaYbF_4$:0.5%Tm^{3+})@CaF_2 core–shell UC nanoparticles [23]. The second limitation arises from the narrow, discrete, and rather weak absorption of doped lanthanide ion owing to the nature of f–f transition. This severely restricts the harvesting capabilities of UC material doped with lanthanide ions.

12.2 Photophysics of RE Ion-Based UC

Principle

Photon up-conversion can be observed when a luminophor is excited with the light of lower energy than the light emitted (anti-Stokes emission) (Fig. 12.2). It is different to the multiphoton excitation process. Multiphoton excitation requires excitation pulses with femtosecond frequency, whereas the strength of the absorption depends on the square of the light intensity. In contrast to that, typical up-conversion mechanisms involve the excitation of an ion in the ground state by one photon and subsequent energy transfer to another ion in the ground state under absorption of one or more photons and de-excitation of the first ion.

The effect of photon UC was first discovered in the late 1950s and studied intensively during the 1960s [18]. Since then, a large variety of applications can be found, such as in detector, bio-imaging, lasers, and photovoltaics. It was applied to photovoltaics first in a GaAs solar cell in 1996 [24]. A comprehensive history and physical fundamentals of photon up-conversion materials have been reviewed by Auzel in 2004 [18].

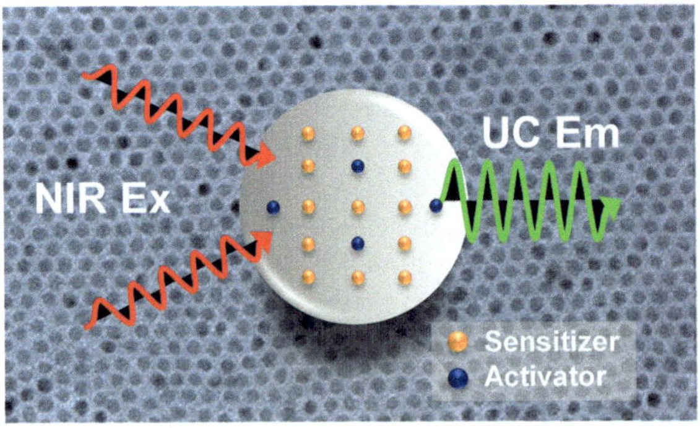

Fig. 12.2 Schematic illustration of NIR excitation and UC emission process in a lanthanide ion-doped UC nanoparticle

Lanthanide Ions

Lanthanides are comprised of a series of elements in the periodic table, containing the 15 Ln elements from lanthanum to lutetium. They possess special electronic configurations of $1s^2 2s^2 2p^6 3s^2 3p^6 3d^{10} 4s^2 4p^6 4d^{10} 4f^n 5s^2 5p^6 5d^m 6s^2$ (n = 0–14, $m = 0$, 1). On the other hand, Ln^{3+} ions are mostly stable. They have an electron configuration in the form of $1s^2 2s^2 2p^6 3s^2 3p^6 3d^{10} 4s^2 4p^6 4d^{10} 4f^n 5s^2 5p^6$, n = 0–14. Due to the dipole-dipole forbidden nature of $4f$-$4f$ transitions as well as the well-shielded effect by the outer $5s^2 5p^6$ electronic state shell, lanthanide ions generally have limited light-absorbing ability due to low extinction coefficient and narrow-band absorption [25]. But meanwhile this results in long-lifetime electronic states, line-like sharp UC emission with high resistance to photobleaching, photochemical degradation, and surrounding crystal field change [26]. Besides, due to the complex electronic configurations, the energy levels of Ln^{3+} are mostly rich and usually exhibit a ladderlike feature (Fig. 12.3) [25]. The ladderlike energy levels and relatively long lifetimes enable a relatively effective UC luminescence.

Main Up-Conversion Mechanisms

The key photophysical processes in an UC of Ln^{3+} ions include the absorption of excitation photons, subsequent energy transfer processes, and finally the release of excitation energy, in particular radiative emission of the up-converted high-energy photons. Different mechanisms have been recognized to be involved in UC either alone or in combination (Fig. 12.4). These mechanisms include, in order of increasing efficiency, two-photon absorption (TPA, Fig. 12.4a), second harmonic generation (SHG, Fig. 12.4b), excited-state absorption (ESA, Fig. 12.4c), and sensitized energy transfer (ETU, Fig. 12.4d). In TPA and SHG, virtual intermediate levels are involved. The two excitation photons must coincide (and also be coherent for SHG, since the emitting level is also virtual). Thus these two processes are very unlikely and very inefficient (with typical efficiency in order of 10^{-11}–10^{-13}).

UC by ground-state absorption/excited-state absorption (ESA, also denoted as GSA/ESA, typical efficiency in the order of 10^{-5}) is the simplest UC mechanism. It only involves one dopant, typically a trivalent lanthanide ion such as erbium Er^{3+}. In ESA, the sequential absorption of two (or more) excitation photons by the dopant leads to the population of the excited level from which the UC emission arises. ETU is the most efficient up-conversion mechanism. It involves non-radiative energy transfers between a sensitizer (e.g., Yb^{3+}) and an activator (e.g., Er^{3+}, Ho^{3+}, Tm^{3+}). The sensitizer usually has a strong absorption coefficient at the excitation wavelength. The sensitizer absorbs one incident photon and gets into the first excited level. Then it relaxes back to a lower energy state (the ground state in the case of Yb^{3+}) by transferring its energy to a neighboring activator, raising the latter to a higher energy state. These energy transfer processes are generally based on electric

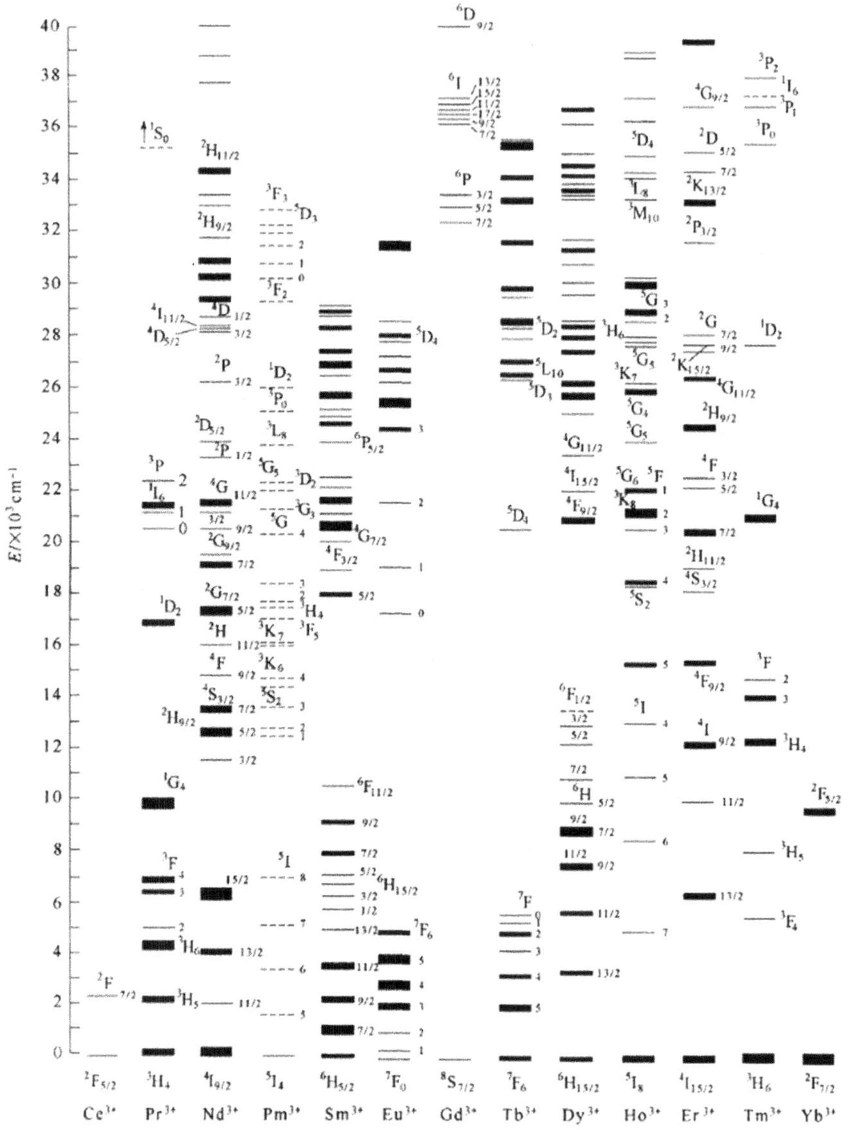

Fig. 12.3 Partial energy-level diagram of lanthanide ions ($Ce^{3+} \rightarrow Yb^{3+}$) doped in a low-symmetry crystal (LaF3). Reproduced from [25] with permission from Tsinghua University Press and Springer-Verlag Berlin Heidelberg 2005

dipole-dipole interactions, with a typical efficiency in the order of 10^{-3} cm^2/W. More recently, as a derivative of ETU mechanism, optical energy migration (EM, Fig. 12.4e) up-conversion was proposed and developed by professor Liu et al.

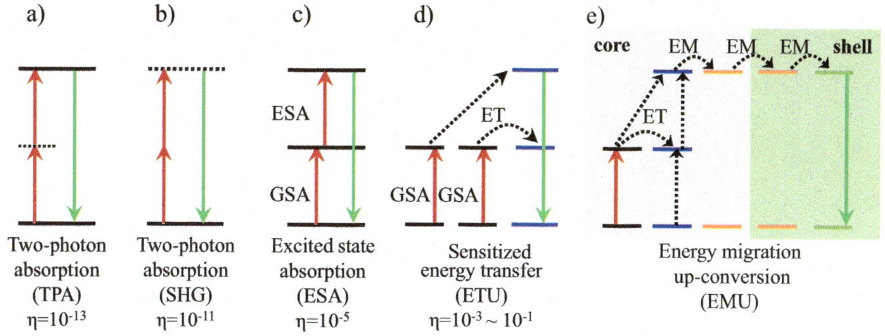

Fig. 12.4 Simplified energy-level diagrams depicting reported photon up-conversion or anti-Stokes processes, including TPA (**a**), SHG (**b**), ESA (**c**), ETU (**d**), and EMU (**e**) mechanisms

[27], in which efficient photon UC can be realized by rational design of core-shell structure with a set of lanthanide ions incorporated into separated layers at precisely defined concentrations. In the EMU mechanism, a sensitizer center excited by GSA first transfers its excitation energy to an accumulator center. The EM then occurs from the high-lying excited state of the accumulator to a migrator center, followed by migration of the excitation energy via the migrator ion through the core-shell interface. Subsequently, the migrated energy is trapped by the activator center in the shell which then emits the UC photons.

In contrary, the ESA and ETU mechanisms are distinctly different from the TPA and SHG optical processes. The key distinction is the involvement of intermediate electronic states in the up-converting system. The UC process is mediated by real intermediate electronic states. These states have long enough lifetime for an excitation to be stored there, and allow a second photon to further excite the phosphor to a higher lying excited state. Since ESA and ETU UC processes involve real energy levels, very high excitation powers are no longer necessary. The excitation power thresholds for these UC are typically about 6–11 orders of magnitude lower than those required for SHG or TPA. It is also worth mentioning that no coherent excitation source is required for them. Given its high-energy conversion efficiencies and low excitation power thresholds, the ESA and ETU UC are particularly relevant for applications. Experimental results indicate that sensitized energy transfer up-conversion is by far the most efficient process for photon UC [18]. Typical energy transfer processes between sensitizer Yb^{3+} and activators Er^{3+} (left), Ho^{3+} (middle), and Tm^{3+} (right), and corresponding UC emissions in nanocrystals, are shown in Fig. 12.5. All the processes presented in Fig. 12.4 are nonlinear two-photon optical processes. The typical emission power density P_{em} depends quadratically on the incident power density P_{inc} with a proportionality coefficient μ such as $P_{em} = \mu P^2{}_{inc}$.

Besides the above UC processes, numerous competing radiative and non-radiative relaxation processes and phonon-assisted energy transfers may also be

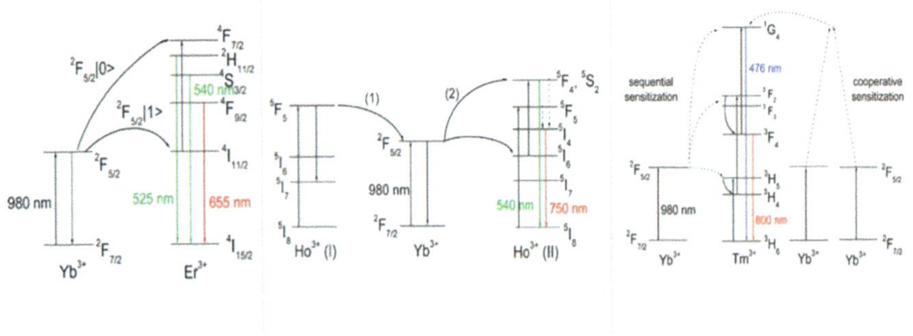

Fig. 12.5 Typical ETU up-conversion processes between sensitizer Yb^{3+} and activators Er^{3+} (left), Ho^{3+} (middle), and Tm^{3+} (right) in nanocrystals. Reproduced from [28]. Copyright 2008 American Chemical Society

involved simultaneously. They all play a vital role in determining the UC emission properties like the overall emission intensity and ratios of different emission bands.

12.3 Principal Factors and Strategies to Enhance UC

Improving the UC light intensity for up-converting phosphors is desired for almost all applications. We recommend references of Boyer and van Veggel [29] and Crosby and Demas [30] for readers to look into absolute quantum yield measurements of UC. Despite the long-recognized potential of Ln-based up-converting materials, current up-converters are still limited by poor UC efficiencies. The Ln ion-based UC could be influenced by various factors, such as the type of dopant ions (ion type, with or without sensitizer), Ln ion concentration, host lattice, UC mechanisms involved, and geometry of up-converters (e.g., bulk, nanoparticle, or core-shell structure). Thus, to improve the UC intensity/efficiency, numerous strategies have been developed for Ln ion-doped UC materials, especially for nanosized materials. These strategies include optimizing the types and concentration of Ln ions, selection and design of host material, geometry design of up-converters (such as nanoparticle, core-shell structure), plasmonic enhancement, surface passivation, and dye sensitization. And they all have been demonstrated effective in UC emission enhancement.

Luminescent Centers

Due to the unique electron configuration of Ln ions, Ln^{3+}-doped materials are particularly suitable for photon UC. Since typical Ln^{3+} ions like Er^{3+}, Ho^{3+},

Tm^{3+}, Nd^{3+}, and Pr^{3+} emit intense UC photons in visible range, these are suitable for photovoltaic/bio-imaging and most other applications. They have most often been incorporated in host materials as luminescent centers (activators). Ln^{3+} ions commonly have multiple spectroscopically active levels except for Yb^{3+} and Ce^{3+} [31]. This provides opportunity and space for manipulation of UC band and color.

Up-converting systems frequently involve more than one type of center. The UC-emitting center is always involved, referred to as activator. However, an efficient UC system usually comprises another type of center that is sensitizer. As partner ions to activators, the sensitizer ions harvest light, and then transfer the absorbed photon energy to activators to promote UC emission. Typically, Yb^{3+} ion has been most widely used as a sensitizer due to its relative intense absorption in the infrared range (centered at 975 and 1550 nm). So far, the literature-reported results indicate that the most efficient infrared-to-visible up-converting ions are Yb^{3+}-Er^{3+}, Yb^{3+}-Ho^{3+}, and Yb^{3+}-Tm^{3+} co-doping systems.

Dopant Concentration Control

The UC emission intensity and spectral properties depend on dopant concentration. Based on the host lattice and the type of rare-earth ions, composition engineering can result in an efficient output of up-conversion. Generally, with high dopant concentrations, quenching effects appear due to cross-relaxation processes and reduce the amount of UC emission. However a low dopant concentration could also result in low UC efficiency due to the long ion-ion distance and insufficient light absorption and/or emitting centers.

The UC efficiency of an ETU process is highly dependent on the concentration of the dopants (Fig. 12.6) [32, 33]. Therefore, in most UC materials the concentration of the dopant ions must be optimized. Unlike the activator ions, the sensitizer Yb^{3+} has only one exclusive excited energy state ($^2F_{5/2}$); thus the Yb^{3+} ions generally do not induce serious cross relaxation and quenching effect. For sensitizer Yb^{3+}, the optimum concentration is around 20–100% for fluoride materials [32], depending on the structure of the nanoparticles, and 10% for oxide host lattices (Fig. 12.7) [34]. For activators, the concentration is determined to be typically about 0.5–1% for Tm^{3+} ions and 2% for both Er^{3+} and Ho^{3+} ions [35]. The optimized activator concentrations work for both fluoride and oxide host matrices. As an exception, quenching effect for higher Yb^{3+} concentration in Yb^{3+}/Er^{3+} (or Ho^{3+})-doped fluoride UC nanoparticles could also be observed [35], which might arise from Yb^{3+}-induced energy migrations from the lattice points inside the particle to the surface quenching sites around the particle.

Besides the overall intensity of UC, the relative intensity ratio of different emission bands can also be manipulated by tuning the doping ion concentration (Figs. 12.6 and 12.7), e.g., by tuning the sensitizer Yb^{3+} concentration and/or sensitizer/activator mole ratio in hosts [32, 34, 36]. Generally, the relative intensity

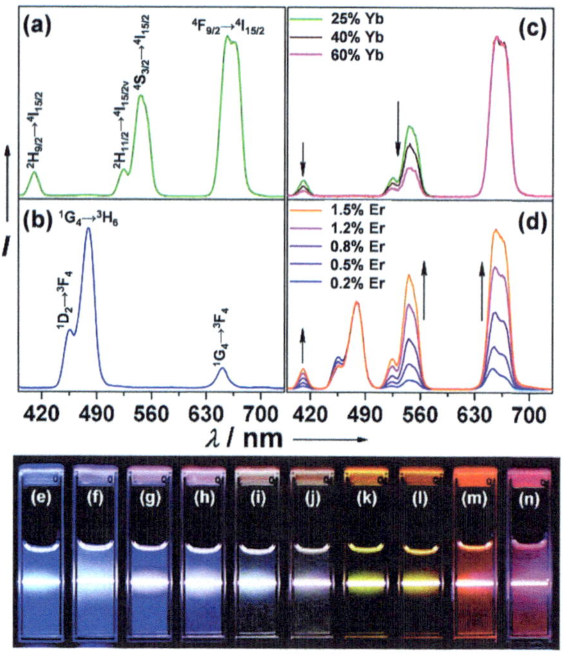

Fig. 12.6 UC spectra of $NaYF_4$:Yb/RE nanoparticles in ethanol solutions (10 mM), with different Yb/RE ratio: (**a**) Yb/Er (18/2 mol%), (**b**) Yb/Tm (20/0.2 mol%), (**c**) Yb/Er (25–60/2 mol%), and (**d**) Yb/Tm/Er (20/0.2/0.2–1.5 mol%). Luminescent photos of corresponding colloidal solutions of (**e**) Yb/Tm (20/0.2 mol%), (**f–j**) Yb/Tm/Er (20/0.2/0.2–1.5 mol%), and (**k–n**) Yb/Er (18–60/2 mol%) doped $NaYF_4$. Excitation: 980 nm, 600 mW laser diode. Reproduced from [32]. Copyright 2008 American Chemical Society

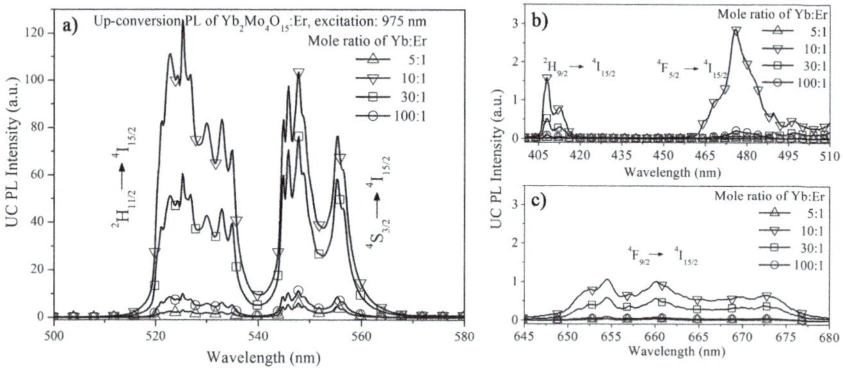

Fig. 12.7 UC spectra of $Yb_{2\text{-}x}Mo_4O_{15}$:$Er_x$, showing the UC intensity change at different bands according to the change of Yb and Er concentration. (**a**) Peaks in the visible range, (**b**) in the violet-blue range, and (**c**) in the red range. Reproduced from [34]. Copyright 2014 WILEY-VCH Verlag GmbH & Co. KGaA, Weinheim

ratio of red-to-green emission increases with the increased Yb^{3+} ion concentrations. Because the higher concentration of Yb^{3+} ions will reduce the distance between Yb^{3+} and Er^{3+}, and promote the back energy transfer from Er^{3+} to Yb^{3+} ions, it suppresses the green ($^2H_{11/2}$, $^4S_{3/2} \rightarrow {}^4I_{15/2}$) emissions and/or enhancement of red emission ($^4F_{9/2} \rightarrow {}^4I_{15/2}$) [32].

Host Lattice

Host lattice plays an important role in enabling efficient UC. First the host material must have a big optical bandgap (usually greater than 6 eV) to ensure transparence of the host lattice to excitation light and UC emissions [37]. This is particularly important for the application in photovoltaic solar cells. Otherwise the incorporated up-converters will compete to absorb the resonant visible solar photons in solar cell application, especially when placed inside the device. To enable efficient UC, the host lattice must provide an asymmetrical crystal field. In principle, a strong asymmetric contribution from the crystal field enhances the optical transition probabilities (absorption and emission) in Ln^{3+} ions [38]. The host lattice can not only influence the transition probabilities of Ln^{3+} ions by lowering the site symmetry but also tune the UC wavelength by shifting the energy levels. Thus varying the asymmetry of the host lattice can be an efficient way to enhance and tune the UC efficiency and properties. In addition, an ideal host lattice for UC must possess low phonon energies [18]. A host lattice with lower maximum phonon energy will present lower non-radiative multiphonon relaxation, thus enabling a higher ratio of excited photons in the UC emission process.

Up to now, Ln ions have been incorporated into all kinds of host materials to achieve photon UC. This includes oxides, phosphates, and vanadates, as well as yttrium fluoride-based and various mixed systems [39]. The broadness and intensity of the absorption spectrum of up-converter can vary from host to host. The choice of the host matrix determines the distance between the emitting ions and the nature of the atoms surrounding the lanthanides and therefore influences the up-conversion efficiency. The internal up-conversion quantum yield rarely exceeds 10–15% in the bulk. Up-converting nanomaterials generally possess lower UC efficiency, not exceeding 2% in quantum yield, due to small size and defects associated with surface quenching [40]. $NaREF_4$ is known to be the optimal host matrix to provide efficient UC due to the low maximal lattice phonon energy (about 350 cm^{-1}) [40, 41]. There are two different phase constructions for $NaREF_4$ nanocrystals, namely the cubic phase (α-phase) and hexagonal phase (β-phase). It is generally adopted that the crystal structure of the *b*-phase favors the emission of the dopant ions rather than the α-phase; therefore Ln^{3+}-doped β-$NaREF_4$ should have a stronger luminescence intensity [42].

Host lattice manipulation can also be realized by tailoring the crystal lattices through ionic doping. Adding certain ions into the crystal lattice is also helpful in reducing the crystal symmetry. A dopant ion with smaller size compared to the oxide or fluoride host atoms, for example Li^+, results in smaller crystal lattice while a relatively large ion will result in lattice expansion. Li^+ doping through wet-chemical method resulted in 30-fold increase of UC luminescence of the $NaYF_4$:Yb-Er solution sample (Fig. 12.8) [43].

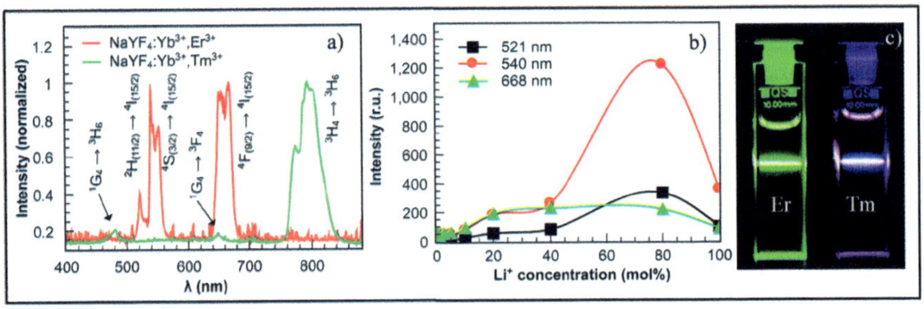

Fig. 12.8 UC spectra of Er and Tm co-doped with Yb in $NaYF_4$ host, under 975 nm excitation (**a**). Influence of Li^+ doping concentration on the UC intensity at different emission bands (**b**). Photographs of $NaYF_4$:Yb-Er and $NaYF_4$:Yb-Tm colloidal sample under 975 nm excitation (**c**). Reproduced from [43]. Copyright 2009 American Chemical Society

Core-Shell Structure Strategies

The excitation energy loss can occur at the surface of up-converting materials due to defect associated with non-radiative relaxation. Especially for those nanosize up-converting materials, this energy loss is generally more pronounced, as the surface-to-volume ratio increases with decreasing size in nanoscale, thus reducing the UC emission/brightness.

At the early stage, with the purpose of stabilizing the up-converting nanoparticles and tailoring the surface chemistry by functional groups in applications, core-shell structures are designed. The up-converting core particles can be coated by a shell of silica or a layer of functional molecules. Inspired by this factor, this strategy is adopted to passivate the surface defects to suppress the surface quenching and increase the UC efficiency (Fig. 12.9) [44, 45]. Core-shell structure designs not only enable us to create efficient UC with desirable excitation and emission properties, but also provide a platform for integrating multiple functionalities that offers new opportunities for various applications. And thus the core-shell structure strategies have increasingly attracted much interest from researchers [46, 47].

Zhang et al. designed and synthesized $NaYF_4$:Yb/Er@$NaGdF_4$ core-shell nanoparticles. The dependence of UC intensity on the $NaGdF_4$ shell thickness was investigated. A limited thickness plateau around 3 nm is determined, corresponding to the critical distance of interaction between surface oscillators and lanthanide dopants [48]. A multilayered core-shell structure was also designed to enhance UC by means of energy migration [27]. $NaYF_4$@$NaYF_4$@$NaYF_4$ core-shell-shell nanoparticles with Yb/Er doped in the inner shell layer were prepared to enhance the ratio of green-to-red emission intensity [49]. And the overall UC enhancement can be attributed to the lowered defect density in the inner shell layer [50]. In addition, nonepitaxial metallic shells were also employed to enhance UC emission, by amplifying excitation flux and in parallel accelerating radiative decay through surface plasmon resonance [51].

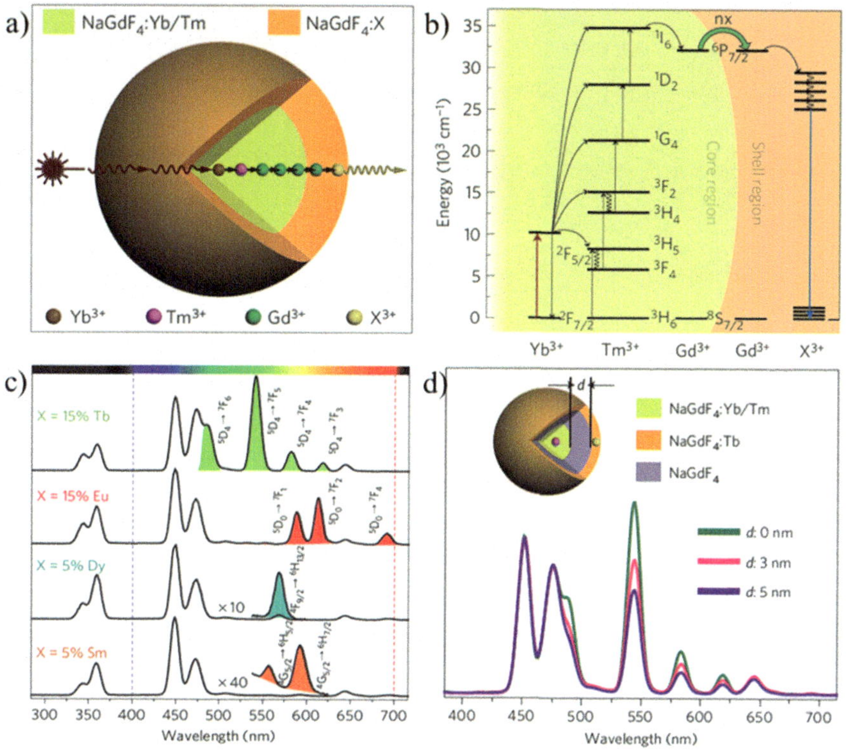

Fig. 12.9 (**a**) Schematic design of a lanthanide-doped $NaGdF_4$@$NaGdF_4$ core-shell nanoparticle for energy migration UC (X: activator ion). (**b**) Proposed energy transfer mechanisms in the core-shell nanoparticle. Note that only partial energy levels of Tm^{3+} and Gd^{3+} are shown for clarity. (**c**) Emission spectra of the as-prepared $NaGdF_4$@$NaGdF_4$ core-shell nanoparticles doped with different activators. (**d**) Schematic design and emission spectra of core-shell-shell nanoparticle with Tm^{3+} and Tb^{3+} separated by a $NaGdF_4$ shell layer of different thicknesses (0, 3, and 5 nm). Reproduced from [27]. Copyright 2011 Springer Nature

Plasmonic Enhancement and Other External Resonators

Localized surface plasmonic resonance is a collective oscillation of free charge carriers at the interface of metallic or self-doped semiconductor nanomaterials and dielectric medium. This has been applied as plasmonic strategy to enhance UC emission in lanthanide-doped nanomaterials.

Noble metal nanomaterials like gold and silver are often used to surround the up-converting rare-earth elements, with an effect to enhance the optical field at the resonance frequencies and enhance the absorption of the rare-earth element. The geometry can be designed in different ways [52]. For instance, by designing gold-shell-coated $NaYF_4$:Er,Yb@SiO_2 nanoparticles [53], UC enhancement factors of ~9.1 and ~6.7 for green and red emission were recorded, respectively, due

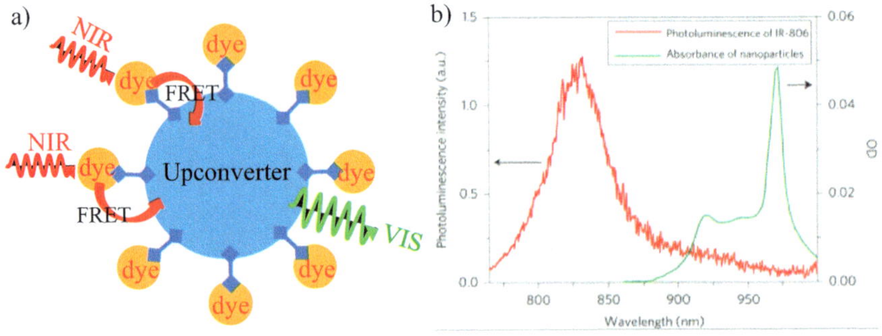

Fig. 12.10 UC enhancement by NIR antenna dye sensitization. (**a**) Schematic illustration of the dye-sensitized nanoparticle. After being absorbed by the antenna dyes, the NIR light energy is transmitted to the Yb^{3+} ion through Förster resonance energy transfer. The Er^{3+} ion then accepts the energy from the excited Yb^{3+} ion, and gives the visible emission. (**b**) Emission (red) and absorption (green) spectra of the cyanine dye (IR-806) and $NaYF_4$:Yb-Er nanoparticles. Redrawn from [55]. Copyright 2012 Springer Nature

to plasmon-enhanced absorption at 980 nm. It is also designed with the metallic material as the inner core, coated with a thin silica layer and then with the UC $NaYF_4$:Er,Yb outer layer. It should be noted that the distance between the emitter and plasmon is critically sensitive. The inner silica layer has to be optimized to achieve efficient plasmonic enhancement.

Other materials like photonic crystals, resonant waveguides [54], and organic dye antenna (Fig. 12.10) [55, 56] have also been used to either enhance the local electric field intensity or increase/broaden the absorption of the UC system, and hence the UC performance. Different to plasmonic enhancement, dye sensitization also spectrally broadens the sensitization effect. More details and discussions about the concepts of nanophotonics for UC enhancements will be present in a separate chapter in this book (by Prof. Won Park, Boulder).

Transition Metal Ions Tuned and Sensitized UC

Transition metals (TM), specifically *d*-block element ions with unfilled *d* orbitals, such as Mn^{2+}, Cr^{3+}, Re^{4+}, V^{3+}, Mo^{3+}, Ni^{2+}, and Ti^{2+}, present UC properties which are generally inefficient and only observed at cryogenic temperatures [18, 57], due to the large electron-phonon coupling effect. However, with TM ions as dopants, the UC emission of Ln^{3+}-doped materials can be effectively manipulated, which benefits from the specific metastable energy levels of Ln^{3+} owing to well-shielded 4f electrons and tunable energy levels of the TM ions. The TM dopants can change the surrounding environment of Ln^{3+} ions and the electronic coupling/transfer with Ln^{3+}. Thus the TM ion doping strategy not only is capable of controlling the morphology/phase of the synthesized material, such as in $MnF_2:Yb^{3+}$-Er^{3+}, but also successfully enhances UC intensity and tunes the single emission band features. Intense UC and a pure single-band emission is obtained in $NaMnF_3:Yb^{3+}$-Er^{3+} [58], $NaYF_4:Yb^{3+}$-Er^{3+} [59, 60], etc. systems, by Mn^{2+} doping. In addition, TM ions exhibit broadband absorption. Thus they can also be used as broadband sensitizers [61, 62]. The contribution of broadband absorption of TM ions was observed in UC materials of $La_3Ga_5GeO_{14}:Cr^{3+}$, Yb^{3+}, Er^{3+} [63], and $La(Ga_{0.5}Sc_{0.5})O_3:Er^{3+},Ni^{2+},Nb^{5+}$ [64].

Optimizing the Excitation Schemes

Besides the above methods focusing on material composites, varying the excitation scheme can be another strategy to enhance/tune UC emission. Without saturation, generally a more intense excitation will result in a stronger UC emission. By changing the excitation pulse parameters such as the width, repetition rate, and peak power, the UC emission properties can be tuned [65]. Moreover, utilizing multiple-wavelength and even broadband light source to replace conventional single-wavelength light source for excitation [66, 67] could be another way to break the UC performance bottlenecks.

12.4 Implementation in Solar Cells

Introduction of UC concept in a solar cell can be an effective approach to compensate the transmission loss in all types of solar cells. In real applications, the UC layer can be incorporated into a photovoltaic device as an external layer or internal layer according to the solar cell category and device configuration (Fig. 12.11). The first proof-of-concept experiment for UC in a solar cell was undertaken by Gibart et al. in 1996 [24]. Since then, considerable progress has been made.

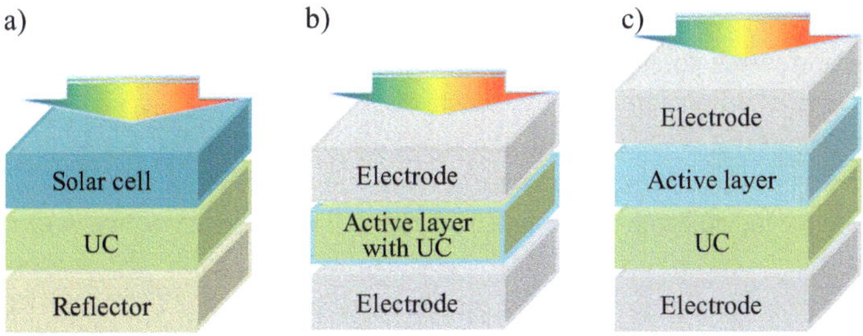

Fig. 12.11 Schematic illustration of implementation of UC in solar cells: (**a**) as an external layer; (**b**) in the active layer; (**c**) as an interfacial/charge-transport layer at the electrode

As an external layer, the UC materials are normally placed at the rear of the device (though the front side or both sides have been investigated as well), without changing the existing designs of the devices, which avoids introducing new parasitic losses to the solar cells. In this case, the resonant solar photons are utilized when passing the solar cell. The nonresonant NIR photons will pass through the cell and reach the UC layer. Then the generated high-energy photons are reflected back into the solar cell and contribute to the current density. In this arrangement, a rear reflecting layer is usually placed behind the UC layer. This helps to reflect the transmitted long-wavelength excitation photons back to the UC layer and direct the up-converted high-energy photons back to the active layer, thus overall enhancing the NIR photon utilization and solar cell performance. In most cases, the back electrode of solar cell is a relatively thick silver/gold metal layer. So this rear-side external setup for UC usually requires that the solar cell features a bifacial functionality, enabling sub-bandgap photons to reach the UC layer and allowing the emitted UC photons to be coupled back into the solar cell.

Besides the advantages mentioned above, certain disadvantages exist in the external setup for UC layer. There exist certain gaps/distance and interfaces between the UC layer, active layer, and back electrode. Of course this will reduce the amount of IR photons that reach the up-conversion layer and also the generated UC photons back to the cell. Besides the rather weak UC efficiencies, these extra losses further complicate the measurement and investigation of the UC effect in a solar cell especially under non-concentrated sun illumination. It is believed that the closer the up-convertor to the active layer the higher the probability for UC photons to be absorbed. All this is a strong argument to integrate the UC as an internal layer inside a solar cell, ideally as another wise functional layer. As an internal layer, it is often placed at the electrode or in the active layer, where it serves as an electronic interface and UC layer, or plays the bifunctional role of scattering and UC in active layer to promote device performance.

Silicon Solar Cells

In 1983 Saxena suggested the application of terbium-doped lanthanum fluoride and thulium-doped calcium tungstate materials in photovoltaic devices [68], but actual measurements with solar cells were not reported. The application of RE-based UC in solar cells was first reported by Gibart et al. in 1996 [24]. A Yb^{3+} and Er^{3+} co-doped UC material was placed at the rear side of an ultrathin GaAs cell. A photo response was observed under excitation with a Ti-sapphire IR laser at 1.39 eV.

For c-Si solar cells, the resonant region of the spectrum is from ultraviolet to 1100 nm. Ln^{3+} ions like Er^{3+}, Ho^{3+}, and Tm^{3+} emit visible UC photons and can be utilized in c-Si solar cells. However the co-doping with Yb^{3+} is not suitable for c-Si solar cells because the Yb^{3+} ions present large absorption coefficient at 980 nm (with $^2F_{5/2}$) which is resonant to the silicon. In 2003, Shalav and coworkers reported the application of RE-based UC material in silicon solar cells. The up-converter

$NaYF_4:Er^{3+}$ was located at the rear side of a bifacial cell. This led to a detectable photon response under excitation of 1500 nm. Then they reported an external quantum efficiency of 2.5% with a bifacial silicon solar cell, by incorporating $NaYF_4:Er^{3+}$ phosphors under 5.1 mW collimated laser excitation at 1523 nm [69]. Trupke et al. [22] placed $NaYF_4:Er^{3+}$ phosphor on the rear side of a bifacial silicon solar cell. EQE of 0.6% was obtained with laser excitation of 1550 nm (2 kWm^{-2}). Other examples on silicon solar cells include Gd_2O_2S:10% Er^{3+} with 4.74% EQE [40], and BaY_2F_8:30%Er^{3+} with 3.6% EQE [70], both at 1511 nm and with an irradiance of 0.1 W/cm^2. Ho^{3+} is also investigated as an emitter, with the advantage of harvesting light in the 1150–1230 nm range and emitting by up-conversion at around 650 and 910 nm. Ho^{3+} was used in up-converting glass ceramics placed on c-Si solar cells [71] and was combined with Yb^{3+} as a sensitizer or with Er^{3+} to broaden the absorption range [72].

In 2010, for the first time Yb^{3+} (18%)/Er^{3+}(2%) co-doped β-$NaYF_4$ was coupled to amorphous silicon solar cells. Due to higher bandgap of amorphous silicon (shorter than 980 nm), Yb^{3+} ions can be used to up-convert NIR photons into visible photons. Current-voltage measurements and spectral response measurements yielded a photocurrent of 10 μAcm^{-2} under illumination of a 980 nm diode laser (10 mW) [73].

Besides external coupling, also internal implementation of UC is applied in silicon solar cells. Lanthanide oxides Ln_2O_3 (Ln = La, Nd, Dy, Sm) are incorporated into solar cells as internal layer to control properties other than luminescence. Conduction between the semiconductor layer and the electrode can be enhanced by introducing the Ln_2O_3 layer and posttreatment [74].

Dye-Sensitized Solar Cells

UC layers can be easily coupled to dye-sensitized solar cells (DSSCs) [75]. Due to the typical device architectures, DSSCs offer unique possibilities for direct integration of up-converters as internal layer into the solar cells. In 2010, Demopoulos et al. [76] reported for the first time the application of Yb^{3+}/Er^{3+}-co-doped LaF_3-TiO_2 to fabricate a "triple-layer" working electrode on FTO in DSSC, where Yb^{3+}/Er^{3+} acts not only as an up-converter, but also as a dopant to improve the electrical properties of the mesoporous TiO_2 matrix. With a 980 nm fiber laser with 2.5 W power supply, open-circuit voltage V_{OC} of 0.40 V and short-circuit current I_{SC} of 0.036 mA were obtained. However, in this situation it is difficult to uniquely identify the contribution from UC effect. This multifunctional role was then explored later in DSSC using nonepitaxial core-shell β-$NaYF_4:Yb^{3+}/Er^{3+}$@SiO2 UCNPs [77] and core-shell-shell β-$NaYF_4:Yb^{3+}/Er^{3+}$@SiO_2@TiO_2 sub-microprisms [78], both producing an impressive relative efficiency increase of 6–11% (Fig. 12.12).

With another method [79], β-$NaYF_4$:20%Yb^{3+},2%Er^{3+} colloidal nanocrystals were intermixed with dye absorber of a DSSC. Under monochromic illumination

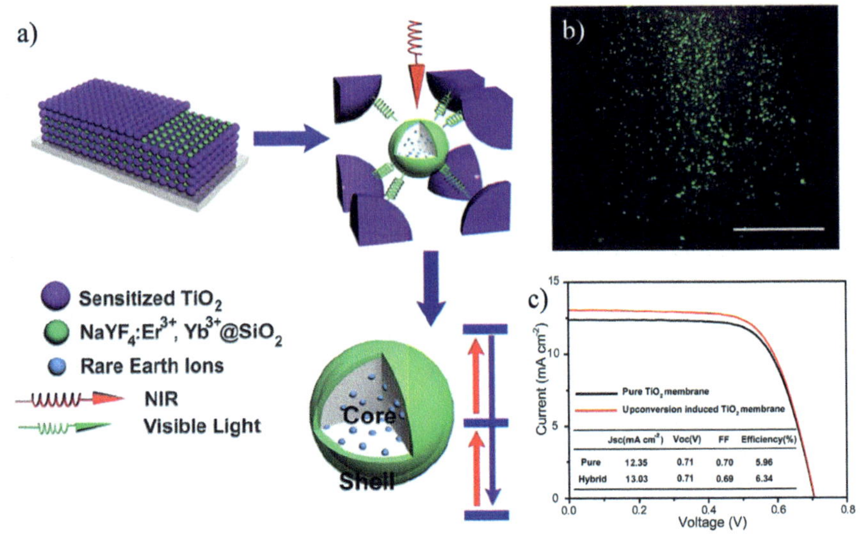

Fig. 12.12 (**a**) Schematic illustration of the working principle of UC-induced photoelectrode in DSSC. (**b**) Luminescence image of the UC-induced TiO_2 porous membrane (photoelectrode with core-shell UCNPs) confirms the existence of core-shell UCNPs in the photoelectrode. (**c**) *J-V* curves of the DSSCs based on TiO_2 porous electrode with or without the core-shell UCNPs. Reproduced from [77] with permission from the Royal Society of Chemistry

with 980 nm and 8 Wcm^{-2} irradiance, an EQE_{UC} of 0.011% was observed. Both fluorescence resonance energy transfer and luminescence-mediated energy transfer are discussed as potential routes for the migration of the energy from the up-converter to the dye.

Organic Solar Cells

Organic photovoltaics has emerged rapidly over the last decades. Due to the advantages like lightweight, flexibility, large-scale production, and low cost, this photovoltaic technology presents promising potential as a new generation of renewable energy source. Most available organic semiconductors exhibit energy bandgaps ≥2 eV, with only a few low-bandgap candidates available [7]. Compared to inorganic solar cells this bandgap is 0.2–0.4 eV larger. Reasons for this are the energetic losses required to split excitons in organic semiconductors, which overall results in a larger difference between the bandgap of the absorber and the maximum achievable built-in voltage of the solar cell [80]. Due to these intrinsic losses, photon conversion technologies are even more important in solar technologies based on this type of excitonic semiconductors. As such, the UC of sub-bandgap photons is more important for organic solar cells even than others.

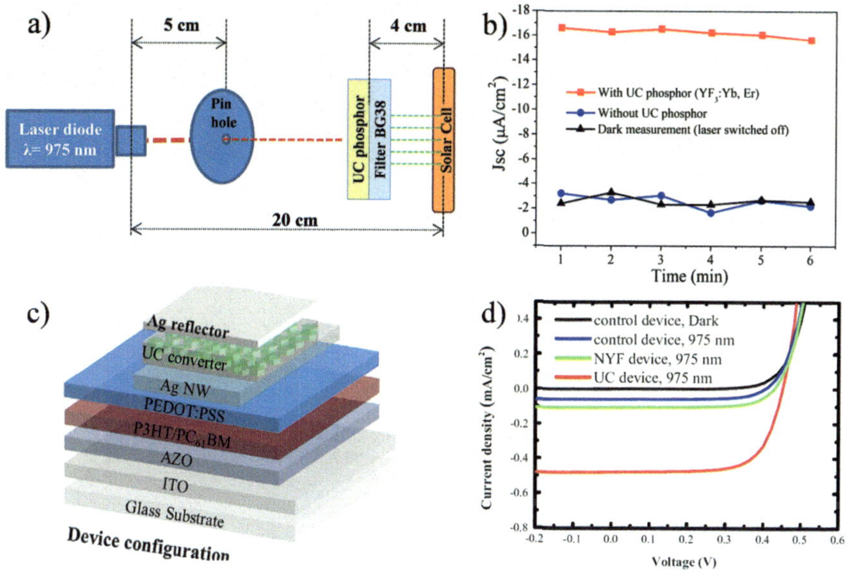

Fig. 12.13 (**a**) Schematic diagram of the setup for UC photocurrent measurement, in which the UC is placed in front of the organic solar cell. (**b**) Short-circuit current density J_{SC} (~16 μA cm^{-2}) recorded for UC/P3HT:PCBM organic solar cell under NIR laser diode excitation (975 nm, ca. 25 mW cm^{-2}). Reproduced from [3]. Copyright 2011 WILEY-VCH Verlag GmbH & Co. KGaA, Weinheim. (**c**) Schematic diagram showing the configuration of semitransparent organic solar cells with a common testing UC-PMMA film (NaYF4:Yb/Er-PMMA). (**d**) *J*-*V* characteristics of four different devices under 975 nm diode laser excitations (ca. 200 mW cm^{-2}): the control device (blue), the NYF device (green), the UC device, and the dark current of control device (dark). Reproduced from [81]. Copyright 2015 Elsevier B.V.

In 2011, a proof of principle to utilize sub-bandgap photons by UC in organic solar cell was demonstrated in our group for the first time (Fig. 12.13). In P3HT:PCBM-based organic solar cell, commercially available UC phosphor $YF_3:Yb^{3+}/E^{3+}$ was incorporated as an external layer on top of the device. A photocurrent density of 16.5 μA cm^{-2} was measured under a laser illumination (ca. 25 mW cm^{-2}, roughly 14 times the power density (1.8 mW cm^{-2}) yielded by the AM 1.5 solar spectrum between 960 and 1000 nm) [3]. After that, by fabricating a composite film with an UC phosphor $NaYF_4$:Yb/Er (20/2 mol%) and PMMA [81], we developed a universal approach/architecture for coupling external UC to all kinds of solar cells (Fig. 12.13). With this, the estimation of UC effect for different types of semiconductor solar cells can be conducted easily. In a semitransparent P3HT:PCBM solar cell, combined with a back reflector layer, the UC-PMMA layer generated a conversion efficiency of 0.1% and a short-circuit current density (J_{SC}) of 450 μA/cm^2 under 975 nm illumination (app. 200 mW/cm^2). Under one sun irradiation, a J_{SC} enhancement of 1.6% was demonstrated for the UC-coupled device which was partially assigned to the UC effect.

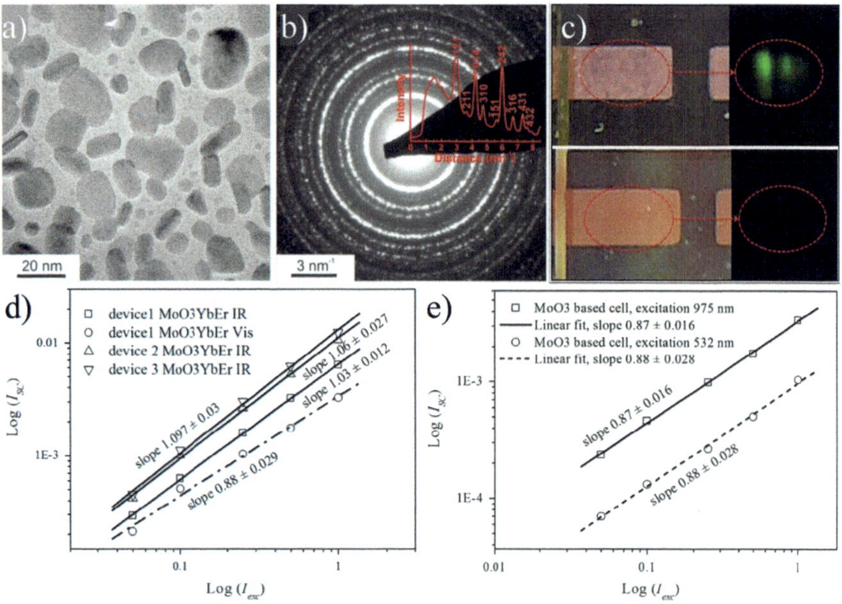

Fig. 12.14 (**a**) TEM image of up-converting metal oxides $Yb_2Mo_4O_{15}$:Er nanocrystals. (**b**) The electron diffraction pattern of the $Yb_2Mo_4O_{15}$:Er nanoparticles. (**c**) Photographs of the UC luminescence in solar cell with or without $Yb_2Mo_4O_{15}$:Er, excited with a laser diode (975 nm). (**d**) Light intensity dependence of the short-circuit current; slopes of Yb/Er-doped cell under NIR or visible excitations. (**e**) The slopes of an un-doped cell under 975 nm and visible (532 nm) excitations. Reproduced from [82]. Copyright 2012 Elsevier B.V.

Further, we demonstrated the multi-functionality of UC material as an internal buffer layer inside organic solar cell. Due to insulating characteristic of the widely used $NaYF_4$ host, it is not suitable to be incorporated inside device as electronic function layer. We developed semiconducting metal oxide-hosted UC phosphor $Yb_2Mo_4O_{15}$:Er to construct the electrode interface. As an internal layer in P3HT:PCBM solar cell, it plays the role as hole selection/extraction layer, and meanwhile as an UC function layer to utilize the sub-bandgap photons in P3HT:PCBM. The UC functionality was proven spectroscopically and demonstrated by analyzing the photocurrent response of the solar cell with/without $Yb_2Mo_4O_{15}:Er^{3+}$ to the excitation intensity (Fig. 12.14). This practically demonstrates that the $Yb_2Mo_4O_{15}:Er^{3+}$ film from a nanoparticle suspension works well as a charge-selective extraction layer in P3HT:PCBM solar cells, and simultaneously as an UC function layer [82].

Guo et al. reported another approach to utilize UC as bifunctional internal materials in an organic solar cell (Fig. 12.15). $NaYF_4:Yb^{3+},Er^{3+}$ nanocrystals were doped into PCDTBT:PCBM, the active layer of the solar cell [83]. The results showed enhanced short-circuit current density while maintaining the open-circuit voltage and fill factor, leading to an increase in power conversion efficiency. In

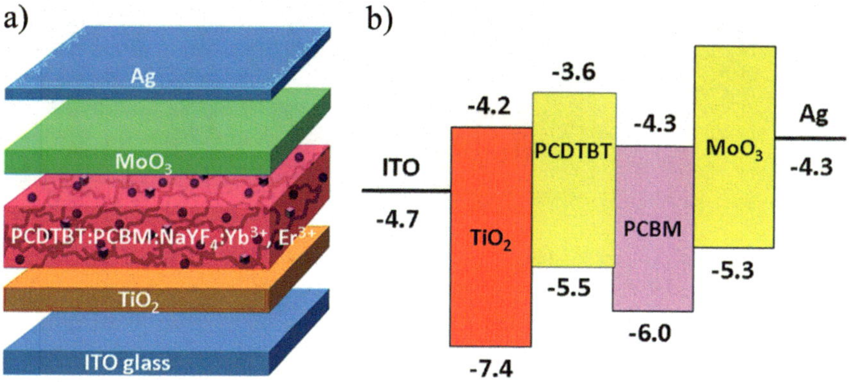

Fig. 12.15 Device structure of an inverted PCDTBT:PCBM polymer solar cell with incorporated $NaYF_4$:Yb^{3+},Er^{3+} UCNPs in the active blend, where the UCNPs play bifunctional roles (**a**), along with the corresponding energy-level structure (**b**). Reproduced from [83]. Copyright 2014 Elsevier B.V.

this architecture, the UC nanocrystals have a dual functionality for light absorption enhancement. On the one hand the photocurrent generated from the UC layer can improve the utilization of solar photons in the NIR range. On the other hand, scattering of $NaYF_4$ nanocrystals can enhance light absorption in the visible region. This approach to integrate UC and scattering functionality into active layer provides a promising and practical pathway for improving visible and NIR absorption in organic solar cells but also requires more extensive experimental characterization in order to distinguish the different contributions.

Perovskite Solar Cells

Integration of UC in perovskite solar cell (PSC) is quite a new research topic. The resonant solar spectral range of common PSCs is around 300–800 nm, which matches well with the emission bands of most typical Ln ions (e.g., Er, Ho, Tm, Nd). Thus PSCs can benefit much from UC by utilizing Ln-doped up-converting materials. Similarly, as an external layer, the up-converter can be attached to the rear of bifacial perovskite solar cells, similar to silicon-based up-converting solar cell devices. Alternatively, as internal UC, the up-convertor can be integrated in a device to function as electrode, such as for UC-TiO_2 (similar as in DSSCs), or mixed in the perovskite layer for light scattering and UC.

In 2016, Chen et al. [84] reported the proof of principle to apply UC $LiYF_4$:Yb^{3+},Er^{3+} single crystal in PSCs as external part. By placing the UC layer in front of the PSCs, PCE is enhanced by relatively 7.9% under the irradiation of simulated sunlight with a concentration factor of 7–8. He et al. [85] reported the utilization of monodisperse $NaYF_4$:Yb/Er up-converting nanoparticles to partly

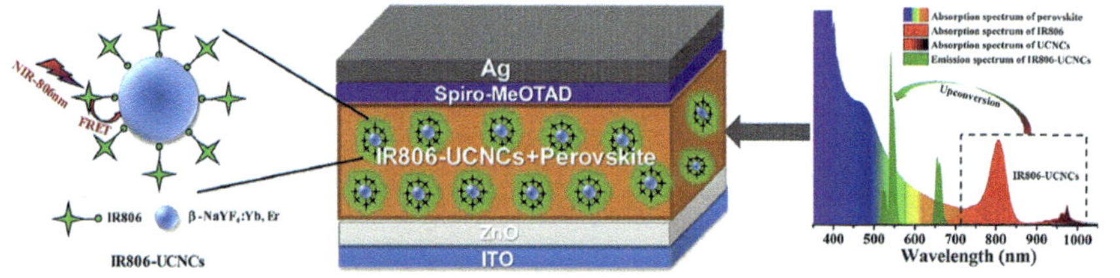

Fig. 12.16 Broadband NIR sunlight harvesting and spectral conversion into visible range to activate perovskite for PSC device. The right part of scheme showing the absorption spectra of the perovskite, IR806 and UCNCs, and emission spectrum of IR806-UCNCs. Reproduced from [88]. Copyright 2017 Elsevier B.V.

replace the m-TiO_2 as the mesoporous layer in PSCs for the first time. The charge-transfer functionality was demonstrated by a reported PCE of 17.8% for the PSC. An absolute efficiency of 0.35% was recorded under illumination of a 980 nm laser (2 W), which confirmed UC functionality. After that, core-shell UC phosphors β-$NaYF_4$:Yb^{3+},Tm^{3+}/$NaYF_4$ [86] and $NaYF_4$:Yb^{3+},Er^{3+}@SiO_2 [87] were designed to reduce surface defects for improving UC and reducing charge recombination, and then incorporated into m-TiO_2 ETL in PSCs to improve light harvesting. As a further internal strategy to modify the active layer, IR806-sensitized UCNPs were incorporated into perovskite layer in PSC (Fig. 12.16). A relative PCE enhancement of 29% was achieved. The absolute PCE enhancement under NIR irradiation (>780 nm) using a solar light simulator was reported to be as high as 0.382% [88].

12.5 Conclusions

Lanthanide ion-doped UC materials have progressed greatly in the material variety, UC efficiency, and applications in various fields. However the exploitation of UC materials and their applications like in photovoltaics are still in the early stage. Currently the relatively low UC quantum yield and narrow and low absorption of Ln ions in the NIR range are the main limiting factors for the application of Ln-doped UC materials. The overall contribution of UC effect under outdoor conditions is still too small for practical application. A more pronounced sub-bandgap photocurrent response will require more efficient UC materials, advanced device architecture, and smart engineering. It can be expected that considerable progress in photovoltaic applications will be made once the limiting factors are overcome.

References

1. L.C. Hirst, N.J. Ekins-Daukes, Fundamental losses in solar cells. Prog. Photovolt. Res. Appl. **19**, 286–293 (2011)
2. W. Shockley, H.J. Queisser, Detailed balance limit of efficiency of p-n junction solar cells. J. Appl. Phys. **32**, 510–519 (1961)
3. H.-Q. Wang, M. Batentschuk, A. Osvet, L. Pinna, C.J. Brabec, Rare-earth ion doped up-conversion materials for photovoltaic applications. Adv. Mater. **23**, 2675–2680 (2011b)
4. J. de Wild, A. Meijerink, J.K. Rath, W.G.J.H.M. van Sark, R.E.I. Schropp, Upconverter solar cells: materials and applications. Energy Environ. Sci. **4**, 4835–4848 (2011)
5. D. Wöhrle, D. Meissner, Organic solar cells. Adv. Mater. **3**, 129–138 (1991)
6. Y. Cai, L. Huo, Y. Sun, Recent advances in wide-bandgap photovoltaic polymers. Adv. Mater. **29**, 1605437 (2017)
7. G. Dennler, M.C. Scharber, C.J. Brabec, Polymer-fullerene bulk-heterojunction solar cells. Adv. Mater. **21**, 1323–1338 (2009)
8. H. Hoppe, N.S. Sariciftci, Organic solar cells: an overview. J. Mater. Res. **19**, 1924–1945 (2011)
9. T. Ameri, G. Dennler, C. Lungenschmied, C.J. Brabec, Organic tandem solar cells: a review. Energy Environ. Sci. **2**, 347–363 (2009)
10. T. Ameri, N. Li, C.J. Brabec, Highly efficient organic tandem solar cells: a follow up review. Energy Environ. Sci. **6**, 2390–2413 (2013)
11. R. Kroon, M. Lenes, J.C. Hummelen, P.W.M. Blom, B. de Boer, Small bandgap polymers for organic solar cells (polymer material development in the last 5 years). Polym. Rev. **48**, 531–582 (2008)
12. T. Dilbeck, K. Hanson, Molecular photon upconversion solar cells using multilayer assemblies: progress and prospects. J. Phys. Chem. Lett. **9**, 5810–5821 (2018)
13. A.J. Nozik, J. Miller, Introduction to solar photon conversion. Chem. Rev. **110**, 6443–6445 (2010)
14. X. Xie, X. Liu, Upconversion goes broadband. Nat. Mater. **11**, 842–843 (2012)
15. C.J. Brabec, S. Gowrisanker, J.J.M. Halls, D. Laird, S. Jia, S.P. Williams, Polymer-fullerene bulk-heterojunction solar cells. Adv. Mater. **22**, 3839–3856 (2010)
16. N. Bloembergen, Solid state infrared quantum counters. Phys. Rev. Lett. **2**, 84–85 (1959)
17. X. Liu, C.-H. Yan, J.A. Capobianco, Photon upconversion nanomaterials. Chem. Soc. Rev. **44**, 1299–1301 (2015)
18. F. Auzel, Upconversion and anti-stokes processes with f and d ions in solids. Chem. Rev. **104**, 139–173 (2004)
19. V. Gray, K. Moth-Poulsen, B. Albinsson, M. Abrahamsson, Towards efficient solid-state triplet–triplet annihilation based photon upconversion: supramolecular, macromolecular and self-assembled systems. Coord. Chem. Rev. **362**, 54–71 (2018)
20. A. Nonat, T. Fix, 6—photon converters for photovoltaics, in *Advanced Micro- and Nanomaterials for Photovoltaics*, ed. by D. Ginley, T. Fix, (Elsevier, 2019), pp. 121–151
21. J.A. Briggs, A.C. Atre, J.A. Dionne, Narrow-bandwidth solar upconversion: case studies of existing systems and generalized fundamental limits. J. Appl. Phys. **113**, 124509 (2013)
22. T. Trupke, A. Shalav, B.S. Richards, P. Würfel, M.A. Green, Efficiency enhancement of solar cells by luminescent up-conversion of sunlight. Sol. Energy Mater. Sol. Cells **90**, 3327–3338 (2006)
23. G. Chen, J. Shen, T.Y. Ohulchanskyy, N.J. Patel, A. Kutikov, Z. Li, J. Song, R.K. Pandey, H. Ågren, P.N. Prasad, G. Han, (α-$NaYbF_4$:Tm^{3+})/CaF_2 core/shell nanoparticles with efficient near-infrared to near-infrared upconversion for high-contrast deep tissue bioimaging. ACS Nano **6**, 8280–8287 (2012)
24. P. Gibart, F. Auzel, J.-C. Guillaume, K. Zahraman, Below band-gap IR response of substrate-free GaAs solar cells using two-photon up-conversion. Jpn. J. Appl. Phys. **35**, 4401–4402 (1996)

25. G. Liu, Electronic energy level structure, in *Spectroscopic Properties of Rare Earths in Optical Materials*, ed. by R. Hull, J. Parisi, R. M. Osgood, H. Warlimont, G. Liu, B. Jacquier, (Springer, Berlin, Heidelberg, 2005), pp. 1–94
26. S.V. Eliseeva, J.C.G. Bunzli, Lanthanide luminescence for functional materials and biosciences. Chem. Soc. Rev. **39**, 189–227 (2010)
27. F. Wang, R. Deng, J. Wang, Q. Wang, Y. Han, H. Zhu, X. Chen, X. Liu, Tuning upconversion through energy migration in core-shell nanoparticles. Nat. Mater. **10**, 968–973 (2011a)
28. O. Ehlert, R. Thomann, M. Darbandi, T. Nann, A four-color colloidal multiplexing nanoparticle system. ACS Nano **2**, 120–124 (2008)
29. J.C. Boyer, F.C.J.M. van Veggel, Absolute quantum yield measurements of colloidal NaYF4: Er3+, Yb3+ upconverting nanoparticles. Nanoscale **2**, 1417–1419 (2010)
30. G.A. Crosby, J.N. Demas, Measurement of photoluminescence quantum yields. Review. J. Phys. Chem. **75**, 991–1024 (1971)
31. Y. Zorenko, V. Gorbenko, T. Zorenko, K. Paprocki, A. Osvet, M. Batentschuk, C. Brabec, A. Fedorov, Enhancement of up-conversion luminescence in Er,Ce doped Y3−xYbxAG single crystalline films. J. Lumin. **169**, 816–821 (2016)
32. F. Wang, X. Liu, Upconversion multicolor fine-tuning: visible to near-infrared emission from lanthanide-doped $NaYF_4$ nanoparticles. J. Am. Chem. Soc. **130**, 5642–5643 (2008)
33. D. Li, H. Ågren, G. Chen, Near infrared harvesting dye-sensitized solar cells enabled by rare-earth upconversion materials. Dalton Trans. **47**, 8526–8537 (2018)
34. H.-Q. Wang, M. Mačković, A. Osvet, I. Litzov, E. Epelbaum, A. Stiegelschmitt, M. Batentschuk, E. Spiecker, C.J. Brabec, A new crystal phase molybdate $Yb_2Mo_4O_{15}$: the synthesis and upconversion properties. Part. Part. Syst. Charact. **32**, 340–346 (2015)
35. K.W. Krämer, D. Biner, G. Frei, H.U. Güdel, M.P. Hehlen, S.R. Lüthi, Hexagonal sodium yttrium fluoride based green and blue emitting upconversion phosphors. Chem. Mater. **16**, 1244–1251 (2004)
36. I. Etchart, Metal Oxides for Efficient Infrared to Visible Upconversion. University of Cambrige Ph.D. Thesis (2010)
37. A. Nadort, J. Zhao, E.M. Goldys, Lanthanide upconversion luminescence at the nanoscale: fundamentals and optical properties. Nanoscale **8**, 13099–13130 (2016)
38. J.H.V. Vleck, The puzzle of rare-earth spectra in solids. J. Phys. Chem. **41**, 67–80 (1937)
39. H.-Q. Wang, T. Nann, Book Chapter: Upconverting nanoparticles, in *Springer Book: Lanthanide Luminescence*, vol. 7, (2010), pp. 115–132
40. J.C. Goldschmidt, S. Fischer, Upconversion for photovoltaics—a review of materials, devices and concepts for performance enhancement. Adv. Opt. Mater. **3**, 510–535 (2015)
41. S. Heer, K. Kömpe, H.-U. Güdel, M. Haase, Highly efficient multicolour upconversion emission in transparent colloids of lanthanide-doped $NaYF_4$ nanocrystals. Adv. Mater. **16**, 2102–2105 (2004)
42. H. Dong, L.-D. Sun, C.-H. Yan, Basic understanding of the lanthanide related upconversion emissions. Nanoscale **5**, 5703–5714 (2013)
43. H.-Q. Wang, T. Nann, Monodisperse upconverting nanocrystals by microwave-assisted synthesis. ACS Nano **3**, 3804–3808 (2009)
44. Z. Li, Y. Zhang, S. Jiang, Multicolor core/shell-structured upconversion fluorescent nanoparticles. Adv. Mater. **20**, 4765–4769 (2008)
45. Y. Wang, L. Tu, J. Zhao, Y. Sun, X. Kong, H. Zhang, Upconversion luminescence of β-$NaYF_4$: Yb^{3+}, Er^{3+}@β-$NaYF_4$ core/shell nanoparticles: excitation power density and surface dependence. J. Phys. Chem. C **113**, 7164–7169 (2009)
46. H.-S. Qian, Y. Zhang, Synthesis of hexagonal-phase core-shell $NaYF_4$ nanocrystals with tunable upconversion fluorescence. Langmuir **24**, 12123–12125 (2008)
47. G.-S. Yi, G.-M. Chow, Water-soluble $NaYF_4$:Yb,Er(Tm)/$NaYF_4$/polymer core/shell/shell nanoparticles with significant enhancement of upconversion fluorescence. Chem. Mater. **19**, 341–343 (2007)
48. F. Zhang, R. Che, X. Li, C. Yao, J. Yang, D. Shen, P. Hu, W. Li, D. Zhao, Direct imaging the upconversion nanocrystal core/shell structure at the subnanometer level: shell thickness dependence in upconverting optical properties. Nano Lett. **12**, 2852–2858 (2012a)

49. X. Li, D. Shen, J. Yang, C. Yao, R. Che, F. Zhang, D. Zhao, Successive layer-by-layer strategy for multi-shell epitaxial growth: shell thickness and doping position dependence in upconverting optical properties. Chem. Mater. **25**, 106–112 (2013)
50. Z. Li, W. Park, G. Zorzetto, J.S. Lemaire, C.J. Summers, Synthesis protocols for δ-doped $NaYF_4$:Yb,Er. Chem. Mater. **26**, 1770–1778 (2014)
51. H. Zhang, Y. Li, I.A. Ivanov, Y. Qu, Y. Huang, X. Duan, Plasmonic modulation of the upconversion fluorescence in $NaYF_4$:Yb/tm hexaplate nanocrystals using gold nanoparticles or nanoshells. Angew. Chem. Int. Ed. **49**, 2865–2868 (2010)
52. S.P. Madsen, J. Christiansen, R.E. Christiansen, J. Vester-Petersen, S.H. Møller, H. Lakhotiya, A. Nazir, E. Eriksen, S. Roesgaard, O. Sigmund, J.S. Lissau, E. Destouesse, M. Madsen, B. Julsgaard, P. Balling, Improving the efficiency of upconversion by light concentration using nanoparticle design. J. Phys. D. Appl. Phys. **53**, 073001 (2019)
53. W. Deng, L. Sudheendra, J. Zhao, J. Fu, D. Jin, I.M. Kennedy, E.M. Goldys, Upconversion in $NaYF_4$:Yb, Er nanoparticles amplified by metal nanostructures. Nanotechnology **22**, 325604 (2011)
54. J.H. Lin, H.Y. Liou, C.-D. Wang, C.-Y. Tseng, C.-T. Lee, C.-C. Ting, H.-C. Kan, C.C. Hsu, Giant enhancement of upconversion fluorescence of $NaYF_4:Yb^{3+},Tm^{3+}$ nanocrystals with resonant waveguide grating substrate. ACS Photonics **2**, 530–536 (2015)
55. W. Zou, C. Visser, J.A. Maduro, M.S. Pshenichnikov, J.C. Hummelen, Broadband dye-sensitized upconversion of near-infrared light. Nat. Photonics **6**, 560–564 (2012)
56. G. Chen, J. Damasco, H. Qiu, W. Shao, T.Y. Ohulchanskyy, R.R. Valiev, X. Wu, G. Han, Y. Wang, C. Yang, H. Ågren, P.N. Prasad, Energy-cascaded upconversion in an organic dye-sensitized core/shell fluoride nanocrystal. Nano Lett. **15**, 7400–7407 (2015a)
57. J.F. Suyver, A. Aebischer, D. Biner, P. Gerner, J. Grimm, S. Heer, K.W. Krämer, C. Reinhard, H.U. Güdel, Novel materials doped with trivalent lanthanides and transition metal ions showing near-infrared to visible photon upconversion. Opt. Mater. **27**, 1111–1130 (2005)
58. Y. Zhang, J.D. Lin, V. Vijayaragavan, K.K. Bhakoo, T.T.Y. Tan, Tuning sub-10 nm single-phase $NaMnF_3$ nanocrystals as ultrasensitive hosts for pure intense fluorescence and excellent T1 magnetic resonance imaging. Chem. Commun. **48**, 10322–10324 (2012b)
59. D. Gao, X. Zhang, W. Gao, Formation of bundle-shaped β-$NaYF_4$ upconversion microtubes via Ostwald ripening. ACS Appl. Mater. Interfaces **5**, 9732–9739 (2013)
60. Z. Wu, M. Lin, S. Liang, Y. Liu, H. Zhang, B. Yang, Hot-injection synthesis of manganese-ion-doped $NaYF_4$:Yb,Er nanocrystals with red up-converting emission and tunable diameter. Part. Part. Syst. Charact. **30**, 311–315 (2013)
61. L. Aboshyan-Sorgho, C. Besnard, P. Pattison, K.R. Kittilstved, A. Aebischer, J.-C.G. Bünzli, A. Hauser, C. Piguet, Near-infrared→visible light upconversion in a molecular trinuclear d-f-d complex. Angew. Chem. Int. Ed. **50**, 4108–4112 (2011)
62. L. Aboshyan-Sorgho, M. Cantuel, S. Petoud, A. Hauser, C. Piguet, Optical sensitization and upconversion in discrete polynuclear chromium-lanthanide complexes. Coord. Chem. Rev. **256**, 1644–1663 (2012)
63. S. Ye, E.H. Song, E. Ma, S.J. Zhang, J. Wang, X.Y. Chen, Q.Y. Zhang, J.R. Qiu, Broadband Cr^{3+}-sensitized upconversion luminescence in $La_3Ga_5GeO_{14}:Cr3^{+},Yb^{3+},Er^{3+}$. Opt. Mater. Express **4**, 638–648 (2014)
64. Y. Takeda, S. Mizuno, H.N. Luitel, T. Tani, A broadband-sensitive upconverter $La(Ga_{0.5}Sc_{0.5})O_3$:Er,Ni,Nb for crystalline silicon solar cells. Appl. Phys. Lett. **108**, 043901 (2016)
65. H. Liu, C.T. Xu, G. Dumlupinar, O.B. Jensen, P.E. Andersen, S. Andersson-Engels, Deep tissue optical imaging of upconverting nanoparticles enabled by exploiting higher intrinsic quantum yield through use of millisecond single pulse excitation with high peak power. Nanoscale **5**, 10034–10040 (2013)
66. Z. Chen, W. Cui, S. Kang, H. Zhang, G. Dong, C. Jiang, S. Zhou, J. Qiu, Fast-slow red upconversion fluorescence modulation from Ho^{3+}-doped glass ceramics upon two-wavelength excitation. Adv. Opt. Mater. **5**, 1600554 (2017)

67. Z. Chen, G. Wu, H. Jia, K. Sharafudeen, W. Dai, X. Zhang, S. Zeng, J. Liu, R. Wei, S. Lv, G. Dong, J. Qiu, Improved up-conversion luminescence from Er^{3+}:LaF_3 nanocrystals embedded in oxyfluoride glass ceramics via simultaneous triwavelength excitation. J. Phys. Chem. C **119**, 24056–24061 (2015c)
68. V. Saxena, *Phosphors for Solar-Cells-tb-Doped Lanthanum Fluoride and th-Doped Calcium Tungstate* (Council Scientific Industrial Research Publ & Info Directorate, New Delhi, 1983), pp. 306–307
69. A. Shalav, B.S. Richards, T. Trupke, K.W. Krämer, H.U. Güdel, Application of $NaYF_4$:Er^{3+} up-converting phosphors for enhanced near-infrared silicon solar cell response. Appl. Phys. Lett. **86**, 013505 (2005)
70. S. Fischer, E. Favilla, M. Tonelli, J.C. Goldschmidt, Record efficient upconverter solar cell devices with optimized bifacial silicon solar cells and monocrystalline BaY_2F_8:30% Er^{3+} upconverter. Sol. Energy Mater. Sol. Cells **136**, 127–134 (2015)
71. F. Lahoz, Ho^{3+}-doped nanophase glass ceramics for efficiency enhancement in silicon solar cells. Opt. Lett. **33**, 2982–2984 (2008)
72. F. Lahoz, C. Pérez-Rodríguez, S.E. Hernández, I.R. Martín, V. Lavín, U.R. Rodríguez-Mendoza, Upconversion mechanisms in rare-earth doped glasses to improve the efficiency of silicon solar cells. Sol. Energy Mater. Sol. Cells **95**, 1671–1677 (2011)
73. J. de Wild, J.K. Rath, A. Meijerink, W.G.J.H.M. van Sark, R.E.I. Schropp, Enhanced near-infrared response of a-Si:H solar cells with β-$NaYF_4$:Yb^{3+} (18%), Er^{3+} (2%) upconversion phosphors. Sol. Energy Mater. Sol. Cells **94**, 2395–2398 (2010)
74. M. Takei, Conductive paste for solar cell, Pat. Nr. JP20100099623 20100423 (A) (2011)
75. C. Miao, T. Liu, Y. Zhu, Q. Dai, W. Xu, L. Xu, S. Xu, Y. Zhao, H. Song, Super-intense white upconversion emission of Yb_2O_3 polycrystals and its application on luminescence converter of dye-sensitized solar cells. Opt. Lett. **38**, 3340–3343 (2013)
76. G.B. Shan, G.P. Demopoulos, Near-infrared sunlight harvesting in dye-sensitized solar cells via the insertion of an upconverter-TiO_2 nanocomposite layer. Adv. Mater. **22**, 4373–4377 (2010)
77. Z. Zhou, J. Wang, F. Nan, C. Bu, Z. Yu, W. Liu, S. Guo, H. Hu, X.-Z. Zhao, Upconversion induced enhancement of dye sensitized solar cells based on core-shell structured β-$NaYF_4$:Er^{3+}, Yb^{3+}@SiO_2 nanoparticles. Nanoscale **6**, 2052–2055 (2014)
78. L. Liang, Y. Liu, C. Bu, K. Guo, W. Sun, N. Huang, T. Peng, B. Sebo, M. Pan, W. Liu, S. Guo, X.-Z. Zhao, Highly uniform, bifunctional core/double-shell-structured β-$NaYF_4$:Er^{3+}, Yb^{3+} @ SiO_2@TiO_2 hexagonal sub-microprisms for high-performance dye sensitized solar cells. Adv. Mater. **25**, 2174–2180 (2013)
79. C. Yuan, G. Chen, P.N. Prasad, T.Y. Ohulchanskyy, Z. Ning, H. Tian, L. Sun, H. Ågren, Use of colloidal upconversion nanocrystals for energy relay solar cell light harvesting in the near-infrared region. J. Mater. Chem. **22**, 16709–16713 (2012)
80. G. Dennler, M.C. Scharber, T. Ameri, P. Denk, K. Forberich, C. Waldauf, C.J. Brabec, Design rules for donors in bulk-heterojunction tandem solar cells towards 15% energy-conversion efficiency. Adv. Mater. **20**, 579–583 (2008)
81. W. Chen, Y. Hou, A. Osvet, F. Guo, P. Kubis, M. Batentschuk, B. Winter, E. Spiecker, K. Forberich, C.J. Brabec, Sub-bandgap photon harvesting for organic solar cells via integrating up-conversion nanophosphors. Org. Electron. **19**, 113–119 (2015b)
82. H.-Q. Wang, T. Stubhan, A. Osvet, I. Litzov, C.J. Brabec, Up-conversion semiconducting MoO_3:Yb/Er nanocomposites as buffer layer in organic solar cells. Sol. Energy Mater. Sol. Cells **105**, 196–201 (2012)
83. W. Guo, K. Zheng, W. Xie, L. Sun, L. Shen, C. Liu, Y. He, Z. Zhang, Efficiency enhancement of inverted polymer solar cells by doping $NaYF_4$:Yb^{3+}, Er^{3+} nanocomposites in PCDTBT:PCBM active layer. Sol. Energy Mater. Sol. Cells **124**, 126–132 (2014)
84. X. Chen, W. Xu, H. Song, C. Chen, H. Xia, Y. Zhu, D. Zhou, S. Cui, Q. Dai, J. Zhang, Highly efficient $LiYF_4$:Yb^{3+}, Er^{3+} upconversion single crystal under solar cell spectrum excitation and photovoltaic application. ACS Appl. Mater. Interfaces **8**, 9071–9079 (2016)

85. M. He, X. Pang, X. Liu, B. Jiang, Y. He, H. Snaith, Z. Lin, Monodisperse dual-functional upconversion nanoparticles enabled near-infrared organolead halide perovskite solar cells. Angew. Chem. Int. Ed. **55**, 4280–4284 (2016)
86. M. Que, W. Que, X. Yin, P. Chen, Y. Yang, J. Hu, B. Yu, Y. Du, Enhanced conversion efficiency in perovskite solar cells by effectively utilizing near infrared light. Nanoscale **8**, 14432–14437 (2016)
87. J. Hu, Y. Qiao, Y. Yang, L. Zhao, W. Liu, S. Li, P. Liu, M. Chen, Enhanced performance of hole-conductor-free perovskite solar cells by utilization of core/shell-structured β-$NaYF_4$:Yb^{3+},Er^{3+}@SiO_2 nanoparticles in ambient air. IEEE J. Photovoltaics **8**, 132–136 (2018)
88. X. Lai, X. Li, X. Lv, Y.-Z. Zheng, F. Meng, X. Tao, Broadband dye-sensitized upconverting nanocrystals enabled near-infrared planar perovskite solar cells. J. Power Sources **372**, 125–133 (2017)

Chapter 13
Nanophotonics for Photon Upconversion Enhancement

Wounjhang Park, Ananda Das, and Kyuyoung Bae

Abstract Lanthanide-based upconversion materials convert low-energy infrared photons into high-energy visible photons with much higher efficiency than two-photon absorption or second harmonic generation. Naturally, they have attracted much attention for potential applications in solar energy harvesting, photocatalysis, security, and biological imaging. Despite the high promise, the intrinsic conversion efficiency of upconversion materials remains low for most applications. The emergent nanophotonic technologies could provide a powerful tool to boost the upconversion efficiency and enable novel applications. In this chapter, we provide an in-depth review of the theoretical foundation for light-matter interaction, dynamics of luminescence upconversion process, and nanophotonic enhancement mechanisms. We then provide a comprehensive survey of recent progress in nanophotonic enhancement of upconversion. The use of plasmonic and dielectric nanostructures has led to 2–3 orders of magnitude improvements in upconversion efficiency. The accelerating pace of progress makes nanophotonically enhanced upconversion a highly promising platform for novel photonic applications.

Keywords Energy transfer · Lanthanide ion · Luminescence upconversion · Nanoparticle · Photonic crystal · Surface plasmon

13.1 Introduction

Luminescence upconversion is a process in which two or more low-energy photons are absorbed, resulting in a higher energy photon emission. Compared to other frequency conversion mechanisms based on nonlinear susceptibility such as high

W. Park (✉) · K. Bae
Department of Electrical, Computer and Energy Engineering, University of Colorado, Boulder, CO, USA
e-mail: won.park@colorado.edu

A. Das
Department of Physics, University of Colorado, Boulder, CO, USA

J. S. Lissau, M. Madsen (eds.), *Emerging Strategies to Reduce Transmission and Thermalization Losses in Solar Cells*, https://doi.org/10.1007/978-3-030-70358-5_13

harmonic generation and optical parametric oscillation, luminescence upconversion is many orders of magnitude more efficient. This is because luminescence upconversion is often accomplished by successive transitions between real states while the high harmonic generation and optical parametric oscillation are second-order transitions involving virtual states. Luminescence upconversion, therefore, does not require phase matching and can be accomplished with an incoherent light source at low intensities. For these reasons, luminescence upconversion materials are ideally suited for a wide range of applications such as lighting, displays, solar energy conversion, biosensing, biomedical imaging, security ink, and photoswitching.

There are multiple mechanisms that can produce upconverted luminescence. Among them, excited-state absorption (ESA) and energy transfer upconversion (ETU) are most commonly observed. ESA involves two successive absorption transitions: first a transition from the ground state to an intermediate state followed by another transition from the intermediate state to a higher excited state from which a radiative transition back to the ground state takes place, emitting a photon [1]. Energy transfer upconversion (ETU), first observed in a Yb^{3+}-Tm^{3+} codoped glass [2], involves two types of optically active ions, sensitizer (or donor) and activator (or acceptor). Low-energy photons are first absorbed by the sensitizers, which then transfer energy to nearby activator ions. When two or more sensitizers transfer energy successively to an activator ion, the activator is excited to a higher excited state from which it can emit a high-energy photon. In addition, luminescence upconversion can arise from cooperative sensitization [3] and cooperative luminescence [4]. In contrast to ESA and ETU, the cooperative processes involve at least one virtual state. In cooperative sensitization, two sensitizers simultaneously transfer energy to an activator via a virtual intermediate state, raising the activator to a real excited state from which upconverted luminescence occurs. Cooperative luminescence is nearly the same as cooperative sensitization except that the emitting state is virtual instead of the intermediate state. In either case, the involvement of virtual states makes them much less efficient than the ESA and ETU processes in which all participating states are real. Much later, a new mechanism for upconversion was discovered and named photon avalanche [5]. In this process, the intermediate state from which ESA takes place is populated by cross relaxation with the terminating state of the luminescence process. The upconverted luminescence originating from the photon avalanche process is typically very weak at low pump power and rapidly increases at high pump power, exhibiting a sharp threshold.

Among the various upconversion mechanisms, ETU is known to be the most efficient and will thus be the main focus of this chapter. It should however be noted that the theoretical framework we present here may be readily extended to other upconversion mechanisms such as ESA and triplet-triple annihilation. Two representative ETU systems are the rare earth ion pairs of Yb^{3+}-Er^{3+} and Yb^{3+}-Tm^{3+} which exhibit green and red luminescence by two-photon ETU and red, blue, and ultraviolet emission by three-photon ETU, respectively. The energy levels and key processes involved in Yb^{3+}-Er^{3+} ETU are shown schematically in Fig. 13.1a. Briefly, the incident light at 980 nm is absorbed by Yb^{3+} ions, which then excites a nearby Er^{3+} ion into the $^4I_{11/2}$ level via Förster energy

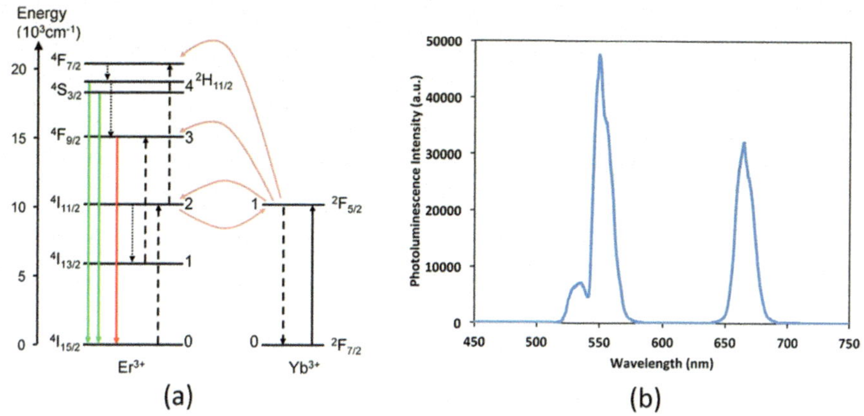

Fig. 13.1 (**a**) Energy levels of Yb^{3+} and Er^{3+} ions relevant to the energy transfer upconversion process. Initial absorption is indicated by the black solid line arrow, subsequent energy transfer processes by the dashed arrows, nonradiative relaxations by dotted arrows, and final upconverted luminescence by the green and red arrows. (**b**) Photoluminescence spectrum of $NaYF_4$:Yb^{3+},Er^{3+} under 980 nm excitation. Reprinted from *Chem. Soc. Rev.***44**, 2940 (2015) [30] with permission from The Royal Society of Chemistry

transfer process. Before the excited Er^{3+} ion decays, another Yb^{3+} ion may transfer energy to the Er^{3+} ion again, raising it to the $^4F_{7/2}$ manifold. The excited Er^{3+} ion then nonradiatively decays into the $^2H_{11/2}$, $^4S_{3/2}$, or $^4F_{9/2}$ levels from which green and red luminescence is emitted. The upconverted luminescence spectrum from $NaYF_4$:Yb^{3+},Er^{3+} nanoparticles under 980 nm excitation is shown in Fig. 13.1b. Like most lanthanide ions, the optical spectra of Yb^{3+}-Er^{3+} and Yb^{3+}-Tm^{3+} are largely insensitive to host materials and thus they exhibit ETU in a wide variety of host materials. While the host materials do not perturb the energy levels significantly, they could have profound impacts on nonradiative transition rates and consequently the ETU efficiency. Since the nonradiative transitions are generally mediated by phonons, host materials with low phonon energy tend to exhibit lower nonradiative transition rates and higher ETU efficiency. For this reason, fluorides which tend to exhibit lower phonon energies than oxides are favored over oxides. The most efficient upconversion material is $NaYF_4$:Yb^{3+},Er^{3+} with energy conversion efficiencies up to ~4% at pump intensities around 1000 W/cm^2 [6].

There has been a tremendous progress in the past decade with the upconversion nanomaterials. Since the first efficient fluoride nanoparticle syntheses were reported in 2004 [7–9], there has been a plethora of refinements and improvements for morphology control, surface treatments, core-shell structures, etc. [10] Despite the remarkable progress, however, the upconversion efficiency is not high enough for truly widespread use. It is therefore critical to find ways to further improve the upconversion efficiency. In this context, nanophotonics provides excellent opportunities that can be pursued concurrently and independently to the continued efforts on improving the synthesis processes and material quality. This book chapter presents the fundamental principles of nanophotonic approaches that can be used for upconversion enhancement, as well as the latest progress reported in the literature.

13.2 Fundamentals of Light-Matter Interaction

In the quantum electrodynamics formalism, the Hamiltonian describing the interaction between electron and photon is given as [11].

$$H_{\text{int}} = -\frac{e}{m}\mathbf{p} \cdot \mathbf{A} - \frac{e^2}{2m}|\mathbf{A}|^2 + e\phi \tag{13.1}$$

where e and m are electronic charge and mass, $\mathbf{p}$ is the canonical momentum of electron, and ϕ and $\mathbf{A}$ are the scalar and vector potential of the electromagnetic field, respectively. The interaction Hamiltonian induces transitions between quantum states. The transition rate, W, is given by Fermi's golden rule:

$$W = \frac{2\pi}{\hbar}\left|\langle\psi_f\left|H_{\text{int}}\right|\psi_i\rangle\right|\rho\left(\omega_f - \omega_i\right) \tag{13.2}$$

where $\hbar\omega_i$ and $\hbar\omega_f$ are energies of the initial and final states, ψ_i and ψ_f, respectively, and ρ is the density of states. Since the optical field varies over a length scale (light wavelength) much greater than the extent of electronic wave functions (Bohn radius), one can take the Taylor expansion to obtain the well-known multipolar Hamiltonian: [11]

$$H_{\text{int}} = -\mathbf{d} \cdot \mathbf{E_0} - \mathbf{m} \cdot \mathbf{B_0} - \left[\left(\overleftrightarrow{\mathbf{Q}}\nabla\right) \cdot \mathbf{E}\right]_0 + \cdots \tag{13.3}$$

where $\mathbf{d}$, $\mathbf{m}$, and $\mathbf{Q}$ are electric dipole, magnetic dipole, and electric quadrupole moments, respectively, and the subscript 0 represents the values evaluated at the position of the atomic nucleus. Magnetic dipole and electric quadrupole terms are generally many orders of magnitude smaller than electric dipole term and are often ignored unless electric dipole transition is forbidden. This leads to the commonly adopted electric dipole approximation in which only the first term in Eq. (13.3) is retained.

Inorganic upconversion materials usually employ lanthanide ions, in which optical transitions take place between $4f^n$ multiplets. The $4f$ shell of lanthanide ions is well shielded from the crystalline environment by the outer shell $5s$ and $5p$ electrons. Thus, the energy levels are largely insensitive to the host material and exhibit weak coupling with phonons. The crystalline environment can, however, make a dramatic impact on the optical transition rates. Optical transitions between $4f^n$ multiplets are subject to the Laporte selection rule. The parity of an electronic configuration is given by $(-1)^{\sum_i l_i}$ where l_i is the angular momentum of individual electron. Thus, all multiplet states of a given $4f^n$ configuration have the same parity and the electric dipole transition between them is forbidden. This is why most optical transitions of lanthanide ions exhibit low-absorption cross sections and long lifetimes. However, it was found that most lanthanide optical transitions in crystals are nonetheless electric dipole transitions. This is possible because an odd parity crystal field may cause mixing between a $4f^n$ state and another state with opposite parity, a $4f^{n-1}5d$ state in most cases, thereby relaxing the parity selection rule. This mixing does not change energy levels in any significant way but can profoundly alter the transition rates. This type of transition mediated by intermediate states is described by the second-order perturbation theory: [12]

$$\langle \psi_i |\mathbf{d}| \psi_f \rangle = \sum_k \frac{\langle \psi_i |\mathbf{d}| \psi_k \rangle \langle \psi_k |V| \psi_f \rangle}{E_f - E_k} + \sum_k \frac{\langle \psi_i |V| \psi_k \rangle \langle \psi_k |\mathbf{d}| \psi_f \rangle}{E_i - E_k} \tag{13.4}$$

where **d** is the electric dipole operator and V is the potential due to the crystal field. It is apparent that both V and the intermediate states ψ_k should have the opposite parity to the initial and final states in order for the matrix elements to be nonzero. For the case of lanthanide transitions, since all of the $4f^n$ multiplets have even parity, both the crystal field, V, and the intermediate states, ψ_k's, must have odd parity in order to observe the second-order electric dipole transitions. The odd parity intermediate states are usually provided by the $4f^{n-1}5d$ states. For odd parity crystal field, the host crystal must offer sites lacking inversion symmetry. For example, an Eu^{3+} ion exhibits many line emissions originating from the $^5D_0$ level. In crystals lacking inversion symmetry, the strongest emission is typically located at around 613 nm due to the $^5D_0 \rightarrow {}^7F_2$ transition, which is a weakly allowed electric dipole transition [13, 14]. If, however, the Eu^{3+} ion is in a crystal with inversion symmetry, e.g., $YBO_3{:}Eu^{3+}$, the crystal field cannot induce mixing with opposite parity states and the $^5D_0 \rightarrow {}^7F_2$ transition is depressed. Instead, the magnetic dipole $^5D_0 \rightarrow {}^7F_1$ transition at 594 nm becomes the most pronounced emission peak [15]. Following Judd [16] and Ofelt [17], we assume that all ψ_k states are degenerate and also $E_i \approx E_f$ so that we may write $E_f - E_k \approx E_i - E_k$ and combine the two sums in Eq. (13.4). Summing over the $2J + 1$ states of the manifolds to which the initial and final states belong and also summing over polarizations, we obtain the well-known expression for the line strength of a transition between J_i and J_f manifolds:

$$S_{\mathrm{ED}}\left(J_i, J_f'\right) = e^2 \sum_{t=2,4,6} \Omega_t \left| \left\langle f^n [\gamma SL] J \left\| U^{(t)} \right\| f^n [\gamma' S' L'] J' \right\rangle \right|^2 \tag{13.5}$$

where the $4f^n$ states are expressed in the Russell-Saunders scheme, $|f^n[\gamma SL]\, J\, J_z\rangle$, $U^{(t)}$ are the tensor operators containing both the crystal field and electric dipole operators, and Ω_t are the intensity parameters which contain radial integrals but are commonly determined experimentally. Despite the seemingly drastic approximations, the Judd-Ofelt theory has been highly successful and correctly describes the weakly allowed nature of lanthanide transitions.

In addition to the direct optical transitions discussed above, any interaction between the optically active ions could strongly affect the overall transition rates. When two optically active ions are close to each other, Coulomb interaction between them may induce transitions, leading to Förster energy transfer process [18]. When the interionic distance, R, is large compared to the ionic radius, the Coulomb interaction can be considered small enough to employ the standard perturbation method. The first-order correction turns out to be zero while the second-order correction gives a term proportional to R^{-6}, which is the van der Waal's energy. The van der Waal's interaction results in a small, often negligible, correction to the electronic energy levels. But it could lead to significant changes in transition rates by enabling energy transfer between ions. Following the time-dependent perturbation theory, the energy transfer rate between ions A and B is given by

$$W_{\mathrm{ET}} = \frac{2\pi}{\hbar}|\langle H_{\mathrm{AB}}\rangle|^2 \int g_{\mathrm{A}}(E)g_{\mathrm{B}}(E)dE \tag{13.6}$$

Here g_{A} and g_{B} are the normalized emission and absorption spectra of the two ions, respectively, and their overlap integral determines the joint density of states of the combined system of ions A and B. The interaction Hamiltonian H_{AB}, if it is purely Coulombic, may be expanded in terms of spherical harmonics to obtain multipolar terms: dipole-dipole ($\propto R^{-6}$), dipole-quadrupole ($\propto R^{-8}$), quadrupole-quadrupole ($\propto R^{-10}$), etc. Magnetic multipolar interaction is many orders of magnitude smaller than the electric counterpart and is usually neglected. The interionic coupling may also be mediated by exchange interaction (Dexter energy transfer process), which exhibits an exponential dependence on the interionic distance and is active only at extremely short distances [19]. Therefore, it is generally believed that the energy transfer among most optically active ions and molecules is usually the Förster type induced by the electric multipolar interaction. Energy transfer is responsible for numerous phenomena commonly observed in luminescent materials such as concentration quenching, cross relaxation, and sensitized luminescence. In upconversion materials, energy transfer is a critical step in one of the most efficient upconversion mechanisms, energy transfer upconversion (ETU), as discussed in the next section.

13.3 Energy Transfer Upconversion

For a complete understanding of the ETU mechanism, we need to set up a system of rate equations containing all involved states and transitions. As an example, we will consider the prototypical Yb^{3+}-Er^{3+} system whose energy-level scheme is shown in Fig. 13.1a. Yb^{3+} ion has 11 electrons in the 4f shell, forming a two-level system with an absorption band centered at 980 nm. In contrast, Er^{3+} ion has a much more complex energy-level structure. The existence of many energy levels and possible transitions between them make Er^{3+} ions highly susceptible to cross relaxation, which results in quenching of luminescence [20]. It is therefore common that the Er^{3+} concentrations are limited to 1–2% in most luminescent materials. Yb^{3+} ions are much more resistant to concentration quenching thanks to the simpler energy-level structure and can be doped more heavily without quenching. Therefore, in a typical ETU material, Yb^{3+} is usually doped heavily at around 20% or higher and Er^{3+} is doped much more lightly at 1–2%. Additionally, Yb^{3+} ion exhibits a higher absorption cross section than Er^{3+} ion. As a result, most of the incident light is absorbed by the Yb^{3+} ions in a Yb^{3+}-Er^{3+}-based ETU material. Upon photon absorption, the Yb^{3+} ion is excited to the $^2F_{5/2}$ level, from which it may decay back to the ground level, $^2F_{7/2}$, through several different channels. It could decay radiatively, emitting an infrared photon, or nonradiatively by multiphonon emission. It is also possible to decay via the Förster energy transfer process, exciting a nearby Er^{3+} ion into the $^4I_{11/2}$ level. The excited state of Yb^{3+} ion ($^2F_{5/2}$) is resonant with the $^4I_{11/2}$ level of Er^{3+} and thus the energy transfer rate, which depends on the overlap integral, as shown in Eq. (13.6), can be very high. If the energy transfer from another Yb^{3+} ion takes place before the excited Er^{3+} ion decays back to the ground state, the Er^{3+} ion is excited to the $^4F_{7/2}$ level and then quickly decays nonradiatively to the $^2H_{11/2}$ and $^4S_{3/2}$ levels from which the green luminescence occurs. A fraction of Er^{3+} ions in the $^4S_{3/2}$ state would decay nonradiatively into the slightly lower $^4F_{9/2}$ level where the red luminescence originates. An additional path for the red emission is for the Er^{3+} ion in $^4I_{11/2}$ level to decay nonradiatively into the $^4I_{13/2}$ level and then get excited to the $^4F_{9/2}$ level by energy transfer from a nearby Yb^{3+} ion. There are other processes that compete with the ETU process, such as radiative and nonradiative decays from $^2F_{5/2}$ of Yb^{3+} and $^4I_{11/2}$ of Er^{3+}, energy transfer among Yb^{3+} ions (often called hopping), back energy transfer from Er^{3+} to Yb^{3+}, and cross relaxation of Er^{3+} ion pairs. Furthermore, there are three-photon upconversion processes, resulting in blue emission and also contributing to the green and red emission. However, it is expected that the three-photon processes make much smaller contributions than the two-photon processes except at very high pump powers. Therefore, the three-photon process may be omitted in the rate equations. Furthermore, the population of the $^4F_{7/2}$ level of Er^{3+} may also be ignored as the relaxation from the $^4F_{7/2}$ level to the lower lying $^2H_{11/2}$ and $^4S_{3/2}$ levels is known to be extremely fast. Also, the $^2H_{11/2}$ and $^4S_{3/2}$ levels are close enough to be considered as a single level. With these simplifications, the complete set of rate equations can be written as [21]

$$\frac{dN_{D1}}{dt} = \sigma \Phi N_{D0} - W_{D10}N_{D1} + c_{Bd2}N_{A2}N_{D0} - c_{Fd2}N_{D1}N_{A0} - c_{d3}N_{D1}N_{A1} - c_{d4}N_{D1}N_{A2} \tag{13.7}$$

$$\frac{dN_{A1}}{dt} = W_{A21}N_{A2} - W_{A10}N_{A1} - c_{d3}N_{D1}N_{A1} \tag{13.8}$$

$$\frac{dN_{A2}}{dt} = c_{Fd2}N_{D1}N_{A0} - c_{Bd2}N_{A2}N_{D0} - c_{d4}N_{D1}N_{A2} - W_{A20}N_{A2} - W_{A21}N_{A2} \tag{13.9}$$

$$\frac{dN_{A3}}{dt} = W_{A43}N_{A4} + c_{d3}N_{A1}N_{D1} - W_{A30}N_{A3} \tag{13.10}$$

$$\frac{dN_{A4}}{dt} = c_{d4}N_{D1}N_{A2} - W_{A40}N_{A4} - W_{A43}N_{A4} \tag{13.11}$$

$$N_D = N_{D0} + N_{D1} \tag{13.12}$$

$$N_A = N_{A0} + N_{A1} + N_{A2} + N_{A3} + N_{A4} \tag{13.13}$$

Here N_i is the density of ions in the energy level i. The subscripts $D1$ and $D0$ represent the $^2F_{5/2}$ and $^2F_{7/2}$ levels of donor (Yb^{3+}), respectively, and $A4$, $A3$, $A2$, $A1$, and $A0$ indicate the $^4S_{3/2}$, $^4F_{9/2}$, $^4I_{11/2}$, $^4I_{13/2}$, and $^4I_{15/2}$ levels of acceptor (Er^{3+}), respectively. W is the transition rate whose subscript indicates the initial and final states of the transition. For example, W_{D10} is the decay rate of donor ion from $D1$ to $D0$ state. c_{d2}, c_{d3}, and c_{d4} are the energy transfer coefficients for the energy transfer processes between the donor and the acceptor in $A2$, $A3$, and $A4$ levels, respectively. The additional subscripts, F and B, in the c_{d2} coefficient indicate the forward (donor to acceptor) and backward (acceptor to donor) energy transfers. Finally, N_D and N_A are the doping densities of donor and acceptor, respectively; σ is the absorption cross section of the donor ion; and Φ is the incident light flux. It is noted that we consider only the backward transfer from the acceptor $^4I_{11/2}$ level to donor as the lifetimes of the higher excited states of the acceptor are so short that backward transfer is negligible. Also, the decay rate, W, should in general include the rates of both radiative and nonradiative decays. In bulk $NaYF_4$:Yb^{3+},Er^{3+} samples, the major nonradiative decay channel is multiphonon emission, but thanks to the small phonon energy of fluoride crystals, the nonradiative decay rate is usually small. Hence, W may generally be replaced with the radiative decay rate. In the photonically enhanced upconversion, however, the upconversion material may be in close proximity to a dissipative material which may offer a new highly efficient nonradiative decay channel. Therefore, nonradiative decay rate not only should be included but could very well dominate. This should be kept in mind in the analysis of nanophotonic enhancement of upconversion.

We now proceed to seek solutions of the rate equations in the weak and strong excitation limits. The details on how to obtain steady-state and time-dependent solutions of the rate equations in the weak and strong excitation limits are given elsewhere [21, 22]. Briefly, in the weak excitation limit, the intermediate energy levels $^2F_{5/2}$ of Yb^{3+} and $^4I_{11/2}$ of Er^{3+} are assumed to be nearly empty while in the strong excitation limit, they are assumed to be nearly full. This approximation allows significant simplifications and offers simple yet revealing analytical solutions. In the weak excitation limit, the green and red photon emission rates (in units of per volume per time) are given as

$$\Phi_{\mathrm{G}} = \frac{W_{A40}}{W_{A4}} \frac{c_{d4} c_{Fd2} N_A N_D}{c_{Bd2} W_{D10}^2} \cdot (\sigma \Phi)^2 \tag{13.14}$$

$$\Phi_{\mathrm{R}} = \left(\frac{W_{A43} c_{d4}}{W_{A4}} + \frac{W_{A21} c_{d3}}{W_{A10}} \right) \frac{c_{Fd2} N_A N_D}{c_{Bd2} W_{D10}^2} \cdot (\sigma \Phi)^2 \tag{13.15}$$

Here the W_{A4} is the total decay rate of energy level $A4$, $W_{A4} = W_{A40} + W_{A43}$. Equations (13.14) and (13.15) show that the upconverted luminescence photon flux depends quadratically on the incident photon flux, which is a consequence of the two-photon process responsible for upconversion. In addition, they show that the upconverted luminescence intensity should increase linearly with the donor and acceptor densities. Also, the upconverted luminescence intensity varies linearly with the energy transfer coefficients, c_{Fd2} and c_{d4}, and inversely with c_{Bd2} and the square of donor decay rate, W_{D10}. In the strong excitation limit, the green and red photon emission rates are given as

$$\Phi_{\mathrm{G}} = \frac{W_{A40}}{W_{A4}} \frac{N_{D0}}{2} \sigma \Phi \tag{13.16}$$

$$\Phi_{\mathrm{R}} = \frac{W_{A43}}{W_{A4}} \frac{N_{D0}}{2} \sigma \Phi \tag{13.17}$$

Equations (13.16) and (13.17) reveal the linear power dependence on the excitation power density. Also, it is important to note that the upconverted luminescence intensity depends only on the donor density and excitation power density and not on any of the energy transfer coefficients since, in the strong excitation limit, the energy transfer rate is so fast that the upconversion is limited by the absorption by the donor ion.

13.4 Photonic Enhancement of Upconversion

Equations (13.14)–(13.17) reveal how a photonic nanostructure can impact the various processes involved in upconversion. The parameters that can be affected by photonic nanostructures are the decay rates (W's), energy transfer coefficients (c's), and absorption cross section, σ. These parameters represent the three different processes involved in ETU: decay (radiative and nonradiative), energy transfer, and absorption. We now discuss how these processes may be influenced by photonic nanostructures, starting with the absorption cross section, σ. Absorption may be calculated classically from Poynting's theorem, which states that the divergence of the Poynting vector specifies the power dissipation in an absorbing medium. Thus, for monochromatic light, the power dissipation is given by

$$P_{\text{abs}} = \frac{1}{2}\int_V \omega\left(\varepsilon''|\mathbf{E}|^2 + \mu''|\mathbf{H}|^2\right) dV \tag{13.18}$$

where ε'' and μ'' are the imaginary parts of permittivity and permeability, respectively. Normally, $\mu'' = 0$ at optical frequencies and we only need to consider the electric part. Now we need to transform Eq. (13.18) into a form appropriate for an optically active ion characterized by its polarizability rather than a continuous absorbing medium specified by the permittivity. For this, we first rewrite the power dissipation in terms of current density, $\mathbf{j}$:

$$P_{\text{abs}} = \frac{1}{2}\int_V \text{Re}\left\{\mathbf{j}^* \cdot \mathbf{E}\right\} dV \tag{13.19}$$

In the dipole approximation, the current density can be expressed in terms of the induced dipole moment, $\mathbf{d}$, as $\mathbf{j} = -i\omega\mathbf{d}\delta(\mathbf{r} - \mathbf{r_0})$, which yields

$$P_{\text{abs}} = \frac{\omega}{2}\,\text{Im}\left\{\mathbf{d}^* \cdot \mathbf{E}(\mathbf{r}_0)\right\} \tag{13.20}$$

where $\mathbf{E}(\mathbf{r}_0)$ is the external electric field at the position of the ion. Since the induced dipole moment is given by the atomic polarizability, α, as $\mathbf{d} = \alpha\mathbf{E}$, the absorbed power may now be written as

$$P_{\text{abs}} = \frac{\omega}{2}\,\text{Im}\{\alpha\}\left|\hat{\mathbf{n}}_d \cdot \mathbf{E}(\mathbf{r}_0)\right|^2 \tag{13.21}$$

where $\hat{\mathbf{n}}_d$ is the unit vector along the direction of the induced dipole moment, $\mathbf{d}$. We can now write the absorption cross section in terms of atomic polarizability, α, and local field, $\mathbf{E}(\mathbf{r}_0)$, as

$$\sigma = \frac{P_{\text{abs}}}{I_{inc}} = c\mu_0\omega\,\text{Im}\{\alpha\}\frac{\left|\hat{\mathbf{n}}_d \cdot \mathbf{E}(\mathbf{r}_0)\right|^2}{|\mathbf{E}_0|^2} \tag{13.22}$$

where $\mathbf{E}_0$ is the incident field. The polarizability is determined by the transition matrix element and thus depends only on the electronic wavefunctions and not on the photonic nanostructures. The local field, however, can be strongly enhanced by a suitably designed photonic nanostructure, thereby impacting the absorption cross section. Specifically, Eq. (13.22) shows absorption cross-section scales as $|\mathbf{E}/\mathbf{E}_0|^2$, that is, linearly with the local intensity. Combined with Eqs. (13.14)–(13.17), we conclude that the upconversion luminescence intensity scales as $|\mathbf{E}/\mathbf{E}_0|^4$ in the weak excitation limit and as $|\mathbf{E}/\mathbf{E}_0|^2$ in the strong excitation limit. It should be noted that the nonlinearity in the weak excitation limit suggests that a modest absorption enhancement could result in a dramatic enhancement of upconversion luminescence.

Next, we discuss decay rates. Photonic nanostructures can affect both radiative and nonradiative decay rates. Radiative decay rates depend on the density of states, as shown in Eq. (13.2). In quantum electrodynamics, we should consider the states of the combined system of atom and photon. Therefore, the density of states in Eq. (13.2) includes the photon states as well as the electronic states. Consequently, the large increase in photon density of states by a nanophotonic structure such as plasmonic nanostructure and photonic crystal can increase the optical transition rate. This is the well-known Purcell effect that has been widely studied in photonics [23, 24]. The theoretical description of Purcell effect can be derived from the Fermi's golden rule. In the dipole approximation, the radiative decay rate is found from Eq. (13.2) using the dipole term of the multipolar Hamiltonian with vacuum electric field:

$$W = \frac{\pi\omega}{3\hbar\varepsilon}|\mathbf{d}|^2\rho\left(\mathbf{r}_0, \omega\right) \tag{13.23}$$

where $\mathbf{r}_0$ is the position of the emitter. Classically, the radiation rate of a dipole emitter can be calculated from the power dissipation rate according to Poynting's theorem:

$$P_{rad} = -\frac{1}{2}\int_V \mathrm{Re}\left\{\mathbf{j}^* \cdot \mathbf{E}\right\} dV \tag{13.24}$$

Expressing the current density in terms of the induced dipole moment, just as before, yields

$$P_{rad} = \frac{\omega}{2}\,\mathrm{Im}\left\{\mathbf{d}^* \cdot \mathbf{E}\left(\mathbf{r}_0\right)\right\} \tag{13.25}$$

Note that although Eq. (13.25) is formally identical to Eq. (13.20), they have different physical meaning. The electric field in Eq. (13.20) represents the incident field being absorbed while the electric field in Eq. (13.25) is the dipole field emanating from the quantum emitter. We can now express the dipole field in terms of the dyadic Green function as

$$P_{rad} = \frac{\omega^3}{2c^2\varepsilon}\left[\mathbf{d} \cdot \mathrm{Im}\left\{\overleftrightarrow{\mathbf{G}}\left(\mathbf{r}_0, \mathbf{r}_0; \omega\right)\right\} \cdot \mathbf{d}\right] \tag{13.26}$$

Comparing Eqs. (13.23) and (13.26) allows us to write the density of states in terms of the dyadic Green function:

$$\rho\left(\mathbf{r}_0, \omega\right) = \frac{6\omega}{\pi c^2}\left[\hat{\mathbf{n}}_{\mathbf{d}} \cdot \mathrm{Im}\left\{\overleftrightarrow{\mathbf{G}}\left(\mathbf{r}_0, \mathbf{r}_0; \omega\right)\right\} \cdot \hat{\mathbf{n}}_{\mathbf{d}}\right] \tag{13.27}$$

It is noted that **d** represents the quantum mechanical transition matrix element in Eq. (13.23) but is a classical dipole in Eq. (13.26). Nevertheless, the classical and quantum mechanical calculations agree when evaluating the enhancement factor, often called the Purcell factor, defined as the ratio of radiation rates in a nanophotonic environment and free space. With Eqs. (13.23)–(13.27), it is now possible to predict the enhancement in radiative decay rate by evaluating the radiative power of a classical point dipole.

Nonradiative decay such as multiphonon emission is not directly impacted by the local field. However, the presence of a lossy medium, e.g., metal, can introduce additional nonradiative decay channels, increasing the overall nonradiative decay rate and consequently resulting in luminescence quenching. The additional nonradiative decay rate may be calculated by evaluating the ohmic loss in the environment for the field emitted by a radiating dipole: [25]

$$\frac{W_{nr}}{W_0} = \frac{1}{P_0}\frac{1}{2}\int_V \mathrm{Re}\left\{\mathbf{j}^* \cdot \mathbf{E}\right\} dV \tag{13.28}$$

Here W_0 is the free-space decay rate, P_0 is the radiation rate of a classical dipole in free space, $P_0 = \omega^4|\mathbf{d}|^2/12\pi\epsilon_0 c^3$, and the integral is performed over the volume of the lossy medium. Also, **E** is the field emitted by the dipole emitter and **j** is the current density induced inside the lossy medium by the dipole field. They can be expressed in terms of the dyadic Green function as follows:

$$\mathbf{j} = \omega\varepsilon''\mathbf{E} \tag{13.29}$$

$$\mathbf{E}(\mathbf{r}) = \frac{\omega^2}{c^2\varepsilon}\overleftrightarrow{\mathbf{G}}(\mathbf{r}, \mathbf{r}_0; \omega) \cdot \mathbf{d} \tag{13.30}$$

Just like the radiative decay rate, the dyadic Green function allows us to estimate the changes in the nonradiative decay rate. The trade-off between the positive effect of Purcell effect and the negative effect of quenching is an important aspect of photonic engineering of quantum emitter. Typically, quenching dominates at short distances while the Purcell effect remains pronounced at larger distances. It is therefore customary to insert a thin (~10 nm) spacer layer between metal surface and quantum emitters.

Finally, we describe the rate of energy transfer. For simplicity, we assume dipole-dipole coupling and estimate the Förster energy transfer rate, which in a dispersive and absorbing medium is given by [26]

$$W_{\mathrm{ET}} = \int d\omega g_A(\omega)\, g_D(\omega)\, \tilde{w}(\omega) \tag{13.31}$$

where g_A and g_D are the free-space normalized absorption and emission spectra of acceptor and donor ions, respectively. The information about the medium is lumped into the transition rate, $\tilde{w}(\omega)$, which can be expressed in terms of the dyadic Green function as follows:

$$\tilde{w}(\omega) = \frac{2\pi}{\hbar^2}\frac{\omega^2}{\varepsilon_0 c^2}\left|\mathbf{d}_A^* \cdot \overleftrightarrow{\mathbf{G}}(\mathbf{r}_A, \mathbf{r}_D, \omega) \cdot \mathbf{d}_D\right|^2 \tag{13.32}$$

Here, $\mathbf{d}_A$ and $\mathbf{d}_D$ are the dipole moments of acceptor and donor ions, respectively; $\mathbf{r}_A$ and $\mathbf{r}_D$ indicate the positions of acceptor and donor ions, respectively.

An earlier study observed enhanced Förster energy transfer rate in optical cavities and attributed it to the enhanced local density of states to which the donor emission rate is directly proportional [27]. Eq. (13.27) shows that the donor emission rate is proportional to the imaginary part of the dyadic Green function $\mathbf{G}(\mathbf{r}_D, \mathbf{r}_D, \omega)$, while the energy transfer rate is related to $\mathbf{G}(\mathbf{r}_D, \mathbf{r}_A, \omega)$ according to Eq. (13.32). For a small donor-acceptor distance, $\mathbf{G}(\mathbf{r}_D, \mathbf{r}_A, \omega)$ may be similar to $\mathbf{G}(\mathbf{r}_D, \mathbf{r}_D, \omega)$. However, the real part of the dyadic Green function often dominates over the imaginary part and thus the energy transfer enhancement is generally quite different from the radiative decay rate enhancement. In fact, it was later shown that the energy transfer enhancement is not related to the density of state enhancement [28, 29]. Theoretical studies indicate that the energy transfer enhancement is generally smaller than the spontaneous emission enhancement, although the enhancement depends strongly on the donor-acceptor distance [26]. Our calculations for a donor-acceptor pair in the vicinity of a silver surface showed that the energy transfer rate can be enhanced at frequencies slightly lower than the surface plasmon frequency and quenched at slightly higher frequencies [21, 30]. Also, both the enhancement factor and the bandwidth over which enhancement is achieved are smaller than the luminescence enhancement.

Now that we have established the quantum electrodynamic foundation for all optical processes involved in ETU, we can start making quantitative predictions on the nanophotonic enhancement of upconversion. Specifically, Eqs. (13.14)–(13.17) can be used to estimate the upconverted luminescence intensity while Eqs. (13.22), (13.25), (13.26), (13.28), and (13.31) allow us to calculate the enhancement factors for absorption cross section, emission rates, and energy transfer coefficients. The expressions for the green emission are a bit simpler as it has only one possible excitation pathway while the red emission has two different excitation pathways as well as possible three-photon processes. We will therefore examine the enhancement of the green luminescence for simplicity and clarity, noting that the analysis can be extended to red luminescence in a straightforward, if tedious, manner. In the strong excitation limit, Eq. (13.16) shows that the photonic enhancement is possible on the radiative efficiency, W_{A40}/W_{A4}, and absorption cross section, σ. The absorption enhancement is described clearly by Eq. (13.22). The behavior of the radiative efficiency is much more complex but with reasonable assumptions detailed elsewhere [21], one can obtain

$$F_{rad} = \left(\frac{W_{A40}}{W_{A4}}\right)_{sp} \Big/ \left(\frac{W_{A40}}{W_{A4}}\right) = \frac{W_{A40} + W_{A43}}{W_{A40} + (W_{A43} + W_{nr})\,/F_P} \tag{13.33}$$

where F_P is the Purcell factor for the green emission. It is noted that the Purcell effect would make a significant impact only for originally inefficient materials. That is, the radiative efficiency is enhanced substantially only when $W_{A43} + W_{nr}$ is much larger than W_{A40}. In the limit of an infinitely large Purcell factor, the radiative efficiency enhancement saturates at $1 + W_{A43}/W_{A40}$. We may therefore conclude that generally the Purcell enhancement of upconverted luminescence would make only a modest contribution. The dominant contribution to the upconversion enhancement therefore comes from absorption enhancement, F_a, which is proportional to the square of local field enhancement. Thus, the overall enhancement factor for the green luminescence intensity in the strong excitation limit is

$$F_{strong} \approx F_a \tag{13.34}$$

In the weak excitation limit, Eq. (13.14) shows that we have to consider the energy transfer coefficients, c_{d4}, c_{Fd2}, and c_{Bd2}, and the infrared decay rate, W_{D10}, in addition to the radiative efficiency and absorption cross section. Since the radiative efficiency should not be affected significantly, the overall enhancement factor, F_{weak}, for the green luminescence intensity in the weak excitation limit can be written as

$$F_{weak} \approx \frac{F_{d4}F_{Fd2}F_a^2}{F_{Bd2}F_{D10}^2} \tag{13.35}$$

where F_{d4}, F_{Fd2}, and F_{Bd2} are the enhancement factors for the energy transfer processes represented by the coefficients c_{d4}, c_{Fd2}, and c_{Bd2}, respectively, and F_{D10} is the enhancement factor for the infrared decay rate, W_{D10}, of the donor ion. W_{D10} can be enhanced by the Purcell effect and also by the additional nonradiative decay rate introduced by metal. Either way any increase in the donor decay rate is detrimental to upconversion, as it reduces the population of the intermediate level from which upconversion takes place. For the energy transfer enhancement, it is reasonable to expect that both the forward and backward transfers are enhanced by the same factor, i.e., $F_{Bd2} = F_{Fd2}$, because the $^2F_{5/2}$ level of Yb^{3+} and $^4I_{11/2}$ level of Er^{3+} are resonant. Thus, the enhancement factor in the weak excitation limit may be simplified to

$$F_{weak} \approx \frac{F_{d4}F_a^2}{F_{D10}^2} \tag{13.36}$$

Based on the analysis given above, we may draw a few conclusions. As stated earlier, even with extremely high Purcell enhancement, the green emission would not be enhanced significantly. Instead, Eqs. (13.34)–(13.36) show that the processes occurring at the excitation wavelength, which are absorption, energy transfer, and donor decay, are important for the overall enhancement of upconversion. It is therefore desirable to tune the photonic resonance to the excitation wavelength rather than the emission wavelength. Furthermore, in the weak excitation limit, the upconversion luminescence intensity is enhanced linearly to energy transfer enhancement and quadratically to absorption enhancement. Thus, absorption enhancement is expected to produce the most dramatic improvement in upconversion.

13.5 Current Status and Outlook

Figure 13.2 shows the enhancement of upconversion luminescence reported during the past decade or so [30, 31]. First, it should be noted that the enhancement factor depends on, among other things, the excitation power, as shown in the theoretical study presented earlier. Unfortunately, some papers, especially the earlier ones, do not provide complete information on the experimental conditions. It is therefore not always possible to meaningfully compare the enhancement factors reported in the literature. Furthermore, nanophotonic structures tend to produce highly localized or spatially varying local field profiles. Upconversion enhancement is therefore highly sensitive to the exact geometry of the nanostructure and the placement of upconversion material, leading to a large variation in reported enhancement factors. Nevertheless, the enhancement factor is one parameter almost universally reported and can be used to obtain some useful insights.

In general, lithographically fabricated nanostructures tend to provide higher enhancement factors. Lithographic techniques generally yield high-quality nanostructures with precise dimensions as prescribed. This in turn leads to strong enhancement of local field and consequently upconversion luminescence. In contrast, colloidal systems often prepared by wet chemistry routes tend to exhibit greater inhomogeneity, leading to broadening and weakening of optical resonance and consequently lower upconversion enhancement. It should, however, be noted that it is not always possible to choose lithographic fabrication to achieve higher enhancement because some applications such as biosensing and biomedical applications require colloidal systems. Therefore, continued progress is required for both lithographically fabricated nanostructures and colloidal nanostructures.

Another interesting point is the apparent shift of focus from emission enhancement to absorption enhancement. Prior to 2015, as shown in Fig. 13.2a, a majority of the nanostructures are resonant at the visible wavelengths, aiming for enhancement of emission via Purcell effect. After 2015, however, more nanostructures target the absorption wavelengths in the infrared region, as shown in Fig. 13.2b. This is presumably due to the realization that the absorption enhancement yields much greater enhancement than emission enhancement, as clearly demonstrated in the theoretical analysis in the previous sections. Indeed, Fig. 13.2 shows larger enhancement factors for nanostructures resonant in the infrared region than those with resonances in the visible.

We now highlight a few notable nanostructures. Das et al. reported a water-dispersible nanostructure targeting the near-infrared region using a metal-insulator-metal (MIM) design (Fig. 13.3a) [32]. The structure is comprised of a layer of

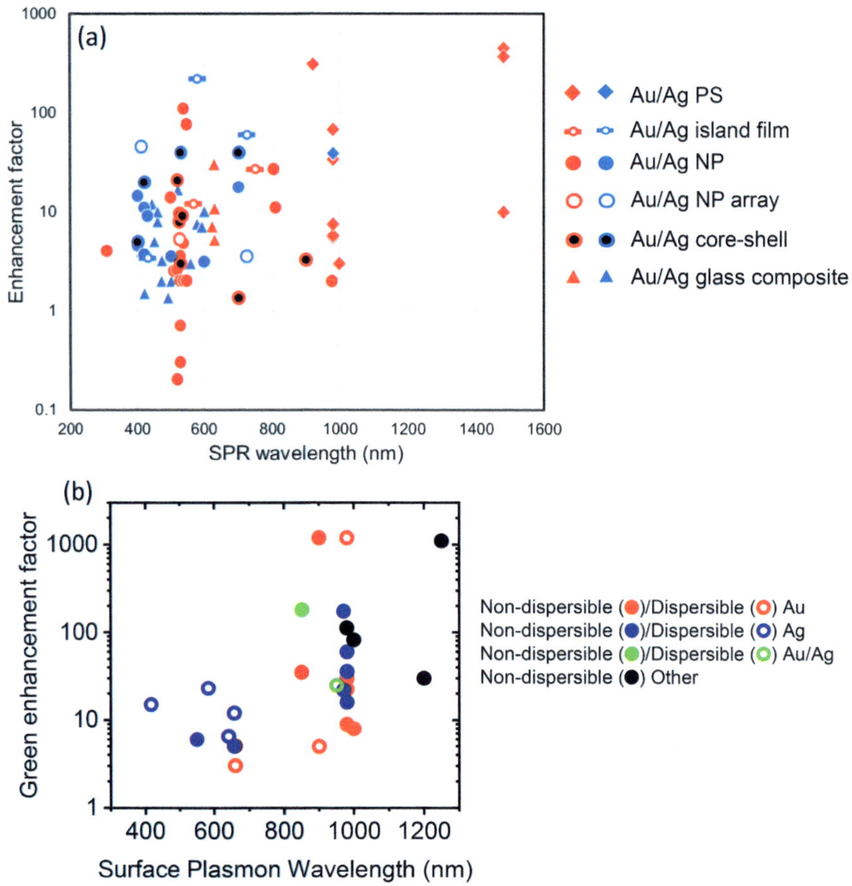

Fig. 13.2 (**a**) Survey of plasmonic enhancement of upconversion before 2015. Gold and silver are represented by red and blue colors, respectively. The various plasmonic geometries include glass composites (triangle); nanoparticles such as nanosphere, nanorod, nanowire, and nanoprism (circle); self-assembled nanoparticle array (ring); core–shell (concentric circle); island film or flat film (horizontal bar with circle); and periodic structure (diamond). NP and PS represent nanoparticle and patterned structure, respectively. Reprinted from *Chem. Soc. Rev.***44**, 2940 (2015) [30] with permission from The Royal Society of Chemistry. (**b**) Survey of plasmonic enhancement of upconversion after 2015. The data is divided by material choice for the nanostructure (red for Au, blue for Ag, green for Au-Ag alloy, and black for other materials) as well as by whether the nanostructure can be dispersed into solution or not (filled circle for non-dispersible nanostructures and unfilled circle for dispersible). Reprinted with permission from *Nanophoton.* 000010151520200159 (2020) [31], licensed under the Creative Commons Attribution 4.0 Public License. BY 4.0

upconversion nanoparticles (UCNPs) sandwiched between two Au disks and uses a gap plasmon resonance to greatly enhance the electric field within the UCNP layer. The MIM nanostructure is unique in that its resonance relies on geometric parameters, specifically the diameter and thickness of the insulating layer, and

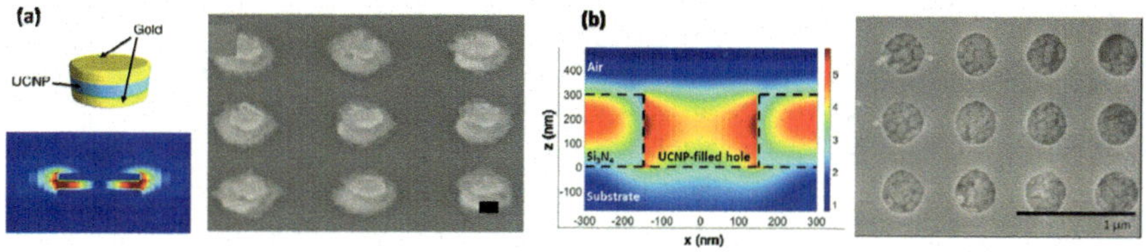

Fig. 13.3 (**a**) Top-left: schematic of metal-insulator-metal (MIM) structure; bottom-left: electric field profile showing strong field enhancement in the insulator region where UCNPs are placed; right: scanning electron micrograph (SEM) of fabricated MIM structures. Scale bar: 100 nm. Reprinted with permission from Nat. Commun. **9**, 4828 (2018) [32], licensed under Creative Commons Attribution 4.0 International License. (**b**) Left: electric field profile showing field enhancement inside the nanoholes; right: SEM of Si_3N_4 photonic crystal structure where the nanoholes are filled with UCNPs by a self-assembly process. Scale bar: 1 μm. Reprinted with permission from ACS Photon. **6**, 1882 (2019) [34], Copyright (2019) American Chemical Society

thus the resonance can be easily tuned to any desired wavelength simply by changing the size of the structure. The plasmon resonance supported by the MIM nanostructure does not depend on the period of the structure so, even when dispersed into solution, a strong localized electric field can be achieved. For fabrication, laser interference lithography was used to produce a large-area, highly uniform nanostructure. Using the MIM designed for a resonance at 980 nm, the authors reported a 1200× enhancement. The MIMs were also lifted off the substrate and dispersed in water by a pattern transfer technique. The water-dispersed MIMs were incubated with cancer cells and imaged using a confocal microscope. Due to their enhanced upconversion, they observed similar upconversion intensities with the MIMs compared to a reference cell culture incubated with free UCNPs at a 1000 times higher concentration. To our knowledge, this remains the largest enhancement achieved using a plasmonic nanostructure at the time of this publication.

Dielectric photonic crystal (PhC) structures with UCNPs provide an excellent alternative to plasmonic designs. Surface plasmon is highly effective in localizing light and creating a strong local electric field, which can be used to enhance upconversion. However, the mode volume where electric field is enhanced tends to be small and thus enhancement over large areas is generally difficult to achieve. Furthermore, metal inevitably quenches luminescence, and thus, the plasmonic nanostructure must be designed carefully to best utilize the plasmonic field while minimizing the effect of quenching. In contrast, PhCs are dielectric nanostructures and thus exhibit less luminescence quenching than metallic nanostructures. Furthermore, by using band edge states, it is possible to excite optical resonances extended over the entire photonic structure, thereby achieving field enhancement over a large area [33]. PhC structures are thus suitable for applications like lighting, displays, and sensors that may require a large active area. Mao et al. incorporated upconversion nanoparticles in a two-dimensional planar PhC structure for upconversion enhancement (Fig. 13.3b) [34]. In this work, a PhC structure consisting of a square array of nanoholes in a Si_3N_4 device layer was fabricated by laser interference lithography. UCNPs were then filled inside the nanoholes using a self-assembly method. The local field enhancement experienced by the UCNPs in the nanoholes resulted in overall enhancement factors of 130× for green and 350× for red luminescence. The local field enhancement in the dielectric PhC structure was not as high as plasmonic nanostructures but the lower field enhancement was partially compensated by the fact that the purely dielectric PhC structure causes much less quenching than the plasmonic structures made of metal.

Continuous-wave (cw) lasering at room temperature in an integrated device platform is an important development with a wide range of potential applications, e.g., quantum photonics, bioimaging, lighting, and sensing. Plasmonic nanocavities with strong light localization provide an excellent device platform for nanolaser development. Recently, Fernandez-Bravo et al. reported room-temperature cw lasering with UCNPs coated on a Ag nanopillar array [35]. The Ag nanopillar array was designed in such a way that the individual plasmon resonance couples together to produce a collective resonance at 660 nm with a very-high-quality factor. After UCNP coating, the entire structure was pumped with a 980 nm laser and the strong coupling between the red upconverted luminescence and the lattice plasmon resulted in lasering. The system exhibited an ultralow threshold power of 70 W/cm^2, orders of magnitude lower than other small lasers. Providing a directional, ultra-stable output at visible frequencies under near-infrared pumping, the new nanolaser opened a new opportunity such as coherent bioimaging.

In closing, we have witnessed a tremendous progress in upconversion enhancement during the past decade, thanks to the combination of deeper understanding of the light-matter interaction in nanostructures and more sophisticated fabrication techniques [30, 31, 36]. The reported enhancement factors have begun exceeding three orders of magnitude. This is particularly exciting for dispersible nanostructures where the design and synthesis constraints make this especially challenging. Also, though not discussed in this review, new exotic materials other than noble metals are emerging as a new tool to enhance upconversion. We anticipate that new strides will be made in the use of dielectric nanostructures where the lack of plasmonic quenching may allow for higher efficiency. We also predict further work into the enhancement of higher order upconversion processes that can generate blue and ultraviolet light. While these are less efficient than the extensively studied two-photon process, the higher order power dependence of these processes should result in a much larger enhancement in upconversion emission.

For applications, we anticipate continued progress in areas encompassing solar energy conversion, lighting, imaging, and sensing, while new applications such as security will be explored. In particular, biological and medical applications have received growing attention thanks to the UCNP's intrinsic advantages such as no background autofluorescence, minimal background tissue damage, and long penetration depth in biological tissues. UCNPs are particularly attractive for in vivo applications where the safe level of laser power is quite small and thus other nonlinear probes such as two-photon dyes are not usable. For widespread use in in vivo settings, the upconversion efficiency must still be improved, calling for further research in nanophotonic approaches. It is also important to develop surface treatment techniques for targeted delivery. The recent development of upconversion laser adds a new dimension to the application of upconversion materials. We may envision upconversion-based coherent imaging and sensing in live animals or humans. With the current pace of developments, it is conceivable that nanophotonically enhanced upconversion materials will soon be placed among the mainstream nanoprobes commonly used in both laboratories and clinics.

References

1. H.W. Leverenz, *An Introduction to Luminescence of Solids* (Dover Publications, 1968)
2. F. Auzel, Comptes Rendus Hebdomadaires Des Seances De L Academie Des Sciences Serie B **263**, 819 (1966)
3. V.V. Ovsyankin, P.P. Feofilov, Sov. Phys. Usp. **15**, 354 (1972)
4. E. Nakazawa, S. Shionoya, Phys. Rev. Lett. **25**, 1710 (1970)
5. J.S. Chivian, W.E. Case, D.D. Eden, Appl. Phys. Lett. **35**, 124 (1979)
6. R.H. Page, K.I. Schaffers, P.A. Waide, J.B. Tassano, S.A. Payne, W.F. Krupke, W.K. Bischel, J. Opt. Soc. Am. B: Opt. Phys. **15**, 996 (1998)
7. S. Heer, K. Kompe, H.U. Gudel, M. Haase, Adv. Mater. **16**, 2102 (2004)
8. K.W. Krämer, D. Biner, G. Frei, H.U. Güdel, M.P. Hehlen, S.R. Lüthi, Chem. Mater. **16**, 1244 (2004)
9. G. Yi, H. Lu, S. Zhao, Y. Ge, W. Yang, D. Chen, L.-H. Guo, Nano Lett. **4**, 2191 (2004)
10. S. Wen, J. Zhou, K. Zheng, A. Bednarkiewicz, X. Liu, D. Jin, Nat. Commun. **9**, 2415 (2018)
11. L. Novotny, B. Hecht, *Principles of Nano-Optics* (Cambridge University Press, 2012)
12. B. Di Bartolo, *Optical Interactions in Solids* (World Scientific, Singapore, 2010)
13. W. Park, C.J. Summers, Y.R. Do, H.G. Yang, J. Mater. Sci. **37**, 4041 (2002)
14. W. Park, R.Y. Lee, C.J. Summers, Y.R. Do, H.G. Yang, Mater. Sci. Eng. B **78**, 28 (2000)
15. J. Zhang, J. Lin, J. Cryst. Growth **271**, 207 (2004)
16. B.R. Judd, Phys. Rev. **127**, 750 (1962)
17. G.S. Ofelt, J. Chem. Phys. **37**, 511 (1962)
18. T. Förster, Ann. Phys. **2**, 55 (1948)
19. D.L. Dexter, J. Chem. Phys. **21**, 836 (1953)
20. X. Yu, C.J. Summers, W. Park, J. Appl. Phys. **111**, 073524 (2012)
21. D. Lu, S.K. Cho, S. Ahn, L. Brun, C.J. Summers, W. Park, ACS Nano **8**, 7780 (2014)
22. D. Lu, C. Mao, S.K. Cho, S. Ahn, W. Park, Sci. Rep. **6**, 18894 (2016)
23. E.M. Purcell, Phys. Rev. **69**, 681 (1946)
24. S. Haroche, D. Kleppner, Phys. Today **42**, 24 (1989)
25. P. Anger, P. Bharadwaj, L. Novotny, Phys. Rev. Lett. **96**, 113002 (2006)
26. H.T. Dung, L. Knoll, D.G. Welsch, Phys. Rev. A **65**, 043813 (2002)
27. P. Andrew, W.L. Barnes, Science **290**, 785 (2000)
28. M. de Dood, J. Knoester, A. Tip, A. Polman, Phys. Rev. B **71**, 115102 (2005)
29. C. Blum, N. Zijlstra, A. Lagendijk, M. Wubs, A.P. Mosk, V. Subramaniam, W.L. Vos, Phys. Rev. Lett. **109**, 203601 (2012)
30. W. Park, D. Lu, S. Ahn, Chem. Soc. Rev. **44**, 2940 (2015)
31. A. Das, K. Bae, W. Park, Nanophotonics **9**, 1359 (2020)
32. A. Das, C. Mao, S. Cho, K. Kim, W. Park, Nat. Commun. **9**, 4828 (2018)
33. K. Min, H. Jung, Y. Park, K.S. Cho, Y.G. Roh, S.W. Hwang, H. Jeon, Nanoscale **9**, 8703 (2017)
34. C. Mao, K. Min, K. Bae, S. Cho, T. Xu, H. Jeon, W. Park, ACS Photonics **6**, 1882 (2019)
35. A. Fernandez-Bravo et al., Nat. Mater. **18**, 1172 (2019)
36. S.P. Madsen et al., J. Phys. D. Appl. Phys. **53** (2020)

Part V
Addressing Thermalisation Losses: Singlet Fission and Quantum Cutting

Chapter 14
Singlet Fission: Mechanisms and Molecular Design

Victor Gray, Leah Weiss, and Akshay Rao

Abstract Exciton multiplication processes provide a means of overcoming thermalization losses in photovoltaic devices. In this chapter we will introduce one promising exciton multiplication process, termed singlet fission, which occurs in carbon-based semiconductors. After introducing the photophysics of organic semiconductors, we discuss the mechanism of singlet fission and the role of spin and electronic structure in the singlet fission process. Based on this mechanistic discussion we will introduce design rules for singlet fission materials related to their energy level alignment, molecular structure and crystal packing.

Keywords Exciton multiplication · Singlet fission · Thermalization losses · Organic semiconductors

Singlet fission is an exciton multiplication process in organic materials that could improve the efficiency of conventional solar cells. In this chapter we first establish a common terminology of photophysical processes and then describe the singlet fission process in detail. We then discuss material considerations for designing singlet fission molecules. The following chapters focus on how singlet fission can be incorporated into solar cells with enhanced efficiency.

V. Gray (✉)
Cavendish Laboratory, Department of Physics, University of Cambridge, Cambridge, UK

Department of Chemistry—Ångström Laboratory, Uppsala University, Uppsala, Sweden
e-mail: victor.gray@kemi.uu.se; vg331@cam.ac.uk

L. Weiss
Pritzker School of Molecular Engineering, University of Chicago, Chicago, IL, USA
e-mail: lrweiss@uchicago.edu

A. Rao
Cavendish Laboratory, Department of Physics, University of Cambridge, Cambridge, UK
e-mail: ar525@cam.ac.uk

J. S. Lissau, M. Madsen (eds.), *Emerging Strategies to Reduce Transmission and Thermalization Losses in Solar Cells*, https://doi.org/10.1007/978-3-030-70358-5_14

14.1 Photophysics of Organic Semiconductors

Here we introduce the nomenclature and theory relevant for singlet fission. For a broader treatment of organic semiconductor physics, the reader is referred to books addressing these topics comprehensively (see for example the books by Köhler and Bässler 2015 [1] and Turro, Ramamurthy, and Scaiano 2010 [2]).

Electrons, Excitons, and Spin

The electronic states of molecular systems are determined by their molecular structure. The electronic structure of organic molecules can be described by sets of molecular orbitals, which are superpositions of atomic orbitals. The class of molecules of interest here, conjugated aromatic molecules, have delocalized π-orbitals with an energy gap between the highest occupied molecular orbital (HOMO) and lowest unoccupied molecular orbital (LUMO) in the optical range.

Electrons each with spin $S = \frac{1}{2}$ then fill these energy levels according to the Pauli exclusion principle, which requires that any two electrons in the same molecular orbital have different spin orientations. This orientation is described by the spin projection quantum number (m) and can take the value of $+1/2$ or $-1/2$ (also termed up ($\uparrow$) and down ($\downarrow$) arrows, respectively).

The total spin S of a multi-electron system is the sum of all electron spin quantum numbers. In most organic semiconductor molecules in the ground state all orbitals are filled maximally and so the total spin is $S = 0$. Spin-zero states are termed *singlet* states. Throughout the text we will refer to singlet states with the notation S_n where the subscript n refers to the energy level considered, e.g., $n = 0$ for the ground state, $n = 1$ for the first excited state, and $n = 2$ for the second excited state.

Electronic transitions in molecules conserve both energy and spin. Photoexcitation of an electron from the HOMO to the LUMO therefore leads to a photoexcited singlet state (S_1), as shown in Fig. 14.1a. The energy gap between the ground and excited S_1 is determined by the HOMO-LUMO gap and any reduction due to the binding energy of the excited state due to electronic interactions, typically on the ~100 meV scale.

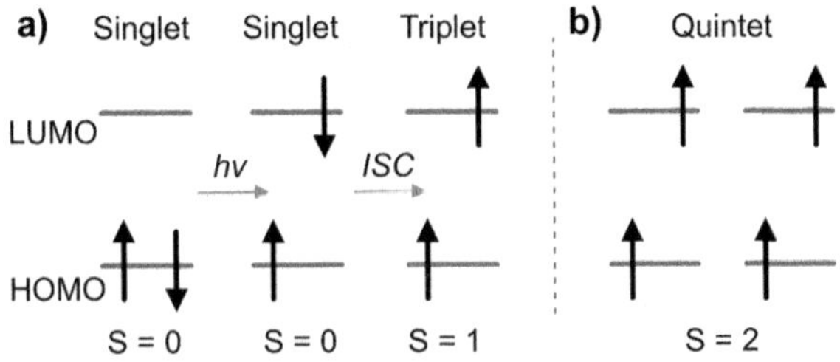

Fig. 14.1 (**a**) Electron spin configuration of singlet and triplet states. (**b**) Electron spin configuration of a quintet state

There are processes that can cause the spin of an electron to flip, most commonly *intersystem crossing* (ISC), which from an excited singlet state then forms a triplet state (T_n); see Fig. 14.1a. Triplet states have an overall spin $S = 1$ with three possible spin projection values for a triplet state, $m = 0, \pm 1$. The terms singlet and triplet refer to the multiplicity of possible spin orientations, where for a state of total spin S the multiplicity is given by $M = 2S + 1$.

A molecule in a photoexcited singlet (or triplet) state will eventually return to the ground state. This relaxation can occur by emission of a photon, or non-radiatively by thermal energy loss associated with vibrations. From a triplet state the change in spin also has to be accounted for through spin-orbit coupling. Many organic molecules stay in the excited singlet state for a few nanoseconds, whereas return from a triplet state is much slower, a consequence of the required spin flip; typically this occurs on a micro-millisecond timescale.

Singlet fission describes the process of triplet exciton pair generation and dissociation from an initially photoexcited singlet exciton [3, 4]. To observe singlet fission in a material it has to occur at a rate faster than normal singlet decay. We now discuss the photophysics of singlet fission and then will detail current design considerations for optimizing singlet fission materials.

14.2 Singlet Fission

Singlet fission was first found in 1965 in anthracene and tetracene crystals [5–8] with a current resurgence of interest when pulsed optical spectroscopy enabled measurement of singlet fission on ultrafast timescales ($\leq$100 fs) in solid-state pentacene with near-unity efficiency [9]. Singlet fission has since been observed in a wide range of physical systems from molecular films [9–11] to dimers [12–14] and solutions [15, 16] with an aim towards integration with photovoltaics (see the following chapter on solar energy applications of singlet fission).

It is useful to consider singlet fission in terms of two steps: (1) the generation of a triplet pair in an overall spin-singlet configuration and (2) the dissociation of the triplet pair to form independent, mobile triplet excitons. This two-step picture is the basis of the first kinetic model of singlet fission, the Johnson-Merrifield model [5, 17]. In this model singlet fission is described by the reaction

$$S_0S_1 \leftrightharpoons {}^1\mathrm{TT} \leftrightharpoons T_1 + T_1$$

where S_0S_1 describes two molecules where one is in the singlet ground state S_0 and the other is in the excited singlet state S_1, ${}^1\mathrm{TT}$ denotes the overall spin-singlet triplet-pair state, and $T_1 + T_1$ denotes noninteracting, dissociated triplet excitons [17–19]. We note that the triplet-pair state ${}^1\mathrm{TT}$ is variably referred to as the multi-exciton (ME) state in the literature—here for consistency we use the nomenclature triplet-pair state and the symbol ${}^M\mathrm{TT}$ for triplet pair with spin multiplicity $M = 1, 3, 5$ for singlet (total spin $S = 0$), triplet ($S = 1$), and quintet ($S = 2$) states, respectively (Fig. 14.1b).

How can singlet fission occur so efficiently? To answer this question we must consider the total spin and energy of the states involved. As described above, conventional triplet formation proceeds via *intersystem crossing* from the excited singlet state S_1 (total spin $S = 0$) to the excited triplet state T_1 (total spin $S = 1$). The two states are separated by the exchange energy (with scale $0.7 - 1.0$ eV in most organic materials). The spin-orbit coupling that links singlet and triplet excited states is relatively weak in pure hydrocarbon molecules (<meV due to the absence of heavy atoms) and so the spin-orbit-mediated transition between these energetically separated states occurs over hundreds of nanoseconds in most organic (hydrocarbon) materials [20].

Singlet fission by contrast occurs between the singlet excited state S_1 and a pair of triplet excitons ^{1}TT in an overall spin-singlet state and so is spin conserving. These two spin states localized on two molecules (labeled a and b) are given explicitly by

$$|S_0S_1\rangle = \frac{1}{2}\,(|\uparrow\downarrow\rangle - |\downarrow\uparrow\rangle)_a(|\uparrow\downarrow\rangle - |\downarrow\uparrow\rangle)_b$$

$$|^1\mathrm{TT}\rangle = \frac{1}{\sqrt{3}}(|\uparrow\uparrow\rangle_a|\downarrow\downarrow\rangle_b + |\uparrow\uparrow\rangle_b|\downarrow\downarrow\rangle_a) - \frac{1}{2\sqrt{3}}(|\uparrow\downarrow\rangle_a + |\downarrow\uparrow\rangle_a)(|\uparrow\downarrow\rangle_b + |\downarrow\uparrow\rangle_b)$$

where the two arrows indicate the spin configuration of the two electrons within a triplet or singlet and the subscript letter indicates the location of the state on molecule a or b. When the total energy of the triplet pair is close to the energy of the S_0S_1 state the process is both spin and energy conserving and so can proceed faster and more efficiently than intersystem crossing.

Triplet-Pair Formation

The rate of fission depends on many interrelated factors, from energy and electronic coupling to entropy and nuclear motion. Models for how these factors determine the rate of triplet-pair formation have been developed from combined experiments and in silico calculations across the broader space of materials. These models continue to be evaluated, refined, and debated as new materials and experimental techniques emerge.

Fission proceeds rapidly in *exothermic* materials like pentacene where the process is downhill in energy and more slowly in systems with an energetic barrier to triplet formation (*endothermic* fission materials). The relationship between the rate, energy splitting, and electronic overlap between the singlet- and triplet-pair states can be understood in the following simplified model [21, 22]. We define the electronic overlap integral $\hat{V} = \left\langle S_0 S_1 | \hat{H} | {}^1\mathrm{TT} \right\rangle$ and the splitting between the triplet pair and singlet state as $\Delta E = E(S_0 S_1) - E(\mathrm{TT})$ where $\hat{H}$ is the relevant interaction. The electronic coupling $\hat{V}$ includes both the direct electronic overlap and super-exchange contributions due to mixing with intermolecular charge-transfer states.

There are two regimes in which singlet fission occurs in which simplified models have been used to understand the rate of singlet fission in terms of qualitative trends across materials. First, in the nonadiabatic regime ($\hat{V} \ll \Delta E$) the rate is proportional to $\left|\hat{V}\right|^2$. Singlet fission has been described in this regime with a Marcus model [21, 22] where the fission rate k_{sf} varies with the coupling and energetic drive as

$$k_{\mathrm{sf}} = \frac{2\pi}{\hbar} \left|\hat{V}\right|^2 \frac{1}{\sqrt{4\pi\lambda kT}} e^{-\frac{(\Delta E+\lambda)^2}{4\lambda kT.}}$$

where λ is the nuclear reorganization energy, k the Boltzmann constant, T the temperature, and $\hbar$ the reduced Planck constant.

When $\hat{V} \gtrsim \Delta E$ the reaction proceeds in the adiabatic regime in which the two manifolds mix as a function of nuclear coordinates. In this regime the fission rate is controlled by the speed of nuclear rearrangement [21]. In its most efficient form this process occurs via a *conical intersection* in which the two potential energy surfaces of the triplet pair and singlet exciton contact at a given nuclear coordinate and fission proceeds preserving vibrational coherence across the reaction [23].

In the above descriptions we have neglected the role of entropy in enhancing the rate of singlet fission. The role of entropy has been a topic of continued investigation, particularly with respect to endothermic singlet fission where entropic drive towards free triplet formation has been proposed as a key determinant of the efficiency. Entropy is proposed to play a role in both the initial formation of the triplet pair and the subsequent formation of free triplet excitons. In triplet-pair generation the effect of entropy depends on the delocalization of the initial S_1 state in a network of chromophores. In solid-state systems, for each delocalized singlet there are multiple localized dimer pairs on which ${}^1\mathrm{TT}$ may form. The entropic gain in this process is then

$$\Delta S = k \ \ln\left[\frac{\Omega\left({}^1\mathrm{TT}\right)}{\Omega\left(S_1\right)}\right]$$

with $\Omega\left({}^1\mathrm{TT}\right)$ and $\Omega(S_1)$ being the number of microstates associated with the triplet pair and singlet exciton, respectively [24]. This effect depends on the availability and interaction of multiple molecules and so depends sensitively on solid-state crystal structure (or oligomeric structure for covalently linked molecules). (Note that an analogous gain can be calculated for the process of free triplet formation.)

Whether adiabatic or nonadiabatic, driven by entropy or energetics, singlet fission has been observed on ultrafast timescales in various materials of interest (as fast as <100 fs, as discussed in the following sections). This process can therefore outcompete singlet exciton decay (> 10 ps) and proceed with ~100% efficiency, depending on available loss channels. We now describe the downstream processes after triplet-pair formation that determine the fate of the triplet pair as it dissociates, diffuses, and decays.

Triplet-Pair Dissociation and Recombination

The efficiency of singlet fission in applications depends not only on the rate of pair formation but also on the efficiency of pair dissociation to form free triplet excitons. Separation of the triplet pair can proceed via Dexter transfer to non-nearest neighbor molecules [25]. Theoretically the Dexter rate of triplet transfer decays exponentially with the energetic barrier and spatial separation of the molecules [26, 27]. The rate at which triplet pairs dissociate therefore depends on a variety of factors including the local crystal structure, the binding energy of the triplet-pair state, and the entropic drive towards free triplet excitons.

Temperature-dependent optical studies have played a key role in understanding the impact of energetics on triplet-pair dissociation [11, 28]. Representative exothermic systems (e.g., pentacene and hexacene as shown in Fig. 14.2) produce free triplet excitons at room temperature, as highlighted in the near-unity triplet formation in pentacene and applications in singlet fission-enhanced photovoltaics [9, 36–38]. In endothermic systems (e.g., tetracene and related derivatives) the triplet-pair formation is temperature independent, while free triplet formation is temperature activated [11, 15]. These results suggest that the triplet pair is roughly *isoenergetic* with the singlet exciton due to inter-triplet interactions while free triplets are higher energy states as expected theoretically from their monomeric energy-level diagrams.

Dissociation of the triplet pair competes with loss processes to determine the overall yield of free triplet excitons. A key loss process is the re-formation of a lone exciton from the fission-generated triplet pair, termed *triplet-triplet annihilation* (TTA) [39, 40]. As in the case of fission, TTA proceeds depending on the constraints of both spin and energetics. In general a pair of spin-1 triplets can form a pair with overall spin $S = 0$, 1, or 2. These states are termed the singlet, triplet, and quintet states of the triplet pair as described in the previous section. TTA proceeds depending on the spin of the triplet pair producing a singlet or triplet exciton with probability proportional to the singlet or triplet content of the triplet-pair spin wave function, respectively [41]. In most molecules the intramolecular quintet state is energetically inaccessible and so only the triplet and singlet decay pathways lead to annihilation [42]. We note that the annihilation of fission-generated geminate triplet pairs can be distinguished from non-geminate triplet-triplet annihilation by their magnetic field effects [5, 40].

Spin and energetics therefore play a key role in the recombination dynamics of triplet pairs. How the overall spin of the pair evolves through time determines the availability of annihilation pathways (through the singlet and triplet channels) and the binding energy of the triplet pair determines how quickly a pair dissociates in the solid state. The spin evolution of the pair is largely determined by the degree of inter-triplet coupling, which sets the strength of the spin-spin exchange interaction between triplets [43–47]. This exchange interaction determines the scale of the energetic separation of pure singlet, triplet, and quintet states and therefore the degree of spin mixing between them. This interaction and subsequent spin mixing process can therefore be tuned by the intermolecular geometry of the pair site where singlet fission and annihilation occur [43].

The lifetime of the triplet state determines the timescale over which dissociation and diffusion can occur (and therefore sets the timescale for harvesting these triplet excitons in devices). The triplet lifetime is determined by the spin-orbit-mediated intersystem crossing from the triplet excited state to the singlet ground state [20]. This process may be either non-radiative (resulting from phonon coupling between the triplet and the surrounding matrix) or radiative (resulting in phosphorescence) [20, 29]. Intersystem crossing of the triplet exciton to the ground state occurs on micro-millisecond timescales in most organic systems with diffusion lengths on the order of tens of nanometers [48–51]. The delicate interplay of fission, dissociation, diffusion, and losses described above determines device efficiencies and therefore motivates interdisciplinary work to define predictive structure-function relationships for the space of molecular materials, as we describe in the next section.

14.3 Materials for Singlet Fission

As described in the section above the singlet fission process is dependent on a number of factors, for example the energetic requirement that $E(S_0S_1) - E(TT) \gtrsim 0$. This requirement is often estimated from the energy of the individual singlet and triplet states so that the requirement is rewritten as $E_{S_1} \gtrsim 2E_{T_1}$. Also the molecular packing and coupling between the involved molecules play an important role in both the formation of the intermediate correlated triplet-pair state and the separation into free triplets. Another important consideration is the application of singlet fission into devices. A promising application is the enhancement of silicon-photovoltaics (Si-PV), which then requires singlet fission materials with energy levels suitable for electron or energy transfer to the Si-PV. Therefore the final device structure will also play a determining role in the choice of singlet fission chromophore.

How these requirements translate into molecular structure is a topic of extensive research. In the following section we will look at different families of chromophores that have displayed singlet fission and the design rules that can be learned from them.

Acenes

Acenes are a class of organic semiconductors consisting of linearly fused benzene rings; see Fig. 14.2. Due to a large-exchange energy the singlet-triplet energy difference tends to be quite large in the acenes (~1eV). As the conjugation is extended in larger acenes the S_1 energy decreases with about 0.6 eV from anthracene to tetracene, and 0.8 eV from tetracene to pentacene [29–34]. Similarly, the T_1 energy decreases by approximately 0.5 eV for the same series. As seen in Fig. 14.2 the energy difference $E_{S_1} - 2E_{T_1}$ changes from significantly *endothermic* (−0.5 eV) anthracene to closer to *isoenergetic* for tetracene (−0.1 to −0.2 eV) and *exothermic* (0.1 eV) for pentacene.

Even though singlet fission in anthracene is significantly endothermic, it was in anthracene that singlet fission was first observed in 1965 [8]. In that case anthracene was excited to a higher vibrational level allowing for the endothermic process to proceed. Other anthracene derivatives, such as 9,10-bis(phenylethynyl)anthracene (BPEA), have recently gained interest as bluer singlet fission materials [52–55]. A few years after anthracene, tetracene was also shown to undergo singlet fission [5–7, 56]. Tetracene is probably the most studied singlet fission material to date and has gained a lot of attention as it is a slightly exothermic singlet fission material (ΔE −0.1 to −0.2 eV [6, 57]) that still displays efficient singlet fission in crystalline films [58]. Entropic gain has been argued as one of the explanations as to why the endothermic process still proceeds efficiently [24]. Surprisingly, the endothermic singlet fission process in tetracene shows no temperature dependence of the triplet formation, arguing for a much smaller endothermicity in the crystalline films as discussed in the previous section [11]. However, the singlet fission dynamics, temperature dependence, and energetics are still a hot topic of debate. To complicate things further, temperature can have a significant effect on the morphology, crystal

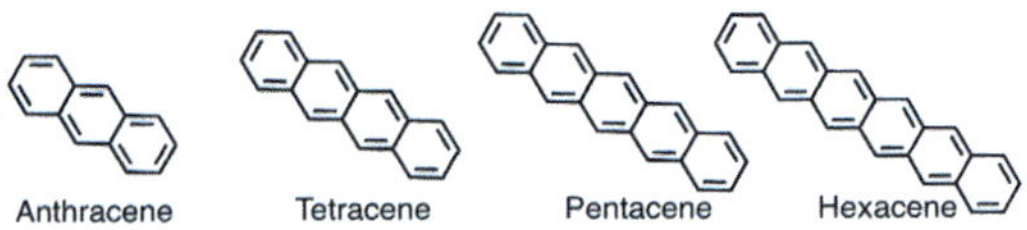

	Anthracene	Tetracene	Pentacene	Hexacene
S_1:	3.16 eV	~ 2.60 eV	1.83 eV	~ 1.5 eV
T_1:	1.83 eV	1.34 eV	0.86 eV	~ 0.5 eV
S_1-$2T_1$:	- 0.5 eV	~ - 0.1 eV	~ 0.1 eV	~ 0.5 eV
k_{SF}^{-1}:	>300 ps	10-100 ps	~100 fs	~ 500 fs

Fig. 14.2 Acene molecules and their singlet (S_1) and triplet (T_1) excited-state energies. Also shown is the singlet fission relevant energy difference $S_1 - 2T_1$, which changes from endothermic to exothermic in moving from anthracene and tetracene to pentacene and hexacene. Experimentally determined singlet fission rates (k_{SF}) are also shown. Note that the energies and rates are rough comparisons as they vary depending on sample morphology and in the case of anthracene also on excitation energy. Values from [3, 29–35]

packing, and vibrational coupling, which all affect the singlet fission dynamics [59–61].

In the series of acenes with increasing size, pentacene is the first acene with exothermic energy alignment and displays ultrafast singlet fission in less than 100 fs [9, 32, 35]. Singlet fission in the even more exothermic hexacene also occurs rapidly, albeit in some reports at a somewhat slower rate than in pentacene [35, 62].

To overcome a significant practical challenge of the unsubstituted acenes, their low solubility, solution-processable versions have been synthesized [63, 64]. These often contain solubilizing acetylene-TIPS groups. Due to the high concentration achievable with TIPS-tetracene and TIPS-pentacene singlet fission has also been observed in solution in these non-dimer chromophores [15, 16]. In fact, the singlet fission process slows down due to the diffusion limit in solution allowing for the direct observation of the intermediate ^{1}TT triplet-pair state [15]. However, it has later been questioned whether the observed intermediate is indeed an intermediate or an excimer trap state [65].

In 1979 Dexter proposed that singlet fission in tetracene could be used to improve the efficiency of silicon solar cells; [66] efforts towards this goal have been pursued [67, 68]. Other approaches include organic and QD solar cells with singlet fission active components and photon multiplication approaches [38, 69–72]. For direct triplet transfer to a PV device (or an emitter material in a photon multiplier device) the triplet level of the singlet fission material has to be higher than the active material bandgap to enable transfer. For a Si-PV with a bandgap of ~1.1 eV the choice of singlet fission material becomes limited to mainly tetracene derivatives as the pentacene triplet energy is too low (~0.9 eV), and higher energy acenes tend to be inefficient singlet fission materials due to the large endothermicity. The small number of practical materials has led to efforts finding efficient singlet fission candidates in other material families.

Biradicaloid Molecules

Radicals are molecules with a singly occupied molecular orbital and molecules with two singly occupied molecular orbitals are known as biradicals. With two unpaired electrons in degenerate orbitals biradicals can have triplet ground states; see Fig. 14.3a. By stabilizing the degenerate orbitals of a biradical, a biradicaloid molecule with a more closed-shell ground-state electronic structure is obtained. Hence biradicaloid molecules are molecules that have some biradical character and some closed-shell character; see Fig. 14.3a.

In 2006 Josef Michl and co-workers proposed that by altering the degree of biradical nature of a molecule, the triplet energy can be tuned, and suitable SF materials obtained [73], as shown schematically in Fig. 14.3a. Later in 2010 they proposed and showed that 1,3-diphenylisobenzofuran undergoes singlet fission with a 200% yield of triplets in solid state [74–76]. The parent biradical for 1,3-diphenylisobenzofuran is the *o*-xylylene biradical with the two unpaired electrons

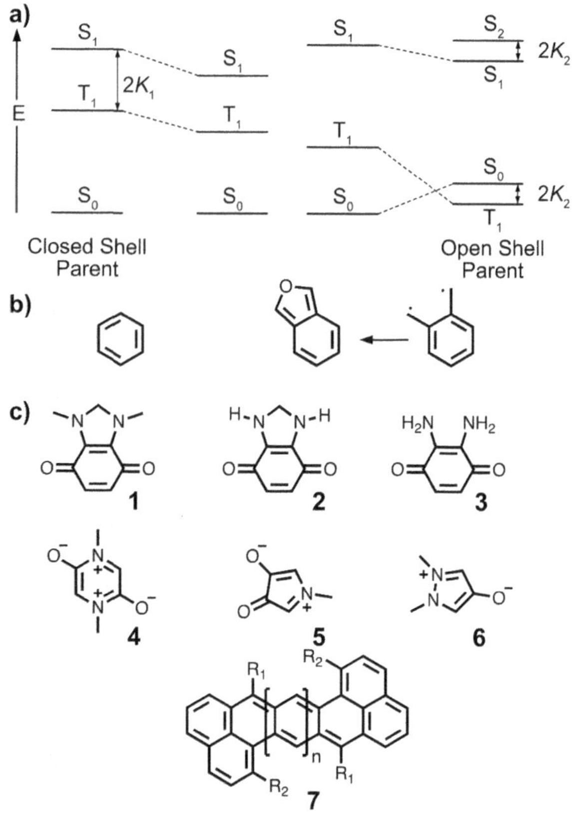

Fig. 14.3 Biradicaloid molecules. (**a**) Energy-level diagram for closed-shell and open-shell molecules illustrating the design principle of biradicaloid molecules as potential singlet fission (SF) material. Adapted from [73] Paci et al., J. Am. Chem. Soc., 2006, **128**, p. 16546. Copyright (2006) American Chemical Society. (**b**) Illustration of closed-shell parent benzene, biradical parent o-xylylene biradical, and biradicaloid molecule isobenzofuran. (**c**) Proposed and studied biradicaloid molecular structures for singlet fission

located in the methyl p-orbitals oriented in the aromatic plane. Envisioning the twisting of the p-orbitals making them parallel to each other and fusing them via an oxygen heteroatom, the biradicaloid isobenzofuran is obtained; see Fig. 14.3b. With a T_1 energy of 1.4 eV and S_1 energy of 2.73 eV 1,3-diphenylisobenzofuran has a similar endothermic energy alignment as tetracene [74]. Lukman et al. demonstrated that other diradicaloid singlet fission materials exist when observing singlet fission in zethrene biradicaloid molecules [77]. These molecules were shown to undergo singlet fission and interestingly also emission from the correlated triplet-pair state (^1TT) increased at lower temperature indicating an energetic barrier for free triplet formation.

So far few other biradicaloid molecules have been explicitly designed and shown to undergo singlet fission. Theoretically, however, Michl and others have proposed a number of small biradicaloid molecules with appropriate energy-level alignment for singlet fission; see Fig. 14.3c [78–82]. In further support of the biradicaloid design rule is the fact that pentacene and further extended acenes can also be considered to possess biradicaloid nature [81, 83].

With an unpaired electron, radicals are highly reactive species, and it is therefore not surprising that molecules with radical character also have a propensity to react easily, rendering them unstable. This is also the case for the biradicaloid molecules; for example both 1,3-diphenylisobenzofuran and, as described above, pentacene, are highly sensitive to oxygen in the presence of light. The limited stability will surely be an important hurdle to overcome for practical application and commercialization of these materials as singlet fission sensitizers for PVs. How this might be done will be discussed in the section **Next-Generation Singlet Fission Molecules**.

Even though stability issues still remain there have been examples of 1,3-diphenylisobenzofuran-based dye-sensitized solar cells (DSSC) [84, 85]. One significant challenge in using surface-anchored singlet fission chromophores in DSSC is the fact that electron injection from the singlet state can outcompete the singlet fission process if the rates are not carefully balanced [84, 85].

Dimers and Oligomers

As discussed in the previous section **Singlet Fission** the singlet fission rate and yield are governed not only by the energetic driving force, but also by the coupling between the involved chromophores. As a result the same material can display completely different singlet fission properties in different morphologies. As a means to overcome issues related to morphology and more directly study the effect of coupling between chromophores dimer systems were developed.

One of the first reports was on bis(tetracene) structures with varying linker structures; see Fig. 14.4a [14]. In these structures singlet fission was quite inefficient, ~3% for the para-phenylene-linked tetracenes and nonexistent in the meta-linked dimer. Given the efficiency of singlet fission in solid-state tetracene, even such low yields provide compelling direct evidence of the importance of molecular coupling between the chromophores involved in singlet fission. The fission yield in other tetracene dimers can be higher, in some cases approaching 200% [13, 86–88].

Beyond tetracene dimers there are many recent reports of pentacene, acene, and diphenylisobenzofuran dimers, displaying various degrees of singlet fission [12, 13, 47, 75, 86, 89–99].

In a recent review Hetzer et al. summarized many studies of different pentacene dimers. An interesting observation was that depending on the position of linkage the mechanism for singlet fission could be altered, from a direct mechanism in 2,2′-linked dimers to a solvent-dependent charge transfer-mediated mechanism in 6,6′-linked dimers; see Fig. 14.5b [101]. In fact, dimers have been crucial in gaining understanding of the two proposed singlet fission pathways; in 2016 Lukman et al. observed an intermediate charge transfer state in 6,6′-linked pentacene dimers, unambiguously answering the question whether a charge-transfer pathway exists [91].

Dimers and larger oligomers/polymers are also of particular interest for practical applications as ideally the singlet fission process is less sensitive to morphological

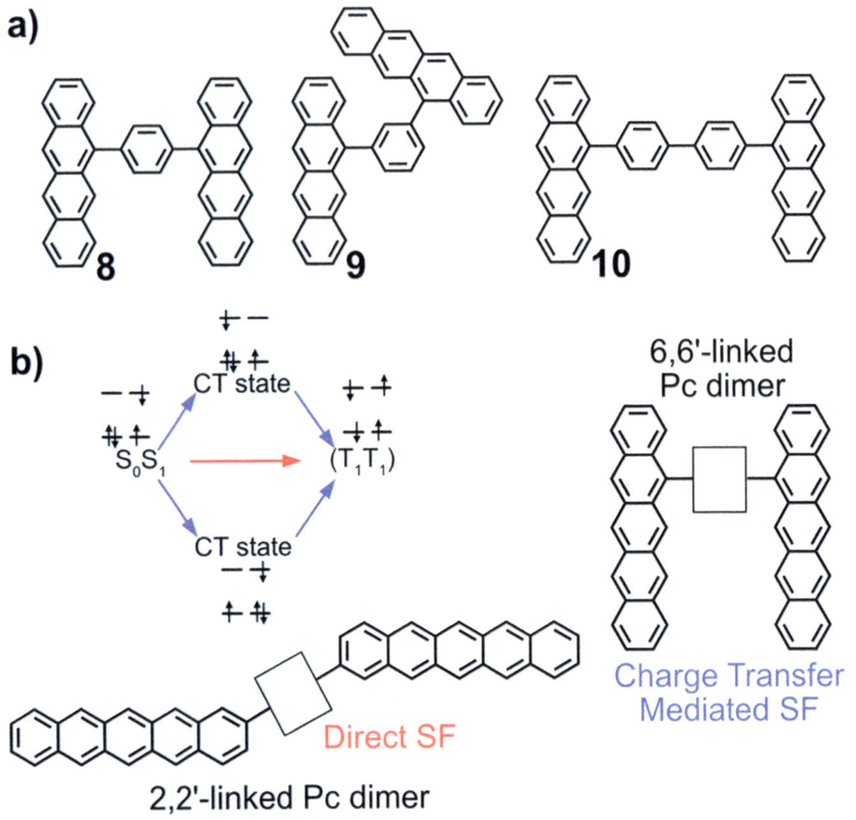

Fig. 14.4 (**a**) Molecular structures of the first singlet fission (SF) tetracene dimers. (**b**) Illustration of the two singlet fission pathways, direct and charge transfer (CT) mediated, and pentacene dimers displaying singlet fission through the two different pathways; squares in the molecular structures indicate that various linker groups have been studied

changes in the solid state, compared to the monomeric counterparts. However, one key issue with dimers for practical applications is the reduced free triplet formation and shorter triplet lifetime, compared to monomer films. Often singlet fission in dimers only leads to the correlated pair state (^{1}TT and subsequent spin-relaxed populations of ^{3,5}TT pairs) in solution as there is nowhere for the triplet to break up and move to; see Fig. 14.5. From the TT state the dimer can return to the singlet state or decay to the ground state directly. Even in larger oligomer and polymer systems free triplets are rarely formed [98, 102–105]. By decoupling the chromophores it is possible to extend the triplet-pair lifetime to 100 s of ns and in some cases also form free triplets; see Fig. 14.5c [47, 90, 100]. The trade-off for extending the triplet-pair lifetime is often a slower rate of singlet fission, due to the weaker coupling.

To address the issue of short-lived triplet-pair states in dimers and oligomers Pun, Sfeir, Campos et al. developed a design principle to achieve ultrafast intramolecular

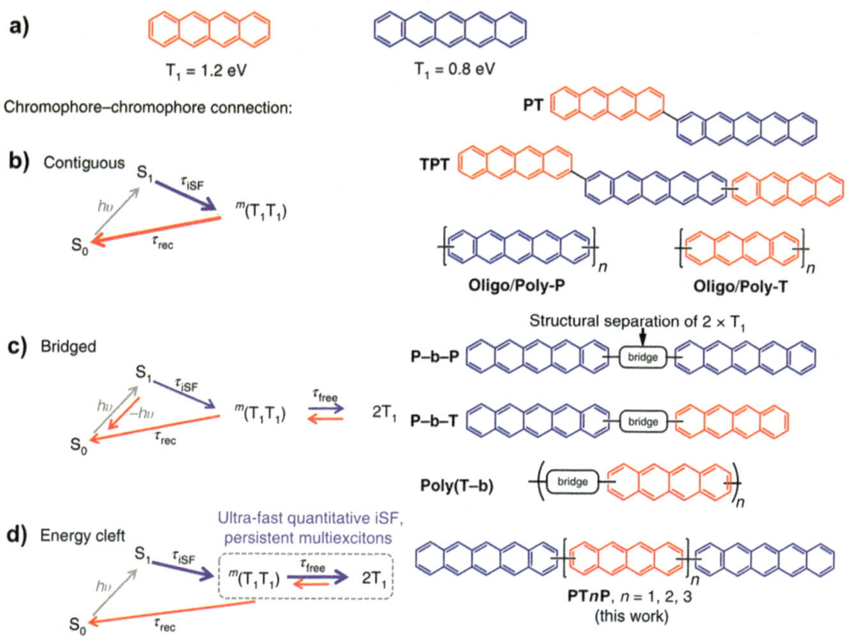

Fig. 14.5 Design strategy for long-lived free triplet generation in singlet fission (SF) dimers and oligomers. Adapted from [100] with permission from Springer Nature

singlet fission while also forming long-lived free triplets; see Fig. 14.5d. The design principle employs a high-energy singlet fission chromophore flanked by lower energy singlet fission chromophores or acceptors, as in the example pentacene-tetracene-pentacene trimer [100]. The energy alignment in these trimers and oligomers drives the separation of the triplets to the two low-energy sites, efficiently separating the triplets spatially; see Fig. 14.5d [100]. The trade-off is the energy loss associated with the triplets moving to the separate low-energy acceptors.

Next-Generation Singlet Fission Molecules

Significant efforts are being put into finding new classes of molecules that can undergo singlet fission. In the quest for new singlet fission materials three key challenges must be met: (1) fulfilling the energy requirement $E_{S_1} \gtrsim 2E_{T_1}$, (2) optimal packing in the solid state, and (3) long-term chemical stability. The first requirement is crucial for achieving efficient and fast singlet fission; however as discussed above for tetracene, some endothermicity can be allowed. Packing of the chromophores can play a crucial role in turning the singlet fission yields and rates, and efforts led by Michl and co-workers towards understanding and tuning

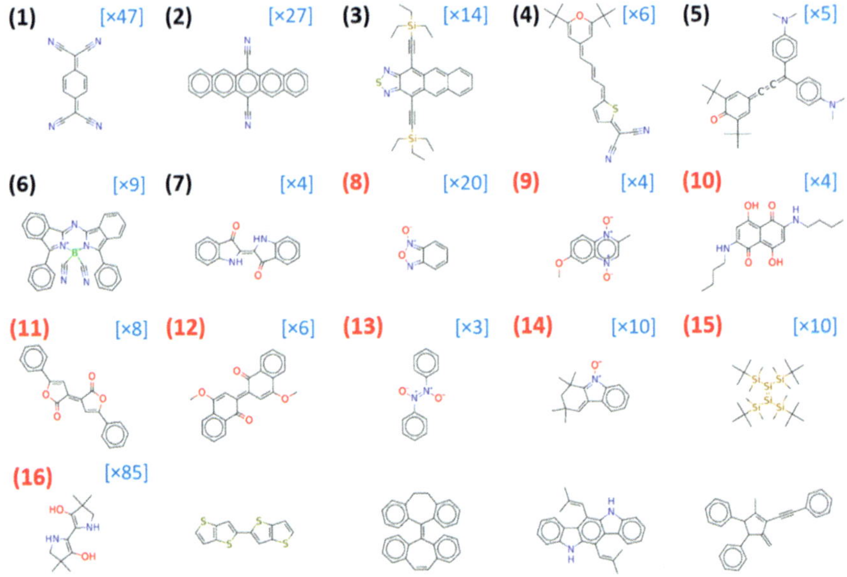

Fig. 14.6 Suggested molecular families of interest as singlet fission (SF) candidates as suggested by Padula et al. Reproduced from [108] with permission of The Royal Society of Chemistry, from Singlet fission molecules among known compounds: finding a few needles in a haystack., Padula, D., Omar, Ö. H., Nematiaram, T. & Troisi, A, Energy Environ. Sci. **12**, 2412–2416 (2019); permission conveyed through Copyright Clearance Center, Inc.

the morphology are aiding the search for singlet fission chromophores [106, 107]. However, if molecular dimers and/or polymers can be developed with high singlet fission yields the packing criteria might be loosened.

Besides the biradicaloid approach discussed above there have been some recent approaches towards finding or designing suitable candidates. Padula et al. screened through 40,000 molecular structures already known in the Cambridge Structural Database. Using the energetic requirements for singlet fission as a criterion calculated energies for the molecular structures were screened and it returned 16 different classes of possible singlet fission molecules [108]. Supporting their method is the fact that seven of the suggested classes have previously been suggested or even observed to undergo singlet fission. Importantly that means that there are nine families of molecules previously not considered for the investigation of singlet fission; see Fig. 14.6.

A second recent approach uses excited-state aromaticity and anti-aromaticity to tune the S_1 and T_1 energies. In a similar diagram to that shown for biradical molecules in Fig. 14.3, the energetic tuning of T_1 can be achieved by tuning the aromatic character of the ground and excited states; see Fig. 14.7. The design strategy is based on that molecules with an aromatic ground state are antiaromatic in the excited state, leading to higher energy T_1 and S_1 states and smaller S_1/T_1 splitting, whereas anti- or nonaromatic ground-state molecules have lower lying T_1

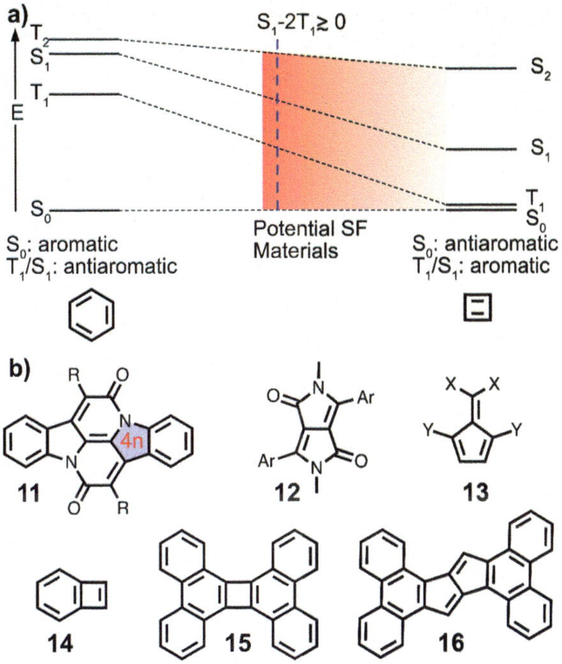

Fig. 14.7 (**a**) Energy-level alignment for aromatic and antiaromatic ground-state molecules indicating a region where possible singlet fission (SF) chromophores can be found. As example of an aromatic ground-state molecule benzene is shown and for an antiaromatic ground state cyclobutadiene is shown. Adapted from [109], El Bakouri et al., J. Am. Chem. Soc., 2020, **142**, p. 5602. (**b**) Molecular structures studied and suggested by Fallon et al. and El Bakouri et al. as possible singlet fission materials with triplet energies stabilized by excited-state aromaticity. Highlighted in molecule 11 is the five-membered 4n electron heterocycle suggested by the authors to increase the excited-state aromaticity [110]

states and a larger S_1/T_1 splitting [111–113]. By tuning the aromatic character of the chromophore an ideal S_1/T_1 splitting can be achieved [109, 110].

Fallon, Budden et al. investigated a family of substituted Cibalackrot derivatives (structure 11 Fig. 14.7) experimentally and expanded the conclusions theoretically by comparing the calculated singlet and triplet energies to the excited-state aromaticity of almost 10,000 candidates. Even though the singlet fission yield was below 100%, electron spin resonance (ESR) was used to distinguish the triplet formation mechanism from normal ISC [110]. Ryerson, Michl, and co-workers studied the parent Cibalackrot molecule, and concluded that triplet formation was high, albeit less than 100%. (We note that without complementary ESR measurements singlet fission could not univocally be designated the triplet formation mechanism.) Fallon et al. concluded, "*In practical terms, the requirement of a [4n] cyclic contribution to a molecule's resonance structures translates into a simple design rule for targeting singlet fission systems*" [110].

Recent work by El Bakouri, Smith, and Ottosson has further investigated this design principle based on Baird's rule on excited-state aromaticity, and proposed many new possible singlet fission chromophores (structures 13–16 in Fig. 14.7), including cyclobutadienes, benzannelated pentalenes, and pentafulvenes [109]. Supporting the conclusion that excited-state (anti)aromaticity is a useful design principle for singlet fission candidates, they point out that the most relevant singlet fission molecules might not necessarily be the ones with most excited-state aromaticity. As can be understood from Fig. 14.7 highly aromatic excited states can result in T_1 energies too low for practical applications [109].

14.4 Outlook

Since the first observation of singlet fission over 50 years ago the understanding has greatly improved, especially thanks to ultrafast spectroscopic techniques that have been able to follow the underlying steps in detail, and time-resolved magnetic field-dependent experiments have shed more light on the role of spin in the singlet fission process. However, there are still questions to be answered and an ongoing debate on certain details relating to the role of entropy and morphology. To gain further insight into the singlet fission process, expanding the current scope to a larger set of molecules will be crucial.

The relationship between molecular structure and singlet fission efficiency is just starting to be explored in more detail and a few recent studies of suggested molecular classes have been proposed. As these classes are scrutinized in more detail the design of a suitable candidate for implementation in solar energy harvesting will hopefully emerge. As will be highlighted in the coming chapter, the potential for singlet fission materials to improve the solar cell efficiencies definitely motivates further studies of the singlet fission process and the underlying material requirements.

References

1. A. Köhler, H. Bässler, *Electronic Processes in Organic Semiconductors* (2015), pp. 389–397. https://doi.org/10.1002/9783527685172
2. N.J. Turro, V. Ramamurthy, J.C. Scaiano, *Modern Molecular Photochemistry of Organic Molecules* (University Science Books, Sausalito, 2010)
3. M.B. Smith, J. Michl, Singlet fission. Chem. Rev. **110**, 6891–6936 (2010)
4. M.B. Smith, J. Michl, Recent advances in singlet fission. Annu. Rev. Phys. Chem. **64**, 361–386 (2013). https://doi.org/10.1146/annurev-physchem-040412-110130.ssion
5. R.E. Merrifield, P. Avakian, R.P. Groff, Fission of singlet excitons into pairs of triplet excitons in tetracene crystals. Chem. Phys. Lett. **3**, 386–388 (1969)
6. N. Geacintov, M. Pope, F. Vogel, Effect of magnetic field on the fluorescence of tetracene crystals: exciton fission. Phys. Rev. Lett. **22**, 593–596 (1969)

7. C.E. Swenberg, W.T. Stacy, Bimolecular radiationless transitions in crystalline tetracene. Chem. Phys. Lett. **2**, 327–328 (1968)
8. S. Singh, W.J. Jones, W. Siebrand, B.P. Stoicheff, W.G. Schneider, Laser generation of excitons and fluorescence in anthracene crystals. J. Chem. Phys. **42**, 330–342 (1965)
9. M.W.B. Wilson et al., Ultrafast dynamics of exciton fission in polycrystalline pentacene. J. Am. Chem. Soc. **133**, 11830–11833 (2011)
10. J.J. Burdett, C.J. Bardeen, The dynamics of singlet fission in crystalline tetracene and covalent analogs. Acc. Chem. Res. **46**, 1312–1320 (2013)
11. M. W. B. Wilson et al., *Temperature-Independent Singlet Exciton Fission in Tetracene* (2013). https://doi.org/10.1021/ja408854u
12. S. Lukman et al., Tuneable singlet exciton fission and triplet-triplet annihilation in an orthogonal pentacene dimer. Adv. Funct. Mater. **25**, 5452–5461 (2015)
13. N.V. Korovina et al., Singlet fission in a covalently linked cofacial alkynyltetracene dimer. J. Am. Chem. Soc. **138**, 617–627 (2016)
14. A.M. Müller, Y.S. Avlasevich, W.W. Schoeller, K. Müllen, C.J. Bardeen, Exciton fission and fusion in bis(tetracene) molecules with different covalent linker structures. J. Am. Chem. Soc. **129**, 14240–14250 (2007)
15. H.L. Stern et al., Identification of a triplet pair intermediate in singlet exciton fission in solution. Proc. Natl. Acad. Sci. **112**, 7656–7661 (2015)
16. B.J. Walker, A.J. Musser, D. Beljonne, R.H. Friend, Singlet exciton fission in solution. Nat. Chem. **5**, 1019–1024 (2013)
17. R.C. Johnson, R.E. Merrifield, Effects of magnetic fields on the mutual annihilation of triplet excitons in anthracene crystals. Phys. Rev. B **1**, 896 (1970). https://doi.org/10.1103/PhysRevB.1.896
18. J.J. Burdett, C.J. Bardeen, Quantum beats in crystalline tetracene delayed fluorescence due to triplet pair coherences produced by direct singlet fission. J. Am. Chem. Soc. **134**, 8597–8607 (2012)
19. R.P. Groff, P. Avakian, R.E. Merrifield, Coexistence of exciton fission and fusion in tetracene crystals. Phys. Rev. B **1**, 815 (1970). https://doi.org/10.1103/PhysRevB.1.815
20. A. Köhler, H. Bässler, Triplet states in organic semiconductors. Mater. Sci. Eng. R: Rep. **66**, 71–109 (2009). https://doi.org/10.1016/j.mser.2009.09.001
21. S.R. Yost et al., A transferable model for singlet-fission kinetics. Nat. Chem. **6**, 492–497 (2014)
22. T.W. Schmidt, A Marcus-Hush perspective on adiabatic singlet fission. J. Chem. Phys. **151**, 054305 (2019)
23. A.J. Musser et al., Evidence for conical intersection dynamics mediating ultrafast singlet exciton fission. Nat. Phys. **11**, 352–357 (2015)
24. W.L. Chan, M. Ligges, X.Y. Zhu, The energy barrier in singlet fission can be overcome through coherent coupling and entropic gain. Nat. Chem. **4**, 840–845 (2012)
25. D.L. Dexter, A theory of sensitized luminescence in solids. J. Chem. Phys. (1953). https://doi.org/10.1063/1.1699044
26. A. Köhler, H. Bässler, What controls triplet exciton transfer in organic semiconductors? J. Mater. Chem. **21**, 4003–4011 (2011)
27. L. Sudha Devi et al., Triplet energy transfer in conjugated polymers. I. Experimental investigation of a weakly disordered compound. Phys. Rev. B Condens. Matter Mater. Phys. **78**, 045210 (2008)
28. H.L. Stern et al., Vibronically coherent ultrafast triplet-pair formation and subsequent thermally activated dissociation control efficient endothermic singlet fission. Nat. Chem. **9**(12), 1205–1212 (2017). https://doi.org/10.1038/nchem.2856
29. S. Reineke, M.A. Baldo, Room temperature triplet state spectroscopy of organic semiconductors. Sci. Rep. **4**, 1–8 (2014)
30. M.R. Padhye, S.P. McGlynn, M. Kasha, Lowest triplet state of anthracene. J. Chem. Phys. **24**, 588–594 (1956)

31. G.N. Lewis, M. Kasha, Phosphorescence and the triplet state. J. Am. Chem. Soc. **66**, 2100–2116 (1944)
32. C. Jundt et al., Exciton dynamics in pentacene thin films studied by pump-probe spectroscopy. Chem. Phys. Lett. **241**, 84–88 (1995)
33. R.H. Friend et al., Excitons and charges at organic semiconductor heterojunctions. Faraday Discuss. **155**, 339–348 (2012)
34. V.K. Thorsmølle et al., Morphology effectively controls singlet-triplet exciton relaxation and charge transport in organic semiconductors. Phys. Rev. Lett. **102**, 3–6 (2009)
35. E. Busby et al., Multiphonon relaxation slows singlet fission in crystalline hexacene. J. Am. Chem. Soc. **136**, 10654–10660 (2014)
36. B. Ehrler, M.W.B. Wilson, A. Rao, R.H. Friend, N.C. Greenham, Singlet exciton fission-sensitized infrared quantum dot solar cells. Nano Lett. **12**, 1053–1057 (2012)
37. J. Lee et al., Singlet exciton fission photovoltaics. Acc. Chem. Res. **46**, 1300–1311 (2013)
38. D.N. Congreve et al., External quantum efficiency above 100% in a singlet-exciton-fission–based organic photovoltaic cell. Science **340**, 334–338 (2013)
39. R.P. Groff, P. Avakian, R.E. Merrifield, Magnetic field dependence of delayed fluorescence from tetracene crystals. J. Lumin. **1–2**, 218–223 (1970). https://doi.org/10.1016/0022-2313(70)90036-0
40. R.C. Johnson, R.E. Merrifield, P. Avakian, R.B. Flippen, Effects of magnetic fields on the mutual annihilation of triplet excitons in molecular crystals. Phys. Rev. Lett. **19**, 285 (1967). https://doi.org/10.1103/PhysRevLett.19.285
41. Y.Y. Cheng et al., On the efficiency limit of triplet–triplet annihilation for photochemical upconversion. Phys. Chem. Chem. Phys. **12**, 66–71 (2010)
42. B. Dick, B. Nickel, Accessibility of the lowest quintet state of organic molecules through triplet-triplet annihilation; an indo ci study. Chem. Phys. **78**, 1–16 (1983)
43. S.L. Bayliss et al., Site-selective measurement of coupled spin pairs in an organic semiconductor. Proc. Natl. Acad. Sci. **115**, 5077–5082 (2018)
44. M. Chen et al., Quintet-triplet mixing determines the fate of the multiexciton state produced by singlet fission in a terrylenediimide dimer at room temperature. Proc. Natl. Acad. Sci. U. S. A. **116**, 8178–8183 (2019)
45. H. Benk, H. Sixl, Theory of two coupled triplet states: application to bicarbene structures. Mol. Phys. **42**, 779–801 (1981)
46. L.R. Weiss et al., Strongly exchange-coupled triplet pairs in an organic semiconductor. Nat. Phys. **13**, 176–181 (2017)
47. M.J.Y. Tayebjee et al., Quintet multiexciton dynamics in singlet fission. Nat. Phys. **13**, 182–188 (2017)
48. R.E. Merrifield, Diffusion and mutual annihilation of triplet excitons in organic crystals. Acc. Chem. Res. **1**, 129–135 (1968). https://doi.org/10.1021/ar50005a001
49. G.M. Akselrod et al., Visualization of exciton transport in ordered and disordered molecular solids. Nat. Commun. **5**, 3646 (2014). https://doi.org/10.1038/ncomms4646
50. M. Born et al., Slow hopping and spin dephasing of Coulombically bound polaron pairs in an organic semiconductor at room temperature. J. Am. Chem. Soc. **1**, 489–564 (2013)
51. Y. Wan et al., Cooperative singlet and triplet exciton transport in tetracene crystals visualized by ultrafast microscopy. Nat. Chem. **7**, 785–792 (2015). https://doi.org/10.1038/nchem.2348
52. Y.J. Bae et al., Singlet fission in 9,10-bis(phenylethynyl)anthracene thin films. J. Am. Chem. Soc. **140**, 15140–15144 (2018)
53. B. Manna, A. Nandi, R. Ghosh, Ultrafast singlet exciton fission dynamics in 9,10-bis(phenylethynyl)anthracene nanoaggregates and thin films. J. Phys. Chem. C **122**, 21047–21055 (2018)
54. C.F. Perkinson et al., Discovery of blue singlet exciton fission molecules via a high-throughput virtual screening and experimental approach. J. Chem. Phys. **151** (2019)
55. S. Jana, A.L. Yapamanu, S. Umapathy, Unraveling structural dynamics in isoenergetic excited S1 and multi-excitonic 1(TT) states of 9,10-bis(phenylethynyl)anthracene (BPEA) in solution: via ultrafast Raman loss spectroscopy. Phys. Chem. Chem. Phys. **21**, 14341–14349

(2019)
56. V. Kocevski, O. Eriksson, C. Gerard, D.D. Sarma, J. Rusz, Influence of dimensionality and interface type on optical and electronic properties of CdS/ZnS core-shell nanocrystals—a first-principles study. J. Chem. Phys. **143** (2015)
57. Y. Tomkiewicz, R.P. Groff, P. Avakian, Spectroscopic approach to energetics of exciton fission and fusion in tetracene crystals. J. Chem. Phys. **54**, 4504–4507 (1971)
58. J.J. Burdett, A.M. Müller, D. Gosztola, C.J. Bardeen, Excited state dynamics in solid and monomeric tetracene: the roles of superradiance and exciton fission. J. Chem. Phys. **133** (2010)
59. G.B. Piland, C.J. Bardeen, How morphology affects singlet fission in crystalline tetracene. J. Phys. Chem. Lett. **6**, 1841–1846 (2015)
60. A.F. Morrison, J.M. Herbert, Evidence for singlet fission driven by vibronic coherence in crystalline tetracene. J. Phys. Chem. Lett. **8**, 1442–1448 (2017)
61. S.L. Bayliss et al., Tuning spin dynamics in crystalline tetracene. J. Phys. Chem. Lett. **10**, 1908–1913 (2019)
62. N.R. Monahan et al., Dynamics of the triplet-pair state reveals the likely coexistence of coherent and incoherent singlet fission in crystalline hexacene. Nat. Chem. **9**, 341–346 (2017)
63. S.A. Odom, S.R. Parkin, J.E. Anthony, Tetracene derivatives as potential red emitters for organic LEDs. Org. Lett. **5**, 4245–4248 (2003)
64. Y.-F. Lim, Y. Shu, S.R. Parkin, J.E. Anthony, G.G. Malliaras, Soluble n-type pentacene derivatives as novel acceptors for organic solar cells. J. Mater. Chem. **19**, 3049 (2009)
65. C.B. Dover et al., Endothermic singlet fission is hindered by excimer formation. Nat. Chem. **10**, 305–310 (2018)
66. D.L. Dexter, Two ideas on energy transfer phenomena: ion-pair effects involving the OH stretching mode, and sensitization of photovoltaic cells. J. Lumin. **18–19**, 779–784 (1979)
67. R. MacQueen et al., Crystalline silicon solar cells with tetracene interlayers: the path to silicon-singlet fission heterojunction devices. Mater. Horiz. **5**, 1065–1075 (2018)
68. M. Einzinger et al., Sensitization of silicon by singlet exciton fission in tetracene. Nature **571**, 90–94 (2019)
69. P.D. Reusswig, D.N. Congreve, N.J. Thompson, M.A. Baldo, Enhanced external quantum efficiency in an organic photovoltaic cell via singlet fission exciton sensitizer. Appl. Phys. Lett. **101**, 113304 (2012)
70. T.C. Wu et al., Singlet fission efficiency in tetracene-based organic solar cells. Appl. Phys. Lett. **104**, 193901 (2014)
71. N.J.L.K. Davis et al., Singlet fission and triplet transfer to PbS quantum dots in TIPS-tetracene carboxylic acid ligands. J. Phys. Chem. Lett. **9**, 1454–1460 (2018)
72. J.R. Allardice et al., Engineering molecular ligand shells on quantum dots for quantitative harvesting of triplet excitons generated by singlet fission. J. Am. Chem. Soc. **141**, 12907–12915 (2019)
73. I. Paci et al., Singlet fission for dye-sensitized solar cells: can a suitable sensitizer be found? J. Am. Chem. Soc. **128**, 16546–16553 (2006)
74. J.C. Johnson, A.J. Nozik, J. Michl, High triplet yield from singlet fission in a thin film of 1,3-diphenylisobenzofuran. J. Am. Chem. Soc. **132**, 16302–16303 (2010)
75. J.C. Johnson et al., Toward designed singlet fission: solution photophysics of two indirectly coupled covalent dimers of 1,3-diphenylisobenzofuran. J. Phys. Chem. B **117**, 4680–4695 (2013)
76. J.N. Schrauben, J.L. Ryerson, J. Michl, J.C. Johnson, Mechanism of singlet fission in thin films of 1,3-diphenylisobenzofuran. J. Am. Chem. Soc. **136**, 7363–7373 (2014)
77. S. Lukman et al., Efficient singlet fission and triplet-pair emission in a family of zethrene diradicaloids. J. Am. Chem. Soc. **139**, 18376–18385 (2017)
78. A. Akdag, Z.K. Havlas, J. Michl, Search for a small chromophore with efficient singlet fission: biradicaloid heterocycles. J. Am. Chem. Soc. **134**, 14624–14631 (2012)
79. J. Wen, Z. Havlas, J. Michl, Captodatively stabilized biradicaloids as chromophores for singlet fission. J. Am. Chem. Soc. **137**, 165–172 (2015)

80. D. López-Carballeira, D. Casanova, F. Ruipérez, Theoretical design of conjugated diradicaloids as singlet fission sensitizers: quinones and methylene derivatives. Phys. Chem. Chem. Phys. **19**, 30227–30238 (2017)
81. T. Minami, M. Nakano, Diradical character view of singlet fission. J. Phys. Chem. Lett. **3**, 145–150 (2012)
82. S. Ito, T. Minami, M. Nakano, Diradical character based design for singlet fission of condensed-ring systems with 4 n πelectrons. J. Phys. Chem. C **116**, 19729–19736 (2012)
83. H. Yang, M. Chen, X. Song, Y. Bu, Structural fluctuation governed dynamic diradical character in pentacene. Phys. Chem. Chem. Phys. **17**, 13904–13914 (2015)
84. J.N. Schrauben et al., Photocurrent enhanced by singlet fission in a dye-sensitized solar cell. ACS Appl. Mater. Interfaces **7**, 2286–2293 (2015)
85. T. Banerjee et al., Diphenylisobenzofuran bound to nanocrystalline metal oxides: excimer formation, singlet fission, electron injection, and low energy sensitization. J. Phys. Chem. C **122**, 28478–28490 (2018)
86. H. Liu et al., A covalently linked tetracene trimer: synthesis and singlet exciton fission property. Org. Lett. **19**, 580–583 (2017)
87. S. Nakamura et al., Quantitative sequential photoenergy conversion process from singlet fission to intermolecular two-electron transfers utilizing tetracene dimer. ACS Energy Lett. **4**, 26–31 (2018). https://doi.org/10.1021/acsenergylett.8b01964
88. Y. Matsui et al., Exergonic intramolecular singlet fission of an adamantane-linked tetracene dyad via twin quintet multiexcitons. J. Phys. Chem. C **123**, 18813–18823 (2019)
89. J. Zirzlmeier et al., Singlet fission in pentacene dimers. Proc. Natl. Acad. Sci. **112**, 5325–5330 (2015)
90. S.N. Sanders et al., Quantitative intramolecular singlet fission in bipentacenes. J. Am. Chem. Soc. **137**, 8965–8972 (2015)
91. S. Lukman et al., Tuning the role of charge-transfer states in intramolecular singlet exciton fission through side-group engineering. Nat. Commun. **7**, 13622 (2016)
92. S.N. Sanders et al., Exciton correlations in intramolecular singlet fission. J. Am. Chem. Soc. **138**, 7289–7297 (2016)
93. S.N. Sanders et al., Intramolecular singlet fission in oligoacene heterodimers. Angew. Chem. Int. Ed. Engl. **55**, 3434–3438 (2016). https://doi.org/10.1002/ange.201510632
94. E.G. Fuemmeler et al., A direct mechanism of ultrafast intramolecular singlet fission in pentacene dimers. ACS Cent. Sci. **2**, 316–324 (2016)
95. E. Kumarasamy et al., Tuning singlet fission in π-bridge-π chromophores. J. Am. Chem. Soc. **139**, 12488–12494 (2017)
96. C.K. Yong et al., The entangled triplet pair state in acene and heteroacene materials. Nat. Commun. **8**, 15953 (2017)
97. J.C. Dean et al., Photophysical characterization and time-resolved spectroscopy of a anthradithiophene dimer: exploring the role of conformation in singlet fission. Phys. Chem. Chem. Phys. **19**, 23162–23175 (2017)
98. A.B. Pun et al., Triplet harvesting from intramolecular singlet fission in polytetracene. Adv. Mater. **29**, 1701416 (2017)
99. R. Montero et al., Singlet fission mediated photophysics of BODIPY dimers. J. Phys. Chem. Lett. **9**, 641–646 (2018)
100. A.B. Pun et al., Ultra-fast intramolecular singlet fission to persistent multiexcitons by molecular design. Nat. Chem. **11**, 821–828 (2019)
101. C. Hetzer, D.M. Guldi, R.R. Tykwinski, Pentacene dimers as a critical tool for the investigation of intramolecular singlet fission. Chem. A Eur. J. **24**, 8245–8257 (2018)
102. E. Busby et al., A design strategy for intramolecular singlet fission mediated by charge-transfer states in donor-acceptor organic materials. Nat. Mater. **14**, 426–433 (2015)
103. S.N. Sanders et al., Singlet fission in polypentacene. Chem **1** (2016)
104. J. Hu et al., New insights into the design of conjugated polymers for intramolecular singlet fission. Nat. Commun. **9** (2018)

105. L.M. Yablon et al., Persistent multiexcitons from polymers with pendent pentacenes. J. Am. Chem. Soc. **141**, 9564–9569 (2019)
106. E.A. Buchanan, J. Michl, Packing guidelines for optimizing singlet fission matrix elements in noncovalent dimers. J. Am. Chem. Soc. **139**, 15572–15575 (2017)
107. A. Zaykov et al., Singlet fission rate: optimized packing of a molecular pair. Ethylene as a model. J. Am. Chem. Soc. **141**, 17729–17743 (2019)
108. D. Padula, Ö.H. Omar, T. Nematiaram, A. Troisi, Singlet fission molecules among known compounds: finding a few needles in a haystack. Energy Environ. Sci. **12**, 2412–2416 (2019)
109. O. El Bakouri, J.R. Smith, H. Ottosson, Strategies for design of potential singlet fission chromophores utilizing a combination of ground state and excited state aromaticity rules. J. Am. Chem. Soc. **142**, 5602–5617 (2020)
110. K.J. Fallon et al., Exploiting excited-state aromaticity to design highly stable singlet fission materials. J. Am. Chem. Soc. **141**, 13867–13876 (2019)
111. M. Rosenberg, C. Dahlstrand, K. Kilså, H. Ottosson, Excited state aromaticity and antiaromaticity: opportunities for photophysical and photochemical rationalizations. Chem. Rev. **114**, 5379–5425 (2014)
112. Y.M. Sung et al., Reversal of Hückel (anti)aromaticity in the lowest triplet states of hexaphyrins and spectroscopic evidence for Baird's rule. Nat. Chem. **7**, 418–422 (2015)
113. J. Oh et al., Unraveling excited-singlet-state aromaticity via vibrational analysis. Chem **3**, 870–880 (2017)

Chapter 15
Singlet Fission Solar Cells

Bruno Ehrler

Abstract The high-energy (blue) part of the solar spectrum is inefficiently converted in conventional solar cells, mainly because the high-energy excitations thermalize to the bandgap before they are extracted. Several strategies have been devised to tackle these thermalization losses, most prominently tandem solar cells. However, these tandem cells require an intricate device design and current matching in case of a series connection. Downconversion via singlet fission and quantum cutting promise a better use of the high-energy photons, avoiding a large fraction of the thermalization losses, but without the intricate fabrication and design constraints of tandem cells. In this chapter we review the progress made towards efficient singlet fission and quantum cutting downconversion. We start with the potential for solar cell integration, reviewing the different integration schemes and their efficiency potential. In the second part we review the progress towards solar cells that utilize singlet fission and quantum cutting from all-organic devices to hybrid two-bandgap devices and fully optical integration. Finally, we lay out the challenges for using these downconversion schemes in commercial solar cells.

Keywords Solar cells · Singlet fission · Exciton conversion · Organic semiconductors

15.1 Introduction

Photovoltaic solar cells convert sunlight into electrical power. These cells are the leading technology for clean renewable energy generation. In 2018, 40% of all newly installed electrical power capacity was solar power [1] and an annual growth rate above 10% for decades has led to a massive reduction in the levelized cost of electricity from photovoltaics [2]. In many locations solar electricity is already the

B. Ehrler (✉)
Center for Nanophotonics, AMOLF, Amsterdam, the Netherlands
e-mail: b.ehrler@amolf.nl

J. S. Lissau, M. Madsen (eds.), *Emerging Strategies to Reduce Transmission and Thermalization Losses in Solar Cells*, https://doi.org/10.1007/978-3-030-70358-5_15

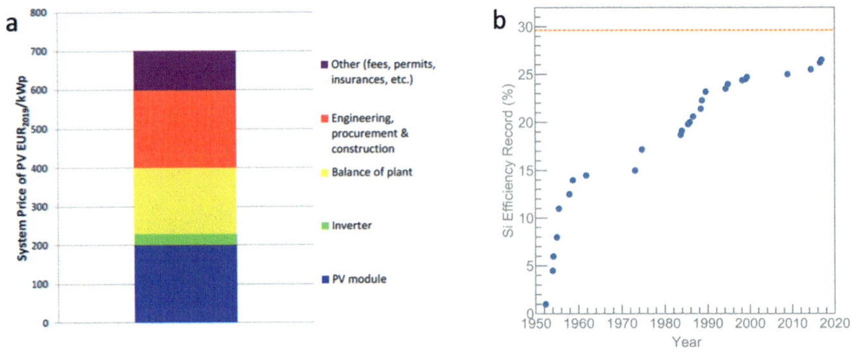

Fig. 15.1 (**a**) Cost breakdown of a utility-scale PV system in a typical competitive market in 2019. Figure reproduced under CC BY 4.0 license from [2]. (**b**) Lab-records for the efficiency of silicon solar cells. The orange line depicts the thermodynamic limit for single-junction silicon solar cells when Auger recombination is considered

cheapest form of electricity generation. However, with an increase in the installed capacity of solar power, the incremental value of the electricity produced decreases. That is to say that if solar energy forms a significant fraction of the power supply, the prices for electricity plummet on a sunny summer day, because of the current grid infrastructure and the lack of storage. In addition, seasonal variation in the available solar irradiation and daily fluctuations mean that we will probably need to install an overcapacity of solar electricity in the future mix of power sources. All these mean that the cost of solar electricity needs to drop further [3].

For state-of-the-art solar installations less than 1/3 of the cost is in the solar module and only about half of that cost is in the solar cell (Fig. 15.1a). The majority of the cost is now in the so-called balance-of-system costs (BOS) [1]. These costs include the land and labor required to install the panels, maintenance, grid connection, inverters, etc.; BOS costs are difficult to reduce by the same margin as the cell costs. The BOS costs are hence a growing fraction of the total installation costs. Soon a further reduction in cell price will have little influence on the cost of a solar installation.

Fortunately, solar cell efficiency reduces almost all the BOS costs simultaneously. For the same power generation fewer solar panels, less land, and lower maintenance costs mean that the cost of a solar installation depends increasingly heavily on the efficiency of the cells. Currently most solar cells used are based on single-junction crystalline silicon. Lab-record efficiencies of these solar cells are close to their efficiency limit (Fig. 15.1b), while panel efficiency is still improving. However, that improvement is coming to an end because also full panels will soon hit their efficiency limit. To further improve the efficiency, radically new solar cells are urgently needed.

Here we will first understand where the efficiency limit for single-junction solar cells comes from and then review the technologies that could overcome the limit, with a focus on singlet fission solar cells.

15.2 The Single-Junction Efficiency Limit

The photons from the sun excite electrons from the conduction band to the valence band of the semiconductor that forms the active layer. The maximum power conversion efficiency of photovoltaics is limited to about 34%, famously called the Shockley-Queisser limit. This limit is based on the principle of detailed balance, and it considers the incoming solar spectrum and the finite solar cell temperature [4]. The solar spectrum is broad, meaning that it covers photons from low-energy, infrared light to high-energy, ultraviolet light. The low-energy photons, with an energy below the bandgap of the active-layer semiconductor, will not excite the electrons across the bandgap, and are hence transmitted through the semiconductor. Photons at the bandgap energy and above will excite the semiconductor. However, any energy above the semiconductor bandgap will be lost because the electrons excited high up in the conduction band will cool quickly to the bandgap. The cooling timescales are typically below a few picoseconds for common solar cell semiconductors, making it near impossible to extract the excess energy. This loss is often called the thermalization loss, even though strictly speaking most of the loss comes from electron cooling and not thermalization. The excited electron gas initially does not have a thermal distribution. Thermalization is the process that this nonequilibrium electron gas undergoes to reach a thermal equilibrium, mostly by electron-electron scattering; afterwards the gas can scatter energy to the lattice, cooling to the band edge. The below-bandgap transmission loss and the high-energy cooling loss are the main contributors to the conversion efficiency limit of solar cells. Many strategies have been proposed to overcome both limitations [5, 6]. So far, the only strategy that is actively employed in solar cells is the use of tandem solar cells.

15.3 Tandem Solar Cells

A tandem solar cell is the combination of two solar cells, typically with different bandgaps. The sunlight passes the solar cell with the large bandgap first, where the high-energy part of the solar spectrum is absorbed, and the remainder is transmitted into the cell with the smaller bandgap. Both parts of the spectrum can thus be converted at the most appropriate electrical potential. The two cells can then be connected in series, the so-called two-terminal tandem cell, or they can be electrically independent and connected by four terminals.

The virtue of tandem solar cells is that they are relatively flexible because each sub-cell can be individually optimized. Especially for the electrically independent four-terminal cells, the two sub-cells can readily be connected to form a tandem cell if the top cell is semitransparent. As a result of the ease of design, tandem solar cells are the only cells that have demonstrated efficiencies above the Shockley-Queisser limit to date [7], and are commercially produced for high-performance applications such as concentrator solar cells and space applications.

However, fabrication of tandem solar cells is complex. The need to fabricate two separate cells requires many fabrication steps, and as a result we have yet to witness a cheap, mass-produced tandem solar cell. The rise of perovskite/silicon tandem solar cells could make this vision a reality. Several start-up companies are actively pursuing this direction with the hope for higher efficiency solar panels, above 30%. Lab efficiencies are already very close to those numbers [7]. Such efforts are very important to push the efficiency of solar panels beyond that of silicon panels, and the prospect of these cells being commercialized within a few years is hugely exciting. However, the added fabrication complexity will likely always make these cells considerably more expensive than the silicon cells alone. Up- and downconversion schemes, as described below, would offer the same efficiency benefit but at much lower complexity, as often only a single additional layer placed on a silicon solar cell would be enough to add additional current. Hence, we consider the downconversion solar cell as the logical next step after the tandem solar cell revolution.

Currently all commercialized tandem solar cells use the two-terminal configuration. A disadvantage of this configuration is the need for current matching. Because each electron must pass both sub-cells, the photocurrent in the tandem cell is limited by the lower current of the two. Under realistic conditions this requirement leads to a decrease in efficiency if the spectrum changes (e.g., during sunrise and sunset) [8]. They are also electrically connected, and to achieve the full efficiency gain at all light intensities, the electrical properties (e.g., fill factor) need to be optimized to a similar level [9]. For a downconversion cell all these considerations are irrelevant because the only electrical part is the underlying (e.g., silicon) solar cell. The downconversion layer only adds additional current to the cell, which otherwise behaves exactly the same as without the downconversion layer.

Efficiency Potential of Tandem Solar Cells

The detailed balance limit calculation performed by Shockley and Queisser has been extended by Vos et al. to tandem solar cells. They calculate a maximum possible efficiency of 42% for two cells [10] using a black body spectrum at 6000 K. For the more realistic AM1.5G spectrum the efficiency limit is slightly higher, at 46% [11]. More junctions lead to an even higher efficiency limit, with an efficiency of 86.8% for an infinite number of junctions [12]. More realistic calculations for example take into account device parameters such as resistances, recombination, and incomplete absorption [13, 14], or environmental parameters such as changes in temperature, spectrum, or irradiance [8, 9, 15]. These models typically show that efficiencies well above 30% should be possible with an efficient silicon base cell.

15.4 Upconversion Solar Cells

Upconversion of low-energy photons by combining the energy of two or more of these to convert it into a particle with energy above the bandgap was proposed to overcome the transmission loss. Traditional upconversion systems based on organic molecules work by a process called triplet-triplet annihilation (TTA), or triplet fusion. This process relies on two triplet excitons combining to form an emissive singlet exciton [16]. The triplet excitons are typically generated in molecules, called sensitizers, containing heavy atoms that absorb the low-energy photons and convert them to triplet excitons via fast intersystem crossing using strong spin-orbit coupling. The triplet excitons are then injected into the TTA material where they undergo fusion. The resulting emissive singlet exciton is then either emitted directly or transferred into separate emitter molecules [17]. However, the organic molecules absorb light that is too high in energy to circumvent losses below the bandgap of common solar cell semiconductors [18].

Recently, upconversion from infrared photons even below the silicon bandgap has been shown by using inorganic II–VI and perovskite quantum dots as sensitizers [19–22]. An excitation in a quantum dot is almost agnostic about the spin state of the exciton because the spin quantum number is not a good quantum number in these highly confined systems [23]. In addition, the spin-singlet and spin-triplet exciton energy in these quantum dots is almost degenerate. As a result, the exciton from a quantum dot can transfer into the triplet exciton state of organic molecules. Low-energy infrared light absorbed in the quantum dots can therefore readily transfer into the triplet exciton state. Several examples of solution systems [19], quantum dot-ligand systems [22], and solid-state bilayers [20] were demonstrated in quick progression. Even bulk perovskites have been successfully employed as triplet sensitizers [24], and presumably the triplet is generated by successive charge transfer in these systems [25].

All of these examples show quantum efficiency well below the theoretical maximum of 50%. Often the power conversion efficiency is even lower because several transfer steps require an energy cascade. In most cases the use for these systems may lie more in enabling new functionality rather than enhancing a solar cell. For example, an infrared sensitizer below the silicon bandgap may be used to broaden the spectral response of silicon-based cameras, and sensitizers that absorb around 800 nm are well suited as light-activated drug delivery systems using the transparency window in biological tissue [26].

For solar cells, upconversion is attractive because any additional photon will add current to the cell, and placing it behind the cell is much less invasive to the cell than a downconversion layer placed in front of the solar cell. On the other hand, to benefit a solar cell, the sensitizer of the upconversion layer needs to be optically thick to absorb all light above its bandgap (and below the solar cell bandgap). This property poses a problem for an upconversion system. The upconverted light is then also likely absorbed by the optically thick sensitizer. For purely organic upconversion systems the photoluminescence of the emitter can be designed to the transparency

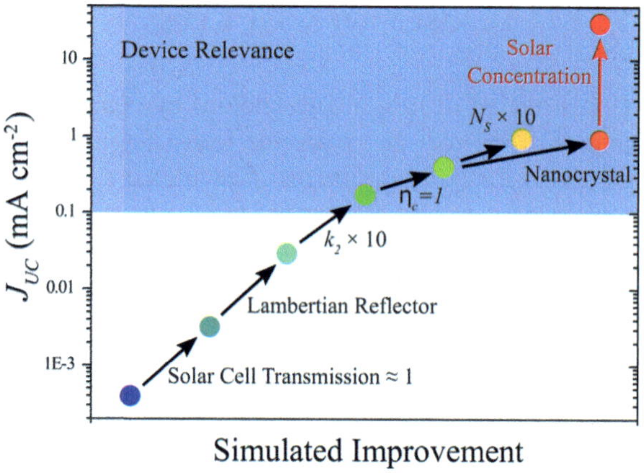

Fig. 15.2 Potential improvement of the current gain from upconversion. k_2 stands for the triplet-triplet annihilation constant, η_c for singlet yield from annihilation, and N_S for the sensitizer concentration. Reprinted with permission from [27]. Copyright 2017 American Chemical Society

window of the sensitizer molecules because organic molecules absorb in discrete states rather than bands. In contrast, the inorganic sensitizers (including quantum dots and perovskites) absorb in continuous bands, which leads to a monotonically increasing absorption with energy. In such systems the emitted light will always be partially reabsorbed. This problem is fundamental and leads to inevitable losses. Still, a solar cell can benefit from such an upconverter, but the maximal current improvement is limited to around 1 mA/cm^2 (Fig. 15.2) [27]. A way to overcome this problem would be electrical injection of the upconverted energy into the solar cell. In such a system the upconverted energy, for example in the form of a singlet exciton, would be direct to an interface between the solar cell and the emitter, and the energy would be transferred by Dexter, Förster, or charge transfer. So far, this approach has attracted only very limited attention [28].

Practically, upconversion only makes sense behind the solar cell, i.e., on the "dark side." The solar cell mass market, based on silicon, is on the verge of a large shift towards bifacial solar cells [29]. For applications of such solar cells in the field, the scattered and reflected light hitting the solar cell from the back is then also absorbed by replacing the opaque back sheet with glass. These solar cells generate up to 20% more power [30] and are hence highly attractive for applications. In these cells, the solar cell does not have a "dark side" and upconversion is not sensible to apply.

15.5 Downconversion Solar Cells

Downconversion, the conversion of high-energy photons into multiples of lower energy, can take many forms. All inorganic semiconductors show multiple exciton generation (MEG), and in quantum dots it can be efficient. Singlet fission, as introduced in previous chapters, occurs only in some organic molecules, but when it occurs it can be very efficient. Quantum cutting using rare-earth atoms has recently caught attention because the integration of rare-earth atoms into perovskite crystals has led to very high photon-to-photon conversion efficiencies. In the following, each process is introduced, and the state of the art of solar cell integration is reviewed.

MEG

MEG is an inverse Auger process by which the energy of a highly excited electron-hole pair is transferred to one or more ground-state electrons to excite them across the bandgap. In most inorganic semiconductors this process has negligible efficiency because fast carrier cooling outcompetes any significant energy transfer to ground-state electrons. However, in semiconductor quantum dots cooling is significantly slowed down because of the "hot phonon bottleneck" [31] so that MEG can compete with cooling. Quantum dots hence show much more efficient MEG than bulk semiconductors with quantum yields of several hundred percent [32], but typically low energy yields because a large energy offset is required.

Efficiency Potential of MEG Solar Cells

The virtue of MEG for solar cells is that the active semiconductor itself undergoes MEG. That means that photons with an energy between the bandgap energy and twice the bandgap energy will produce one electron-hole pair, but any photon with a higher energy has a chance to produce more. In the ideal case the quantum yield could be two from a photon energy of twice the bandgap, three from a photon energy three times the bandgap, etc.; in this ideal case the efficiency limit is 44.4% [33]. However, in all real devices, the increase in quantum yield is typically gradually increasing from about 2.5 to 3 times the bandgap energy [34]. The high threshold means that the benefit of MEG quickly disappears because the number of photons in the solar spectrum decreases drastically at an energy above 3.5 eV.

MEG Solar Cells

MEG for photovoltaics has only been seriously explored in solar cells made from colloidal quantum dots, in most cases from II–VI semiconductors [35] and recently from perovskite quantum dots [36, 37]. The quantum dots are fabricated by synthetic procedures, most often the hot injection method, where two precursors are reacted at elevated temperature to form colloidally stable, monodisperse quantum dots [38]. The quantum dot cores are stabilized by ligands, often organic molecules. To fabricate a solar cell, the quantum dots are spin-coated onto a substrate, and the ligands are exchanged (either in solution or in the film) for short, conductive ligands [39]. While MEG was shown spectroscopically in a multitude of different quantum dot samples early on [40], it took until 2011 to demonstrate an external quantum efficiency above 100%, unambiguous proof of MEG contributing to a solar cell; see

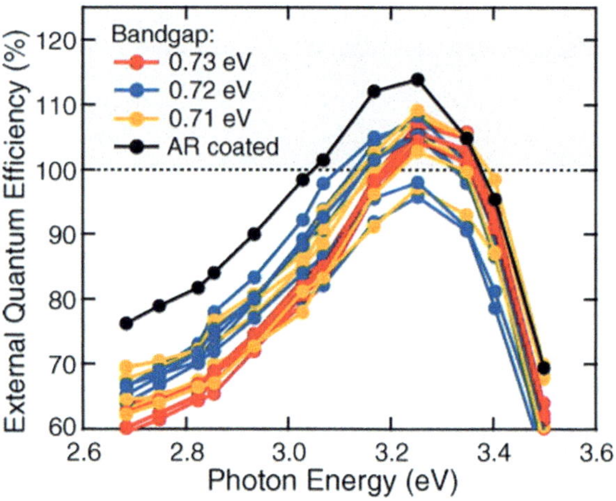

Fig. 15.3 External quantum efficiency of a quantum dot solar cell in the blue region of the spectrum, where, at an energy multiple times the bandgap energy, the quantum efficiency exceeds unity thanks to multiple-exciton generation. Reprinted with permission from [41]

Fig. 15.3 [41]. This progression shows how difficult it is to extract charge carriers from the short-lived multiple excitons. Further developments on quantum dots [42] and nanorods [43] further pushed the quantum efficiency to values well above 120%. However, in all these examples the power conversion efficiency remained low. For a high quantum efficiency the bandgap of the quantum dot needs to be low because MEG efficiency scales as the ratio of the photon energy to the bandgap. Efficient quantum dot solar cells are typically fabricated from quantum dots of much larger bandgap energy, where synthetic protocols are well established [44, 45]. It is notable that none of the high-efficiency quantum dot solar cell shows signs of MEG in the external quantum efficiency spectrum, at least not in the range typically considered (wavelength > 400 nm).

For MEG to improve the power conversion efficiency of a solar cell, two aspects of the process need to improve: a stepwise onset of MEG at twice the bandgap and a high MEG efficiency close to the theoretical maximum.

Quantum Cutting

Quantum cutting is another flavor of downconversion. Here, the excited state in a semiconductor is transferred into two long-lived excited states in a rare-earth impurity, e.g., ytterbium. In these atoms some forbidden transitions are very efficient emitters of photons. Because they are forbidden, they cannot directly be populated with photons because of the selection rules. Thus, typically an intermediate (trap) state is populated which then populates the two forbidden transitions.

The most recent development is the sensitization of perovskite quantum dots [46, 47] and thin films [48] with rare-earth atoms. The most common examples are $CsPb(Cl_xBr_{1-x})_3$, where x can be between 0 and 1. These systems show very high

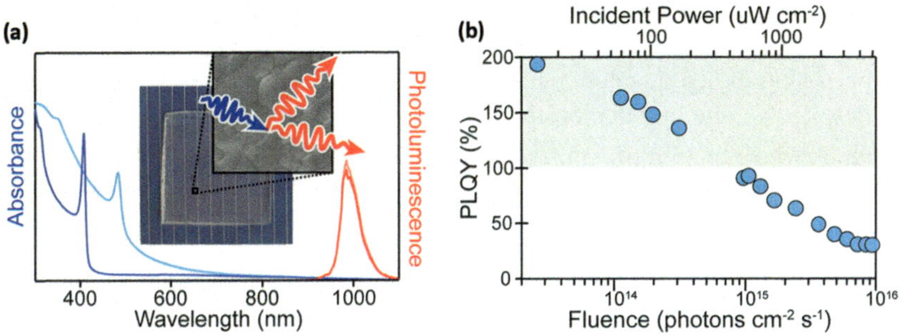

Fig. 15.4 (**a**) Illustration of the quantum cutting process in ytterbium-doped perovskite nanocrystals. The absorbed photons in the blue are transferred into lower energy, but forbidden, transitions in the ytterbium ions. (**b**) Quantum yield of the process, which is almost two at low fluence and reduces at higher fluence due to power saturation. The integrated solar flux of a 1 sun AM1.5G spectrum above 1.1 eV is 2.75 × 1017 photons/cm^2/s. Adapted with permission from 48. Copyright 2018 American Chemical Society

quantum efficiency close to 200% (Fig. 15.4). Two Yb^{3+} ions can replace three Pb^{2+} ions in the perovskite crystal forming a Yb–V_{Pb}–Yb complex, where the lead vacancy V_{Pb} is required for charge balance. This complex ensures that the ytterbium atoms are forming pairwise in the perovskite lattice, presumably supporting the quantum cutting.

One unresolved issue with the quantum cutting system is the saturation at high intensity. When an exciton is in close proximity to an excited ytterbium atom it can undergo Auger recombination, limiting the overall efficiency. The saturation is an issue even at comparably low intensity because of the very long (millisecond) lifetime of the ytterbium excited state. Typically, the efficiency drops already well below 1 sun intensities [47, 48]. For application in solar cells a higher threshold for saturation is necessary.

Efficiency Potential of Quantum Cutting Solar Cells

The ideal limit for the efficiency of a quantum cutting solar cell is the same as that calculated for singlet fission. In the idealized form the two processes are identical. However, in practice the efficiency of a quantum cutting solar cell depends heavily on the breadth of the emission peak, the underlying solar cell considered, and the threshold for saturation. The efficiency limit for conventional silicon cells in different locations yields potential relative efficiency gains between 10 and 20%, which comprises a highly significant advance [49]. A fraction of the gain comes from the poor blue response of some silicon solar cells, and the other fraction of the gain comes from the quantum cutting. The poor blue response means that even under saturation, the quantum cutting layer can be beneficial. The saturation problem is further alleviated by the fact that the natural solar intensity is below 1 sun for a large fraction of the time.

Quantum Cutting Solar Cells

The perovskite-based quantum cutting systems are largely still in the early stages. Yet, despite this, the first demonstrations of solar cell improvements with quantum cutting layers show promise. A Si solar cell was improved from 18.1 to 21.5% by adding a quantum cutting layer [46]. In another example, CIGS and Si cells were improved by similar margins [50] and real-life examples like phone charging were used to demonstrate the practical use.

Much like an optical singlet fission downconverter, quantum cutting relies on the in-coupling of photons from the quantum cutting layer into the solar cell. To a large degree in-coupling can be achieved by the refractive index contrast between the solar cell underneath and the air above the quantum cutting layer, and nanostructures could further improve the coupling efficiency.

15.6 Singlet Fission Solar Cells

Singlet fission is the conversion of a singlet exciton into two lower energy triplet excitons. The mechanistic details of the process are still subject of active study [51–53] and are treated in the earlier chapters of the book. Here we focus on the use of singlet fission for solar cells.

Singlet fission, like other downconversion schemes, and like tandem solar cells, can reduce the so-called thermalization losses in photovoltaic energy conversion. The light absorbed in a singlet fission material excites one of the chromophores to form a spin-singlet exciton. This excited state quickly converts into two triplet excitons of lower energy. Thus, a single photon can generate two lower energy excited states. In a solar cell, a singlet fission layer placed in front of a low-bandgap cell can double the photocurrent from the high-energy part of the solar spectrum (see Fig. 15.5).

The product of singlet fission is a triplet exciton which allows for a variety of methods to extract the exciton energy in a solar cell. Below we will introduce the different device schemes and the experimental progress on each of the schemes. Initially most singlet fission solar cells used organic materials or a combination of organic materials and quantum dots. This chapter reviews some of these developments before focusing on the use of singlet fission to increase the efficiency of highly developed solar cells, mainly silicon.

Singlet fission was probably operational in solar cells before it was recognized. Pentacene solar cells showed unusually high external quantum efficiency of 69% [54] with pentacene/C60 heterojunctions and even >80% with pentacene/PTCDI-C13H27 blends [55], which was only later identified originating from the carrier multiplication thanks to singlet fission. First internal quantum efficiencies (IQEs) >100% were reported for photodetectors made from alternating layers of pentacene and C_{60} to aid charge extraction. In these devices optical modeling suggested an IQE of 128% for the photons absorbed in pentacene, and the high value was

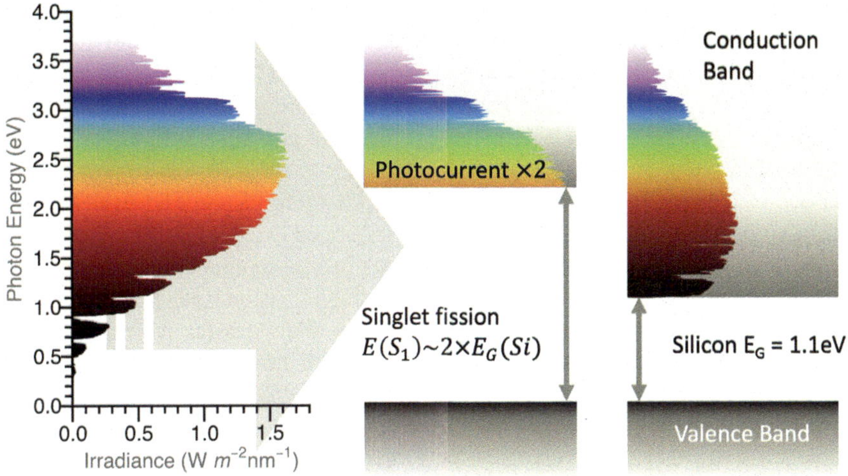

Fig. 15.5 The solar spectrum (shown on the left-hand side) is incident on a singlet fission solar cell. The singlet fission layer absorbs the high-energy light and generates two excitons, potentially leading to twice the photocurrent. The underlying (silicon) solar cell absorbs the low-energy part of the spectrum and converts it in a conventional fashion

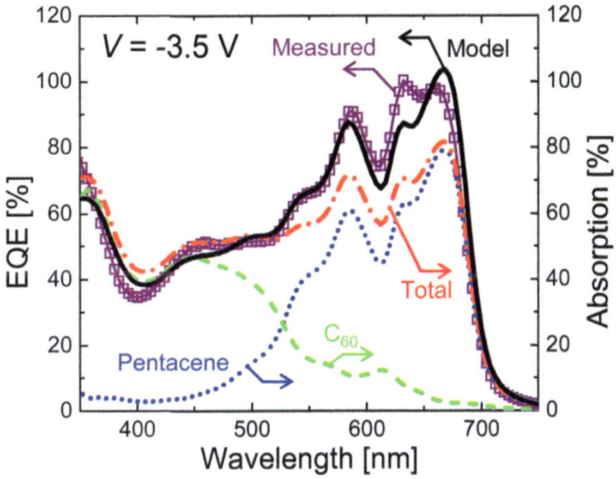

Fig. 15.6 External quantum efficiency (EQE) of a pentacene photodetector at negative bias, compared to the absorption of light in the individual layers. The internal quantum efficiency is estimated at 128% due to singlet fission. Reprinted from [56], with the permission of AIP Publishing

assigned to singlet fission, quantitatively by measuring the magnetic field-dependent photocurrent; see Fig. 15.6 [56]. These initial examples all used pentacene and C_{60} for the organic heterojunction. Later, other organic acceptor molecules like PDI derivatives [57] and copper phthalocyanine [58], and other singlet fission molecules, were also used (tetracene [58–60], TIPS-pentacene [61]).

For combination of most purely organic molecules the bandgaps of the singlet fission material and the acceptor material are similar such that a large fraction of the low-energy infrared photons remains unused. A notable exception is an example by Jadhav et al., where copper phthalocyanine [58] was inserted in between a tetracene/C_{60} heterojunction to absorb the low-energy photons. The first examples of low-bandgap acceptors with a bandgap on the order of the triplet exciton energy were PbS and PbSe quantum dots [57, 62, 63]. The virtue of these quantum dots is that their bandgap depends on their size such that it can be exactly tuned to the triplet exciton energy. They are also very tuneable in terms of processability and passivation. Short organic ligands are sufficient to passivate the surface electronically while still allowing for charge transport and triplet exciton dissociation. These nanocrystals have been developed extensively for quantum dot-based solar cells [64], and the same techniques for ligand exchange and deposition were used for the hybrid quantum dot/organic singlet fission solar cells.

Mechanistically, the triplet exciton seems to first transfer its energy to the quantum dots, after which the hole is transferred back into the organic material. Two studies showed this mechanism at work in tetracene/PbS [65] and pentacene/PbSe [66] heterojunctions. If the quantum dot bandgap is significantly larger than the triplet exciton energy, charge generation can still take place [61], which must mean that charge transfer from the triplet exciton takes place directly at the interface in these systems.

While most singlet fission solar cells to date use linear acenes such as pentacene and tetracene [67], these are probably not the ideal candidates for future singlet fission solar cells. As outline below, large-scale applications of singlet fission solar cells will probably need coupling to a highly efficient solar cell material like silicon or low-bandgap perovskites. This coupling requires suitable singlet and triplet energy, and they also need to be highly stable and easy to process. Pentacene has a low triplet energy of only 0.86 eV [68], which is too low for most applications with highly efficient solar cells (with the possible exception of charge transfer into silicon, see below). While tetracene has a singlet and triplet energy suitable for sensitization of silicon, it is unstable and requires careful processing. Some advances can be made with tetracene derivatives such as TIPS-tetracene [61], but in the long run more intrinsically stable molecules need to be found. Promising candidates are perylene diimides [69, 70] and terrylene derivatives [71], which have both shown high triplet yield and high stability. To date, however, no efficient solar cell with triplet yield from these materials has been demonstrated.

Acenes need careful processing because the triplet yield depends heavily on the arrangement of the molecules in the crystal. A route to circumvent that requirement is to develop individual molecules that perform singlet fission intrinsically (intramolecular singlet fission, iSF), for example singlet fission dimer molecules [72], oligomers [73, 74], or polymers [75, 76]. A key consideration for these molecules is that the triplet pair needs to be able to dissociate, and that the triplet excitons then need to be mobile enough to reach the heterojunction. The crystalline packing of the simple acenes allows for relatively efficient charge and exciton transport, while that may not always be the case with larger, more complex molecules.

Next to the development of the singlet fission material, the interface between the organic material and the underlying low-bandgap solar cell is a key component in the drive towards highly efficient singlet fission solar cells. This interface will be highly specific to the solar cell materials considered, the singlet fission material, the device geometry, and the mechanism of harvesting the triplet exciton energy. For simplicity, here we will demonstrate the various challenges with the dominant solar cell material on the market, silicon. Other solar cell materials, in particular low-bandgap halide perovskites, are also promising candidates and maybe somewhat simpler to couple to singlet fission materials, mostly because the surface passivation appears to be more straightforward. However, these materials are not nearly as mature and commercially relevant as silicon, and the bandgap is typically higher, reducing the benefit of singlet fission.

The holy grail is then to fabricate an efficient silicon solar cell whose power conversion efficiency is enhanced by the photocurrent contribution from triplet excitons.

Such an efficient singlet fission-sensitized silicon solar cell has not yet been realized, but many important steps have recently been taken in this rapidly developing field. Below we consider four different ways to harvest the triplet exciton energy in a silicon solar cell and outline the progress and potential for each of them.

How to Harvest Triplet Energy for a Silicon Solar Cell

There are at least four fundamentally different ways to harvest triplet exciton energy for Si solar cells. In all four ways the singlet fission material is placed in front of the Si solar cell to absorb the high-energy photons and convert them into triplet excitons. Once the triplet excitons are generated, their energy can be extracted:

(a) By dissociation of triplet excitons and subsequent charge transfer (CT)
(b) By energy transfer from triplet excitons into quantum dots (QDs) and subsequent Förster resonant energy transfer (FRET) into Si
(c) Subsequent radiative transfer (RT)
(d) Direct Dexter energy transfer (DET) of the triplet excitons into Si

Figure 15.7 illustrates the harvesting of triplets via these four routes, each of which is described in detail below. It is of fundamental interest under which conditions triplet excitons will follow each of the four processes (a)–(d). There are also very different constraints and loss mechanisms for all four pathways.

Charge Transfer at the Tetracene/c-Si Interface

The separation (dissociation) of excitons at a heterojunction is the most common way to utilize excitons in organic solar cells. Excitons are coulombically bound electron-hole pairs, and extracting the charges requires overcoming the coulomb binding energy. The triplet exciton in the organic material diffuses to the interface and dissociates by donating the electron (or hole) into the acceptor. This architecture works well for heterojunctions between tetracene or pentacene with C_{60} [60, 77] and

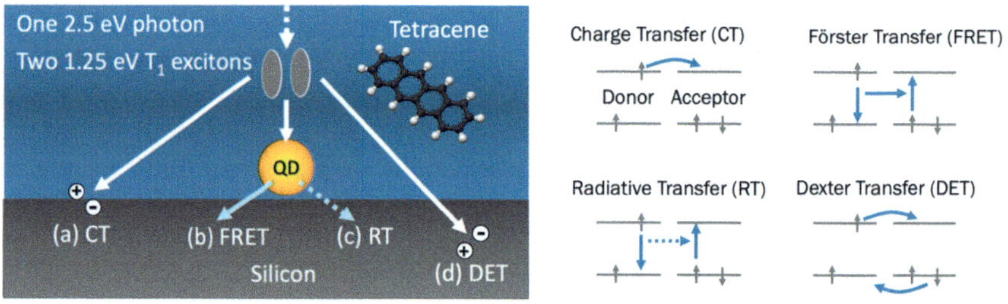

Fig. 15.7 Four possible mechanisms to transfer the energy of triplet excitons into Si, illustrated with state diagrams (left) and a real-space diagram (center). (**a**) Charge transfer (CT), (**b**) Förster resonant energy transfer (FRET), (**c**) radiative transfer (RT), and (**d**) Dexter energy transfer

pentacene with CuPc [58] and between TIPS-pentacene and PbS quantum dots [61] with a remarkable internal quantum efficiency of 175–200%. However, the power conversion efficiency of examples to date is not competitive, with none of them reaching above 5% [61]. For the combination with organic acceptors, both materials only absorb a small fraction of the solar spectrum, the bandgap of the acceptor is much larger than the triplet energy, and the voltage losses are high because of non-radiative recombination. For the combination with inorganic acceptors, most examples use PbS or PbSe quantum dots as acceptors [57, 62, 63]. They have a tuneable bandgap, and the acceptor level can hence be tuned to be exactly at the ionization energy of the triplet excitons. Unfortunately, these quantum dot solar cells do not show competitive efficiency either. A recent example with a metal halide $MAPBI_3$ perovskite acceptor shows charge transfer into the perovskite [78]. This is potentially a very promising route if high-efficiency, low-bandgap perovskites are used.

If the dissociation of the exciton could be achieved at the organic/Si interface with high efficiency, a simple heterojunction organic/n-Si solar cell based on PEDOT:PSS/Si solar cells [79] would be possible, but with a much higher current due to singlet fission. This architecture has some major upsides regarding the efficiency potential, as discussed below. However, the architecture also requires all the holes (or electrons) photogenerated in Si to pass the organic layer. This high charge density will increase the overall series resistance of the device, and requires organic semiconductors with high conductivity, or nanostructured organic layers. To avoid this loss, the silicon top contact could remain on the Si cell itself and collect the holes from both silicon and singlet fission material, while the electrons from the organic are donated to the silicon and extracted at the bottom of the cell. Another complication in both schemes may be the large charge carrier density close to the surface of silicon when a large density of triplet excitons donates electrons into silicon. The silicon surface typically has higher rates of non-radiative recombination [80], especially because thick passivation layers could not be used if electron transfer is required. Thirdly, transfer of singlet excitons should be avoided, because this would lead to a loss in energy. To prevent singlet transfer into silicon, either singlet fission needs to kinetically outcompete the transfer of singlets into silicon or a barrier layer that only accepts triplets (e.g., by spin or energetic selection) would need to be inserted.

Förster Energy Transfer (FRET) via a Quantum Dot Interlayer

The dominant transfer mechanism for singlet excitons is Förster resonant energy transfer (FRET). FRET is rather long range (10 nm) and very efficient in organic systems. It would hence be convenient if the triplet energy could be transferred into silicon via FRET. However, for triplet excitons FRET is unlikely, because it relies on optical dipole-dipole coupling, and therefore requires radiative decay. Usually triplet excitons do not decay radiatively because the decay to the ground state is spin forbidden, and intersystem crossing is weak in the organic semiconductors used [81]. However, the triplet energy can be transferred into quantum dots with >90% efficiency [65, 66], for example PbS and PbSe, where the excited state is emissive because the spin state is not a good quantum number in the quantum dots [23] and the intersystem crossing rate is very high in PbS/PbSe. Thus, an excited quantum dot could allow for radiative coupling and hence FRET into Si. FRET from quantum dots into silicon has also been observed, yet with large-bandgap quantum dots that would not be suitable for triplet excitons generated in tetracene (1.25 eV) or pentacene (0.86 eV).

FRET proceeds via virtual photon exchange and hence needs an emissive donor material with its emission overlapping with the absorption of the acceptor. Si is an indirect-bandgap semiconductor, with a low-absorption cross section, slowing down FRET from quantum dots into Si. The calculated Förster radius for point dipoles indeed shows that the quantum dots need to be within about 1 nm from the Si surface for efficient transfer [82]. Further, the Stokes shift within the quantum dots reduces the available power conversion efficiency. As a result, it is hard to achieve high overall efficiency with FRET transfer into silicon.

Radiative Transfer: The "Photon Multiplier"

The energy transfer into quantum dots opens a second pathway for harvesting triplet excitons via energy transfer. They can emit this energy simply by emitting a photon, which is subsequently absorbed by the Si solar cell. This downconverter, which resembles the quantum cutting scheme in function, is often referred to as a photon multiplier. As with the quantum cutting layer, one main challenge is the coupling of emitted photons into the solar cell underneath. A priori, randomly oriented quantum emitters radiate in all directions. Several strategies could be used to increase directionality. The placement of the emitters next to a high-index material would already direct a large fraction of the radiation towards that material. If they were placed directly on top of the silicon surface, the index contrast alone would allow high directionality (fraction of photons emitted towards the silicon cell). However, one of the virtues of the photon multiplier would be the simple implementation: it can be placed onto any low-bandgap solar cell as an add-on film; even existing panels could be upgraded to higher efficiency with such a film. This means, however, that it would be highly desirable to find schemes that direct photon emission while not requiring the close proximity to the silicon surface. Instead of the high-index material close to the emitter, once could also modify the local density of optical states on one side of the emitter using nanostructures [83]. These can be very efficient in directing the (narrowband) emission, but fabrication would need to be very simple and cheap, and hence likely rely on self-assembly techniques. Another approach could be to align the emitting dipole, for example using quantum rods as emitters that align into a favorable direction for emission. Together with a simple scheme to break the up-down symmetry, this could be a low-cost approach to a downconversion foil with directional emission.

Another, more practical challenge is to find quantum dots that emit with high efficiency. While II–VI quantum dots often show unity emission efficiency [84], these have a bandgap that is far too large for accepting triplet exciton energy from materials useful for solar cells. The lower bandgap III–V quantum dots can be highly emissive, but only in solution or solid solution [85]. Low-bandgap quantum dots deposited into the solid state show lower efficiency. This drop in efficiency presumably comes from energy transfer across quantum dots which increases the chance for trapping, and from changes to the quantum dot surface during deposition. Potentially a solid solution could provide much more favorable conditions, ideally with the matrix material doubling as the singlet fission material [86]. In this case the number of quantum dots should be tuned such that they are minimally affecting the absorption of the film, while still every triplet exciton finds a quantum dot within one diffusion length. For common materials such as tetracene, the diffusion length is long enough so that the quantum dot absorption is not significant [87].

The ability to fabricate an efficient photon multiplier would mean that it could be added to any low-bandgap solar cell as an add-on layer. Ultimately this combination even works without physical contact with the bottom cell. The challenges are that all steps (light absorption, singlet fission, energy transfer to the quantum dots, quantum dot emission, and light directivity) need to work with near-unity efficiency for the entire process to increase overall efficiency [87].

Dexter Energy Transfer via Surface Functionalization

The most promising, and probably the most challenging, way to use the triplet exciton energy in a Si solar cell is the direct energy transfer via Dexter energy transfer (DET, also termed "exchange-mediated energy transfer"). DET is described by the correlated transfer of two electrons, one from the excited state of the donor to the excited state of the acceptor, and the other ground-state electron from the acceptor to the ground state of the donor [88, 89]. DET does not require spin-flip, and is thus the dominating transfer for triplet excited states in materials without appreciable spin-orbit coupling (such as common singlet fission organic semiconductors). It also does not depend on the oscillator strength, which could be beneficial in the realization with the indirect-bandgap silicon. The simplicity of a Dexter transfer architecture is enticing: an organic singlet fission layer is placed on a silicon solar cell which then simply transfers its energy into the silicon cell. Silicon cells have been optimized to perfection to generate charges from the excited states, so once the excited state is in silicon, charge generation should be efficient. Also, it should work with a range of silicon solar cell geometries. Solar cells with interdigitated back contacts are ideal because then the entire surface is free for triplets to inject into silicon. But also silicon solar cells with contacts on the top would work, because the contacts are spaced millimeters apart and the triplet diffusion length is typically only a few tens [90] or hundreds of nanometer [60, 91] so that almost no triplet excitons are quenched at the contacts.

There are, however, many challenges. The DET rate depends exponentially on the donor-acceptor distance. The absolute transfer rate further depends on the extent and overlap of the molecular wave functions, and these wave functions typically decay exponentially with distance. Hence, the donor molecules need to be within 1–2 nm from the Si surface to allow for efficient transfer. The wave function for tetracene has a spatial extent between 0.5 and 3.5 nm, depending on the orientation, highlighting short distances involved and the need for the control of molecular orientation [92, 93]. Achieving efficient silicon passivation at that thickness is a challenge on its own. The recombination on the surface of silicon needs to be especially good since the injection of triplet exciton energy happens at the surface. Hence, the steady-state carrier density will be much higher compared to a silicon solar cell operating on its own. One way to partially avoid this trade-off could be a system with point contacts for triplet injection. A thick-surface passivation could have holes spaced roughly one triplet diffusion length apart, so that most charges in silicon see a well-passivated surface before they diffuse into the bulk. This scheme would work best for organic molecules with a very long triplet diffusion length.

Despite these difficulties, significant progress towards the DET architecture has recently been made. MacQueen et al. deposited tetracene on top of crystalline silicon solar cells and showed that the photocurrent of these solar cells potentially shows a small contribution of triplet excitons [94]. However, in this work the results are consistent with no triplet contribution as well, and it was not clear if any transfer would proceed via charge or energy transfer. Unambiguous evidence was only published later by Einzinger et al. In this breakthrough, the surface of a back-contacted silicon solar cell was passivated with HfO_xN_y and tetracene deposited on top [95]. The solar cells showed a signature of photocurrent from tetracene in the EQE, and a magnetic field-dependent photoluminescence contribution of the silicon PL when tetracene was excited (Fig. 15.8). This evidence clearly showed that triplet excitons were not only transferred into silicon, but also contributed to the photocurrent. The operation of the system appears to depend very strongly on the properties of the interlayer; for example, it only works efficiently for an 8 Å thick HfO_xN_y layer, while 6 Å and 10 Å are both significantly worse. The system also

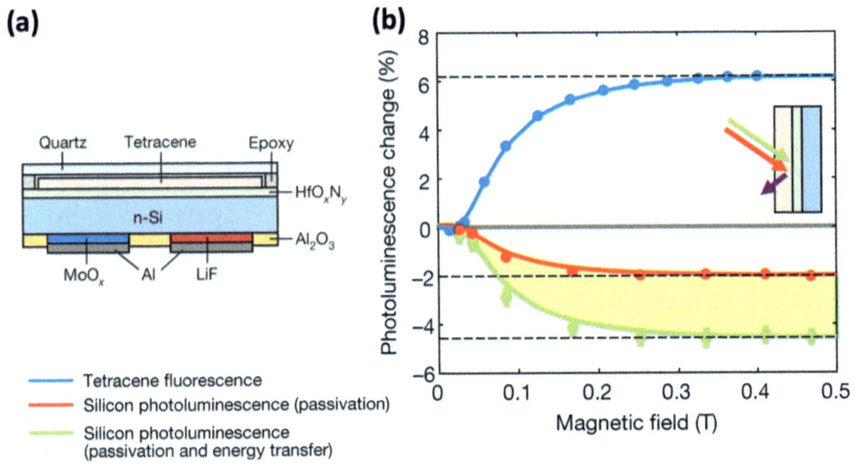

Fig. 15.8 (**a**) Device structure of a silicon solar cell sensitized by singlet fission. This cell uses tetracene with a HfOxNy interlayer. (**b**) The magnetic field dependence unambiguously shows the contribution of singlet fission to the photocurrent. Adopted with permission from [95]

warrants further investigation on the mechanism. While triplets presumably undergo Dexter transfer, it is also possible that the photocurrent is generated via subsequent charge transfer.

Efficiency Limit of Singlet Fission Solar Cells

While singlet fission solar cells were proposed in the 1970s [6], it took until 2006 to simulate the efficiency potential of singlet fission solar cells. In a seminal paper, the efficiency of singlet fission solar cells was compared to other carrier multiplication schemes. The calculations were based on detailed balance, and assumed an ideal solar cell and ideal singlet fission. This means that all light about the bandgaps of the materials included is absorbed while all light below the bandgap is transmitted. Further all recombination of photogenerated carriers is radiative, and there are no losses at the contacts. The authors also assumed that the bandgap of the singlet fission material needs to be twice the bandgap of the underlying solar cell to make sure that the triplet energy is at least as large as the bandgap of the solar cell. Under these assumptions a singlet fission solar cell could achieve about 44% power conversion efficiency (PCE) [33]. This calculation led to a revival of the interest in singlet fission solar cell as it was published at a time when silicon solar cells were hitting their efficiency limit.

In 2012 Tayebjee et al. published a revised efficiency limit that realized that the singlet fission process allows for an entropic gain during singlet fission [96]. This additional energy gain lifts the efficiency limit for singlet fission solar cells to about

45.9%. In the same paper the authors have calculated the efficiency limit for a more realistic case including the exciton binding energy which results in a reduction in the efficiency limit to 42.8% (37.8%) for an exciton binding energy of 0.3 eV (0.5 eV). Modern organic solar cells with non-fullerene acceptors require hardly any driving energy to dissociate the excitons. This observation calls into question the need for an energy-loss step to dissociate the exciton.

More realistic models to predict the efficiency potential of singlet fission solar cells necessarily need to include more assumptions about the device geometry. An example is the calculation of the efficiency potential of a singlet fission-based photon multiplier. This scheme relates closely to the radiative transfer schemes described above. Futscher et al. calculated an efficiency enhancement with such a photon multiplier of up to 4.2% absolute [87]. In their model, a record-efficiency silicon solar cell was used and each step in the conversion from one high-energy photon to two lower energy photons (absorption, singlet fission, transmission and transfer into an emitter, directional emission, and reabsorption) was assumed to be highly efficient. While these could be achieved with much optimization, it also shows how difficult it is to improve upon a highly optimized silicon cell.

There are interesting differences in the efficiency potential between the different realizations of singlet fission solar cells described above. For example, the energy transfer geometry does not suffer from some of the loss channels present in the photon multiplier geometry. The quantum dot emitter in the photon multiplier always induces an additional Stokes shift, which translates into a direct loss of energy. Also, the emission of the photon multiplier needs to be absorbed by the underlying solar cell, and at the edge of the bandgap the EQE of most silicon solar cells is well below 1. For the energy transfer geometry, these losses are not present, and the available efficiency benefit is hence roughly doubled compared to the photon multiplier.

An even more efficient cell could be achieved with the charge transfer geometry. In the DET architecture, the triplet exciton energy needs to be at least as large as the silicon bandgap. The charge transfer geometry is different. At that heterojunction between the singlet fission donor and the silicon acceptor, triplet ionization energy needs to be shallower than the silicon conduction band energy. But that is the only requirement. That means that the HOMO of the singlet fission material can be above the valence band energy of silicon. The smaller the diagonal bandgap (difference between the singlet fission HOMO and the silicon conduction band) the larger the possible absorption in the singlet fission material. This relation comes from the fact that the smaller diagonal bandgap allows for a smaller triplet energy, which in turn allows for a smaller singlet exciton energy, which determines the absorption edge of the organic material. On the flipside, the smaller diagonal bandgap also reduces the maximal photovoltage one can achieve. Thus, a trade-off between current gain and voltage loss leads to an ideal triplet energy that is below the silicon bandgap. Overall the efficiency potential in these cells is higher than in any other singlet fission solar cell, potentially increasing the silicon solar cell efficiency by more than 10% absolute ACS Energy Lett. 2021, 6, 8, 2800–2808. However, the device is significantly more complex than the energy transfer or photon multiplier geometry;

additional transparent contacts on top of the singlet fission material potentially reduce absorption and add fabrication complexity, and all the current from the silicon solar cell needs to pass through the organic material which adds to the series resistance. A potential route around that problem is a very thin singlet fission layer paired with a sensitizer [97] that is strongly absorbing, and has excellent carrier transport. One could think of a perovskite layer of suitable bandgap, for example [98].

15.7 Perspective on Downconversion and Tandem Solar Cells

Tandem solar cells and downconversion cells such as singlet fission solar cells are both solving the same problem: the thermalization losses for high-energy photons in single-bandgap solar cells. They do so, however, in fundamentally different ways. The tandem solar cell essentially consists of two individual cells that are connected in series, or electrically independent as a four-terminal variant. Since all commercial tandem cell applications so far are two-terminal cells we will focus on this implementation here. The serial connection of the two sub-cells leads to an addition of the photovoltage, but the current of both sub-cells needs to match (Fig. 15.9a). In contrast, singlet fission cells add current to another solar cell, so it remains a single-junction cell overall. This means that current matching is not required, and the electrical properties of the cell are mostly unaffected (except in some implementations such as the charge transfer architecture discussed above). This leads to an addition of the photocurrent from the singlet fission layer and the underlying solar cell (Fig. 15.9b). Instead, the voltage is matched, in the sense that the energy of a triplet exciton needs to match or exceed the bandgap of the underlying solar cell.

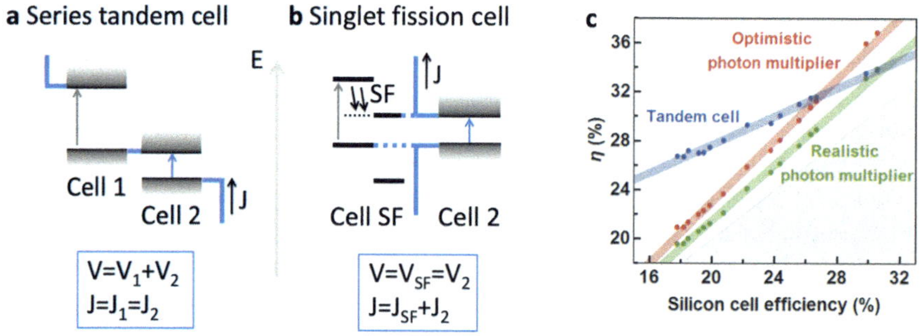

Fig. 15.9 Schematic comparison of (**a**) a series-connected tandem solar cell and (**b**) a singlet fission solar cell. (**c**) Comparison of the efficiency potential of a singlet fission photon multiplier and a perovskite silicon tandem solar cell based on different silicon base cells. Used with permission from [87]

It is interesting that the singlet fission solar cells and the tandem solar cells tackle the same problem, but by challenging different assumptions made in the Shockley-Queisser limit. The singlet fission solar cell breaks with the assumption that each photon only generates one electron-hole pair. Tandem cell on the other hand breaks with the assumption that only one bandgap is used to absorb the light from the sun. While the ultimate efficiency limit is the same for both, the differences in their operation lead to differences in how the efficiency increases for realistic solar cells. One example compares the photon multiplier solar cells to perovskite/silicon tandem solar cells (Fig. 15.9c) [87]. The efficiency of an implementation that makes realistic but optimistic assumptions about the efficiency of all processes in the cell shows that both tandem solar cells and singlet fission solar cells can significantly increase the silicon efficiency alone. However, the tandem solar cell increases a bad silicon cell much more than an efficient one. In contrast, the singlet fission implementation increases all solar cells by a similar factor. The reason for this difference is that the tandem cell needs to work together electrically, and the additional resistance and fill factor losses from the perovskite cell reduce the benefit if the underlying silicon cell is heavily optimized. The singlet fission cell simply adds to the photocurrent without affecting other parameters, thereby increasing the efficiency by an almost constant factor, independent of the efficiency of the underlying solar cell.

Tandem cells are a well-established technology, and many products exist for niche applications. These are typically III–V semiconductor combinations that serve for the very-high-efficiency, high-cost markets such as space applications.

More recently, lead-halide perovskite semiconductors have emerged as a top cell material that is made from abundant constituents and is easy and cheap to fabricate. The commercialization of affordable perovskite/silicon tandem solar cells is now well underway, and promises a revolution in solar cell technology [99]. Efficiencies of 30% and more in mass-produced solar cells are realistic with these tandem cells [9, 15]. While challenges regarding the stability and potentially the toxicity of these materials still need answering, the efforts are globally ramping up and investment is pouring in to accelerate the transition to tandem solar cells. In principle they could be employed wherever conventional silicon cells were employed before, even though spectral changes, especially in bifacial silicon cells, are somewhat more complex [100, 101].

On the other hand, singlet fission-sensitized silicon solar cells are much further away from commercial applications. The first prototypes in the lab are slowly emerging now, and the efficiencies are well below the silicon-only efficiency. It is almost certain that the market introduction of singlet fission solar cells is much further away than that of tandem solar cells, and that it is also significantly less likely and therefore more risky.

One remarkable feature of singlet fission solar cells is that in most realizations they do not change the electrical properties of the solar cell. That means that the singlet fission layer only adds additional current, but does not change the resistances or recombination of the underlying solar cell. This is in stark contrast to the tandem solar cell, where two cells are connected (in the two-terminal configuration), and

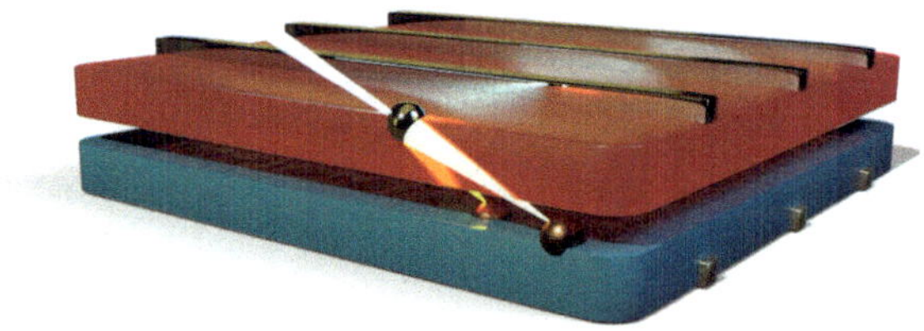

Fig. 15.10 Schematic of a singlet fission solar cell implementation

their current needs to be matched, but also their resistances should be similarly optimized to benefit from the tandem configuration [9]. This difference means that a singlet fission layer adds more to a good solar cell compared to a worse cell, because it just raises the current density [87]. The perovskite/silicon tandem cells on the other hand make more sense for worse bottom cells because it is likely that optimizing the perovskite cell to the same level as the best silicon solar cell will take decades from now.

So, why work on singlet fission solar cells? The reasons are threefold: simplicity, flexibility, and efficiency.

(a) Simplicity: The implementation of singlet fission into a silicon solar cell depends on the exact mechanism (see above), but the simplest implementation adds only a single layer of material to the silicon solar cell (Fig. 15.10). That layer adds additional current to the silicon solar cell so the electronics of the silicon cell remain mostly untouched. In tandem solar cells one either has to design the cell for current matching or needs to fabricate a four-terminal cell, significantly increasing fabrication complexity.
(b) Flexibility: The singlet fission layer could be placed on a silicon solar cell independent of its geometry. Silicon cells with conventional geometries as well as back-contacted silicon cells would work, and mono- as well as bifacial cells could easily be fitted with a singlet fission layer. In contrast, a current-matched bifacial tandem is almost not possible, because even in a single panel, each cell receives dramatically different backlight conditions [102]. Another flexibility that singlet fission materials allow for is chemical flexibility. Many materials have been synthesized recently, including small molecules, dimers, and polymers with a range of active molecules, so that a wide range of target properties could be accessed. This tuneability means that various needs in terms of processing, color, and energy levels can be fulfilled.
(c) Efficiency: The efficiency of a two-junction tandem solar cell based on perovskites and silicon is realistically limited to about 32% [15]. While such an efficiency would be a tremendous success, and would outshine all previous affordable solar cells, it is still well below the numbers calculated for the optimistic case of a singlet fission solar cell. In case of charge transfer, with entropic gain and relatively optimistic assumptions (but all realistically achievable), efficiencies of around 35% could be reached ACS Energy Lett. 2021, 6, 8, 2800–2808.

In that light, one should not ask whether to work towards tandem cells *or* singlet fission cells, but rather think about a natural order where tandem cells will be introduced to the market first, and singlet fission cells are one candidate for the following generation that reduces complexity and increases efficiency beyond tandem solar cells.

References

1. IRENA. *Renewable Capacity Statistics 2020. International Renewable Energy Agency* (2020)
2. A. Jäger-Waldau, *PV Status Report 2019, EUR 29938. Publications Office of the European Union* (2019). https://doi.org/10.2760/326629
3. Energy Watch Group. *Global Energy System based on 100% Renewable Energy. Energy Watch Group* (2019)
4. W. Shockley, H.J. Queisser, Detailed balance limit of efficiency of p-n junction solar cells. J. Appl. Phys. **32**, 510–519 (1961)
5. Edmond D Jackson, *Solar energy converter* (U.S. Patent No. 2,949,498) (1960)
6. D.L. Dexter, Two ideas on energy transfer phenomena: ion-pair effects involving the OH stretching mode, and sensitization of photovoltaic cells. J. Lumin. **18–19**, 779–784 (1979)
7. M.A. Green et al., Solar cell efficiency tables (Version 55). Prog. Photovolt. Res. Appl. **28**, 3–15 (2020)
8. H. Liu et al., The realistic energy yield potential of GaAs-on-Si tandem solar cells: a theoretical case study. Opt. Express **23**, A382 (2015)
9. M.H.M.H. Futscher, B. Ehrler, Modeling the performance limitations and prospects of perovskite/Si tandem solar cells under realistic operating conditions. ACS Energy Lett. **2**, 2089–2095 (2017)
10. A.D. Vos, Detailed balance limit of the efficiency of tandem solar cells. J. Phys. D. Appl. Phys. **13**, 839 (1980)
11. S.P. Bremner, M.Y. Levy, C.B. Honsberg, Analysis of tandem solar cell efficiencies under AM1.5G spectrum using a rapid flux calculation method. Prog. Photovoltaics Res. Appl. **16**, 225–233 (2008)
12. A.S. Brown, M.A. Green, Detailed balance limit for the series constrained two terminal tandem solar cell, in *Physica E: Low-Dimensional Systems and Nanostructures*, vol. 14, (North-Holland, 2002), pp. 96–100
13. T.P. White, N.N. Lal, K.R. Catchpole, Tandem solar cells based on high-efficiency c-Si bottom cells: top cell requirements for >30% efficiency. IEEE J. Photovoltaics **4**, 208–214 (2014)
14. M.H. Futscher, B. Ehrler, Efficiency limit of perovskite/Si tandem solar cells. ACS Energy Lett. **1**, 863–868 (2016)
15. M.T. Hörantner, H.J. Snaith, Predicting and optimising the energy yield of perovskite-on-silicon tandem solar cells under real world conditions. Energy Environ. Sci. **10**, 1983–1993 (2017)
16. C.A. Parker, C.G. Hatchard, Delayed fluorescence from solutions of anthracene and phenanthrene. Proc. Chem. Soc. (1962). https://doi.org/10.1098/rspa.1962.0197
17. T.N. Singh-Rachford, F.N. Castellano, Photon upconversion based on sensitized triplet-triplet annihilation. Coord. Chem. Rev. **254**, 2560–2573 (2010)
18. T. Dilbeck, K. Hanson, Molecular photon upconversion solar cells using multilayer assemblies: progress and prospects. J. Phys. Chem. Lett. **9**, 5810–5821 (2018)
19. Z. Huang et al., Hybrid molecule-nanocrystal photon upconversion across the visible and near-infrared. Nano Lett. **15**, 5552–5557 (2015)
20. M. Wu et al., Solid-state infrared-to-visible upconversion sensitized by colloidal nanocrystals. Nat. Photonics **10**, 31–34 (2015)

21. M. Mahboub, Z. Huang, M.L. Tang, Efficient infrared-to-visible upconversion with subsolar irradiance. Nano Lett. **16**, 7169–7175 (2016). https://doi.org/10.1021/acs.nanolett.6b03503
22. C. Mongin, S. Garakyaraghi, N. Razgoniaeva, M. Zamkov, F.N. Castellano, Direct observation of triplet energy transfer from semiconductor nanocrystals. Science **351**, 369–372 (2016)
23. K.I.M. Jeoncho, C.Y. Wong, G.D. Scholes, Exciton fine structure and spin relaxation in semiconductor colloidal quantum dots. Acc. Chem. Res. **42**, 1037–1046 (2009). https://doi.org/10.1021/ar8002046
24. L. Nienhaus et al., Triplet-sensitization by lead halide perovskite thin films for near-infrared-to-visible upconversion. ACS Energy Lett. **4**, 888–895 (2019). https://doi.org/10.1021/acsenergylett.9b00283
25. S. Wieghold et al., Triplet sensitization by Lead halide perovskite thin films for efficient solid-state photon upconversion at subsolar fluxes. Matter **1**, 705–719 (2019). https://doi.org/10.1016/j.matt.2019.05.026
26. J.V. Frangioni, In vivo near-infrared fluorescence imaging. Curr. Opin. Chem. Biol. **7**, 626–634 (2003)
27. L. Frazer, J.K. Gallaher, T.W. Schmidt, Optimizing the efficiency of solar photon upconversion. ACS Energy Lett. **2**, 1346–1354 (2017). https://doi.org/10.1021/acsenergylett.7b00237
28. K.M. Felter et al., Solid-state infrared upconversion in perylene diimides followed by direct Electron injection. ACS Energy Lett. **5**, 124–129 (2020). https://doi.org/10.1021/acsenergylett.9b02361
29. M. Gaetan, O. Sinead, R. Manoel, Global market outlook. Eur. Photovolt. Ind. Assoc. **60** (2018). https://doi.org/10.1787/key_energ_stat-2014-en
30. H. Park, S. Chang, S. Park, W.K. Kim, Outdoor performance test of bifacial n-type silicon photovoltaic modules. Sustainability **11** (2019)
31. A.J. Nozik et al., Semiconductor quantum dots and quantum dot arrays and applications of multiple exciton generation to third-generation photovoltaic solar cells. Chem. Rev. **110**, 6873–6890 (2010)
32. R.J. Ellingson et al., Highly efficient multiple exciton generation in colloidal PbSe and PbS quantum dots. Nano Lett. **5**, 865–871 (2005)
33. M.M.C. Hanna, A.A.J. Nozik, Solar conversion efficiency of photovoltaic and photoelectrolysis cells with carrier multiplication absorbers. J. Appl. Phys. **100**, 74510 (2006)
34. C. Smith, D. Binks, Multiple exciton generation in colloidal nanocrystals. Nanomaterials **4**, 19–45 (2013)
35. J.M. Luther et al., Multiple exciton generation in films of electronically coupled PbSe quantum dots. Nano Lett. **7**, 1779–1784 (2007)
36. M. Li et al., Low threshold and efficient multiple exciton generation in halide perovskite nanocrystals. Nat. Commun. **9**, 4197 (2018). https://doi.org/10.1038/s41467-018-06596-1
37. C. de Weerd et al., Efficient carrier multiplication in CsPbI3 perovskite nanocrystals. Nat. Commun. **9**, 4199 (2018). https://doi.org/10.1038/s41467-018-06721-0
38. M.L. Steigerwald, L.E. Brus, Synthesis, stabilization, and electronic structure of quantum semiconductor nanoclusters. Annu. Rev. Mater. Sci. **19**, 471–495 (1989)
39. I.J.I. Kramer, E.H. Sargent, Colloidal quantum dot photovoltaics: a path forward. ACS Nano **5**, 8506–8514 (2011)
40. A. Luque, A. Martí, A.J. Nozik, Solar cells based on quantum dots: multiple exciton generation and intermediate bands. MRS Bull. **32**, 236–241 (2007)
41. O.E. Semonin et al., Peak external photocurrent quantum efficiency exceeding 100% via MEG in a quantum dot solar cell. Science **334**, 1530–1533 (2011)
42. M.L.M.L. Böhm et al., Lead telluride quantum dot solar cells displaying external quantum efficiencies exceeding 120%. Nano Lett. **15**, 7987–7993 (2015)
43. N.J.L.K. Davis et al., Multiple-exciton generation in lead selenide nanorod solar cells with external quantum efficiencies exceeding 120%. Nat. Commun. **6**, 8259 (2015)
44. M. Biondi et al., A chemically orthogonal hole transport layer for efficient colloidal quantum dot solar cells. Adv. Mater. **32**, 1906199 (2020). https://doi.org/10.1002/adma.201906199

45. J. Yuan et al., Metal halide perovskites in quantum dot solar cells: progress and prospects. Joule **4**, 1160–1185 (2020). https://doi.org/10.1016/j.joule.2020.04.006
46. D. Zhou et al., Cerium and ytterbium codoped halide perovskite quantum dots: a novel and efficient downconverter for improving the performance of silicon solar cells. Adv. Mater. **29** (2017)
47. T.J. Milstein, D.M. Kroupa, D.R. Gamelin, Picosecond quantum cutting generates photoluminescence quantum yields over 100% in ytterbium-doped CsPbCl3 nanocrystals. Nano Lett. **18**, 3792–3799 (2018). https://doi.org/10.1021/acs.nanolett.8b01066
48. D.M. Kroupa, J.Y. Roh, T.J. Milstein, S.E. Creutz, D.R. Gamelin, Quantum-cutting ytterbium-doped CsPb(Cl1-xBrx)3 perovskite thin films with photoluminescence quantum yields over 190%. ACS Energy Lett. **3**, 2390–2395 (2018). https://doi.org/10.1021/acsenergylett.8b01528
49. M.J. Crane, D.M. Kroupa, D.R. Gamelin, Detailed-balance analysis of Yb3+:CsPb(Cl1-: XBrx)3 quantum-cutting layers for high-efficiency photovoltaics under real-world conditions. Energy Environ. Sci. (2019). https://doi.org/10.1039/c9ee01493d
50. D. Zhou et al., Impact of host composition, codoping, or tridoping on quantum-cutting emission of ytterbium in halide perovskite quantum dots and solar cell applications. Nano Lett. **19**, 6904–6913 (2019). https://doi.org/10.1021/acs.nanolett.9b02139
51. G.B. Piland, C.J. Bardeen, How morphology affects singlet fission in crystalline tetracene. J. Phys. Chem. Lett. **6**, 1841–1846 (2015)
52. A.J. Musser et al., Evidence for conical intersection dynamics mediating ultrafast singlet exciton fission. Nat. Phys. **11**, 352–357 (2015)
53. A.B. Kolomeisky, X. Feng, A.I. Krylov, A simple kinetic model for singlet fission: a role of electronic and entropic contributions to macroscopic rates. J. Phys. Chem. C **118**, 5188–5195 (2014)
54. S. Yoo et al., Analysis of improved photovoltaic properties of pentacene/C60 organic solar cells: effects of exciton blocking layer thickness and thermal annealing. Solid State Electron. **51**, 1367–1375 (2007)
55. A.K. Pandey, S. Dabos-Seignon, J.-M. Nunzi, Pentacene: PTCDI-C_{13}H_{27} molecular blends efficiently harvest light for solar cell applications. Appl. Phys. Lett. **89**, 113506 (2006)
56. J. Lee, P.J. Jadhav, M.A. Baldo, High efficiency organic multilayer photodetectors based on singlet exciton fission. Appl. Phys. Lett. **95**, 33301 (2009)
57. P.J. Jadhav et al., Triplet exciton dissociation in singlet exciton fission photovoltaics. Adv. Mater. **24**, 6169–6174 (2012)
58. P.P.J. Jadhav, A. Mohanty, J.M. Sussman, J. Lee, M.A. Baldo, Singlet exciton fission in nanostructured organic solar cells. Nano Lett. **11**, 1495–1498 (2011)
59. P.S. Abthagir et al., Studies of tetracene- and pentacene-based organic thin-film transistors fabricated by the neutral cluster beam deposition method. J. Phys. Chem. B **109**, 23918–23924 (2005)
60. T.C. Wu et al., Singlet fission efficiency in tetracene-based organic solar cells. Appl. Phys. Lett. **104**, 193901 (2014)
61. L. Yang et al., Solution-Processable singlet fission photovoltaic devices. Nano Lett. **15**, 354–358 (2014)
62. B. Ehrler, M.W.B. Wilson, A. Rao, R.H. Friend, N.C. Greenham, Singlet exciton fission-sensitized infrared quantum dot solar cells. Nano Lett. **12**, 1053–1057 (2012)
63. B. Ehrler et al., In situ measurement of exciton energy in hybrid singlet-fission solar cells. Nat. Commun. **3**, 1019 (2012)
64. M. Yuan et al., Colloidal quantum dot solids for solution-processed solar cells. Nat. Energy **1**, 16016 (2016)
65. N.J. Thompson et al., Energy harvesting of non-emissive triplet excitons in tetracene by emissive PbS nanocrystals. Nat. Mater. **13**, 1039–1043 (2014)
66. M. Tabachnyk et al., Resonant energy transfer of triplet excitons from pentacene to PbSe nanocrystals. Nat. Mater. **13**, 1033–1038 (2014)
67. J. Xia et al., Singlet fission: progress and prospects in solar cells. Adv. Mater. **29** (2017)

68. C. Jundt et al., Exciton dynamics in pentacene thin films studied by pump-probe spectroscopy. Chem. Phys. Lett. **241**, 84–88 (1995)
69. N. Renaud, F.C. Grozema, Intermolecular vibrational modes speed up singlet fission in perylenediimide crystals. J. Phys. Chem. Lett. **6**, 360–365 (2015)
70. S.W. Eaton et al., Singlet exciton fission in polycrystalline thin films of a slip-stacked perylenediimide. J. Am. Chem. Soc. **135**, 14701–14712 (2013)
71. S.W. Eaton et al., Singlet exciton fission in thin films of tert-butyl-substituted terrylenes. J. Phys. Chem. A **119**, 4151–4161 (2015)
72. S.N. Sanders et al., Quantitative intramolecular singlet fission in bipentacenes. J. Am. Chem. Soc. **137**, 8965–8972 (2015)
73. N. Alagna et al., Singlet fission in tetraaza-TIPS-pentacene oligomers: from fs excitation to μs triplet decay via the biexcitonic state. J. Phys. Chem. B **123**, 10780–10793 (2019)
74. N.V. Korovina, C.H. Chang, J.C. Johnson, Spatial separation of triplet excitons drives endothermic singlet fission. Nat. Chem. **12**, 391–398 (2020)
75. E. Busby et al., A design strategy for intramolecular singlet fission mediated by charge-transfer states in donor-acceptor organic materials. Nat. Mater. **14**, 426–433 (2015)
76. A.J. Musser et al., Activated singlet exciton fission in a semiconducting polymer. J. Am. Chem. Soc. **135**, 12747–12754 (2013)
77. D.N. Congreve et al., External quantum efficiency above 100% in a singlet-exciton-fission based organic photovoltaic cell. Science **340**, 334–337 (2013)
78. D. Guo et al., Charge transfer dynamics in a singlet fission organic molecule and organometal perovskite bilayer structure. J. Mater. Chem. A **8**, 5572–5579 (2020)
79. S. Jäckle et al., Junction formation and current transport mechanisms in hybrid n-Si/PEDOT:PSS solar cells. Sci. Rep. **5**, 13008 (2015)
80. S.W. Glunz, F. Feldmann, SiO2 surface passivation layers—a key technology for silicon solar cells. Sol. Energy Mater. Sol. Cells **185**, 260–269 (2018). https://doi.org/10.1016/j.solmat.2018.04.029
81. S. Reineke, M.A. Baldo, Room temperature triplet state spectroscopy of organic semiconductors. Sci. Rep. **4**, 3797 (2014)
82. S.W. Tabernig, B. Daiber, T. Wang, B. Ehrler, Enhancing silicon solar cells with singlet fission: the case for Förster resonant energy transfer using a quantum dot intermediate. J. Photonics Energy **8** (2018)
83. D.K.G. de Boer et al., Directional sideward emission from luminescent plasmonic nanostructures. Opt. Express **24**, A388–A396 (2016). https://doi.org/10.1364/oe.24.00a388
84. D.A. Hanifi et al., Redefining near-unity luminescence in quantum dots with photothermal threshold quantum yield. Science **363**, 1199–1202 (2019). https://doi.org/10.1126/science.aat3803
85. S. Pradhan et al., High-efficiency colloidal quantum dot infrared light-emitting diodes via engineering at the supra-nanocrystalline level. Nat. Nanotechnol. **14**, 72–79 (2019). https://doi.org/10.1038/s41565-018-0312-y
86. N.J.L.K. Davis et al., Singlet fission and triplet transfer to PbS quantum dots in TIPS-tetracene carboxylic acid ligands. J. Phys. Chem. Lett. **9**, 1454–1460 (2018). https://doi.org/10.1021/acs.jpclett.8b00099
87. M.H. Futscher, A. Rao, B. Ehrler, The potential of singlet fission photon multipliers as an alternative to silicon-based tandem solar cells. ACS Energy Lett. **3**, 2587–2592 (2018)
88. D.L. Dexter, A theory of sensitized luminescence in solids. J. Chem. Phys. **21**, 836 (1953)
89. G.D. Scholes, Long-range resonance energy transfer in molecular systems. Annu. Rev. Phys. Chem. **54**, 57–87 (2003)
90. M. Tabachnyk, B. Ehrler, S. Bayliss, R.H. Friend, N.C. Greenham, Triplet diffusion in singlet exciton fission sensitized pentacene solar cells. Appl. Phys. Lett. **103**, 153302 (2013)
91. M. Pope, N.E. Geacintov, D. Saperstein, F. Vogel, Calculation of the diffusion length, diffusion coefficient and lifetime of triplet excitons in crystalline tetracene. J. Lumin. **1–2**, 224–230 (1970). https://doi.org/10.1016/0022-2313(70)90037-2

92. F. Plasser, S.A. Mewes, A. Dreuw, L. González, Detailed wave function analysis for multireference methods: implementation in the Molcas program package and applications to tetracene. J. Chem. Theory Comput. **13**, 5343–5353 (2017). https://doi.org/10.1021/acs.jctc.7b00718
93. S.L. Bayliss et al., Geminate and nongeminate recombination of triplet excitons formed by singlet fission. Phys. Rev. Lett. **112**, 238701 (2014)
94. R.W. MacQueen et al., Crystalline silicon solar cells with tetracene interlayers: the path to silicon-singlet fission heterojunction devices. Mater. Horiz. **5**, 1065–1075 (2018)
95. M. Einzinger et al., Sensitization of silicon by singlet exciton fission in tetracene. Nature **571**, 90–94 (2019). https://doi.org/10.1038/s41586-019-1339-4
96. M.J.Y. Tayebjee, A.A. Gray-Weale, T.W. Schmidt, Thermodynamic limit of exciton fission solar cell efficiency. J. Phys. Chem. Lett. **3**, 2749–2754 (2012)
97. P.D. Reusswig, D.N. Congreve, N.J. Thompson, M.A. Baldo, Enhanced external quantum efficiency in an organic photovoltaic cell via singlet fission exciton sensitizer. Appl. Phys. Lett. **101**, 113304 (2012)
98. H. Lu, X. Chen, J.E. Anthony, J.C. Johnson, M.C. Beard, Sensitizing singlet fission with perovskite nanocrystals. J. Am. Chem. Soc. **141**, 4919–4927 (2019)
99. Y. Hu, L. Song, Y. Chen, W. Huang, Two-terminal perovskites tandem solar cells: recent advances and perspectives. Solar RRL **3**, 1900080 (2019). https://doi.org/10.1002/solr.201900080
100. R. Asadpour, R.V.K. Chavali, M. Ryyan Khan, M.A. Alam, Bifacial Si heterojunction-perovskite organic-inorganic tandem to produce highly efficient ($\eta T_* \sim 33\%$) solar cell. Appl. Phys. Lett. (2015). https://doi.org/10.1063/1.4922375
101. J. Lehr et al., Energy yield of bifacial textured perovskite/silicon tandem photovoltaic modules. Sol. Energy Mater. Sol. Cells **208**, 110367 (2020)
102. G. Coletti et al., Bifacial four-terminal perovskite/silicon tandem solar cells and modules. ACS Energy Lett. **5**, 1676–1680 (2020)

Index

J. S. Lissau, M. Madsen (eds.), *Emerging Strategies to Reduce Transmission and Thermalization Losses in Solar Cells*, https://doi.org/10.1007/978-3-030-70358-5

E

Q

Printed by Printforce, the Netherlands